AF323767

CALLIANASSOIDEA
OF THE WORLD
(DECAPODA, THALASSINIDEA)

Published in this series:

CRM 001 - *Stephan G. Bullard*	Larvae of anomuran and brachyuran crabs of North Carolina
CRM 002 - *Spyros Sfenthourakis et al.*	The biology of terrestrial isopods, V
CRM 003 - *Tomislav Karanovic*	Subterranean Copepoda from arid Western Australia
CRM 004 - *Katsushi Sakai*	Callianassoidea of the world (Decapoda, Thalassinidea)

In preparation (provisional titles):

CRM 005 - *Kim Larsen*	Deep-sea Tanaidacea (Peracarida) from the Gulf of Mexico
CRM 00x - *Darren C. Yeo & Peter K. L. Ng*	Southeast Asian freshwater crabs
CRM 00y - *Frank D. Ferrari & Hans-U. Dahms*	Copepod development
CRM 00z - *Chang-tai Shih*	Marine Calanoida of the China seas
CRM 00p - *Katsushi Sakai*	Upogebiidae of the world (Decapoda, Thalassinidea)

Author's address:

Prof. Dr. K. Sakai, Biological Laboratory, Shikoku University, Tokushima 771-1192, Japan; e-mail: ksakai@shikoku-u.ac.jp

Manuscript first received 2 June 2003; final version accepted 22 April 2004.

Cover: Neocallichirus mortenseni sp. nov., see p. 171.

CALLIANASSOIDEA OF THE WORLD
(DECAPODA, THALASSINIDEA)

BY

Katsushi Sakai

Crustaceana Monographs, 4

BRILL

LEIDEN · BOSTON

This book is printed on acid-free paper.

Library of Congress Cataloging-in-Publication Data

The Library of Congress Cataloging-in-Publication Data is available from the Publisher.

ISBN 90 04 14211 8

CONTENTS

PREFACE

To Prof. Dr. L. B. Holthuis, Nationaal Natuurhistorisch Museum, Leiden, this work is gratefully dedicated by the author as a slight token of respect.

The taxonomic composition of the superfamily Callianassoidea Dana, 1852 (with the exception, however, of the possibly also included Axiidae Huxley, 1879 and Callianideidae Kossmann, 1880) is revised. Three families, 12 subfamilies, 20 genera, and 218 species, including 13 new species are assigned to the superfamily Callianassoidea. This includes the family Callianassidae Dana, 1852, with the following subfamilies and genera: Callianassinae, *Callianassa*; Callichirinae, *Callichirus*, *Glypturus*, *Lepidophthalmus*, *Michaelcallianassa*, *Neocallichirus*, and *Podocallichirus*; Eucalliacinae, *Calliax* and *Paraglypturus*; Calliapaguropinae, *Calliapagurops*; Anacalliacinae, *Anacalliax*; Lipkecallianassinae n. subfam., *Lipkecallianassa*; Bathycalliacinae, *Bathycalliax*; and Paracalliacinae n. subfam., *Paracalliax*; the family Gourretiidae Sakai, 1999 with: Gourretiinae, *Gourretia*, *Laurentgourretia*, and *Paragourretia*; Callianopsinae, *Callianopsis* and Pseudogourretiinae n. subfam., *Pseudogourretia* n. gen.; and the family Ctenochelidae Manning & Felder, 1991 with: Ctenochelinae, *Ctenocheles*.

Two genera, *Necallianassa* and *Pseudobiffarius*, are synonymized with *Callianassa*, *Grynaminna* is synonymized with *Podocallichirus*, and *Dawsonius* is synonymized with *Callianopsis*[*]. *Corallichirus bayeri* is recognized as a junior synonym of *Gourretia assimilis*.

KATSUSHI SAKAI

---

[*]) But see "Note added in proof" on p. 245.

INTRODUCTION

Our systematic understanding of the Callianassidae s.l. has been established by Borradaile (1903), De Man (1928a, b), De Saint Laurent (1973, 1979), Poore & Griffin (1979), Ngoc-Ho (1991, 1994, 1995, 2003), Felder & Manning (1994, 1995, 1997), and Sakai (1999a, c). Manning & Felder (1986, 1989, 1991, 1992, 1995) included two families in their classification of the infraorder Thalassinidea, viz., Callianassidae and Ctenochelidae. Both Poore (1994) and Tudge et al. (2000) applied a heuristic, computer-aided phylogenetic analysis to the families of the Thalassinidea, and revised the classification that had been followed by many scientists since Borradaile (1903). Their investigation appears, at least on the surface, to have adopted a rigorous empirical approach, whereas the present interpretation is rather based on the traditional taxonomic approach of taxonomists who have an extensive knowledge of the group they are studying. In fact, the cladistic analysis yielded a result that is incompatible with the taxa recognized at present, and has, consequently, caused some confusion. From a study of the morphology of the gastric mill (see below), the infraorder Thalassinidea appears to be composed of two superfamilies, Thalassinoidea and Callianassoidea, as De Saint Laurent (1979: 1396) already surmised, based on external morphology, where she mentioned that in the Callianassoidea P2 is chelate, whereas in the Thalassinoidea, the P2 is simple. In the present work, then, 3 families, 12 subfamilies, and 20 genera in the superfamily Callianassoidea from the infraorder Thalassinidea are revised, and the family Ctenochelidae is divided into two families, Ctenochelidae s.s. and Gourretiidae (cf. Sakai, 2004). I had already published an account of the systematics of the family Callianassidae (cf. Sakai, 1999c), but the present situation, as described above, has necessitated me to compose a comprehensive review of the taxa of the Callianassoidea and their classificatory framework, by detailing my views and by adding new genera and species.

In June 1978, Dr. Torben Wolff of the Zoologisk Museum, University of Copenhagen, sent me a collection of unidentified Indo-Pacific Thalassinidea. The axiid component of this collection, which included one new genus and six new species, has been reported upon already (Sakai, 1992a). The callianassid part is documented in the present paper, within the framework of an updated revision of the family Callianassidae. The collections of Callianassidae are based principally on the material collected during the "Galathea" Expedition 1950-1952; during Dr. Th. Mortensen's expeditions, including the Java–South

African Expedition 1929-1930, and the Pacific expeditions 1913-1930; the Danish expedition to the Kei Islands 1922; Dr. G. Thorson's Persian [= Iranian] expedition; the Danish "Dana" Expedition 1928-1930; the Thai-Danish expedition 1966; and other, smaller collections in Copenhagen. In addition to those collections, one new species, *Callianassa costaricensis* sp. nov., which was brought back by the German exploring ship, R/V "Victor Hensen" from the Golfo de Nicoya, Bahia Herradura, Costa Rica, Pacific side, is described. Two genera, *Lipkecallianassa* and *Michaelcallianassa*, and 17 species of Callianassidae from the Andaman Sea (Sakai, 2002: 461) are also included in the present work.

Abbreviations used. — A1, antennule or antenna 1; A2, antenna or antenna 2; AB1, abdominal length including telson; Cl, carapace length; Mxp3, maxilliped 3; P, pereiopod; Plp, pleopod; Tl, total length, from the tip of the carapace to the end of the telson, measured by attaching a thread. In referring to material examined, measurements of Tl/Cl are given in mm, without further indication. LACM, Los Angeles County Museum, Los Angeles, California; MNHNP, Muséum nationale d'Histoire naturelle, Paris; NHMW, Naturhistorisches Museum, Vienna; NMW, National Museum of New Zealand, Wellington; NSMT, National Science Museum, Tokyo; PMBC, Phuket Marine Biological Center, Phuket; QMB, Queensland Museum, Brisbane; SIC, Stazione Idrobiologica di Chioggia, Dipartimento di Biologia, Università degli Studi di Padova, Padova; SMF, Forschungsinstitut Senckenberg, Frankfurt am Main; SMNH, Swedish Museum of Natural History or Natur Historiska Riksmuseet, Stockholm; ZLUA, Zoological Laboratory, University of Athens, Athens; ZMUC, Zoological Museum, Copenhagen.

SYSTEMATIC ACCOUNT OF THE CLASSIFICATION OF THE INFRAORDER THALASSINIDEA LATREILLE, 1831

Twenty-nine species, including 9 new species from the collection of the Zoological Museum, University of Copenhagen, one new species from the collection of the Forschungsinstitut Senckenberg, Frankfurt a. M., one new species from Costa Rica, and one new species from Guinea, are herein assigned to the Callianassoidea. As a result, three families with 12 subfamilies, including three new subfamilies, 20 genera, and 218 spp. are presented in this revision of the world's Callianassoidea (table I): the Callianassidae: Callianassinae with *Callianassa*; Callichirinae with *Callichirus*, *Glypturus*, *Lepidophthalmus*, *Michaelcallianassa*, *Neocallichirus*, and *Podocallichirus*; Eucalliacinae with

TABLE I

Families, subfamilies, genera, and species in the superfamily Callianassoidea, and their numbers

Subfamilies	Genera	Number of species
CALLIANASSIDAE		
Callianassinae	*Callianassa*	90
Callichirinae	*Callichirus*	5
	Glypturus	15
	Lepidophthalmus	13
	Michaelcallianassa	1
	Neocallichirus	37
	Podocallichirus	8
Eucalliacinae	*Calliax*	14
	Paraglypturus	1
Calliapaguropinae	*Calliapagurops*	2
Anacalliacinae	*Anacalliax*	3
Lipkecallianassinae	*Lipkecallianassa*	1
Bathycalliacinae	*Bathycalliax*	1
Paracalliacinae	*Paracalliax*	1
GOURRETIIDAE		
Gourretiinae	*Gourretia*	10
	Laurentgourretia	1
	Paragourretia	2
Callianopsinae	*Callianopsis*	3
Pseudogourretiinae	*Pseudogourretia*	1
CTENOCHELIDAE		
Ctenochelinae	*Ctenocheles*	9
TOTALS	20	218

Calliax and *Paraglypturus*; Calliapaguropinae with *Calliapagurops*; Anacalliacinae with *Anacalliax*; Lipkecallianassinae n. subfam. with *Lipkecallianassa*; Bathycalliacinae with *Bathycalliax*; and Paracalliacinae n. subfam. with *Paracalliax*. Then in the Gourretiidae: the Gourretiinae with *Gourretia, Laurentgourretia,* and *Paragourretia*; the Callianopsinae with *Callianopsis*; and the Pseudogourretiinae n. subfam. with *Pseudogourretia* n. gen.; and finally in the Ctenochelidae: the Ctenochelinae with *Ctenocheles*.

Following the anatomical study of the gastric mill of some Penaeidae by Kubo (1949: 157), a comparative analysis of this structure and an application of the resulting characters to the taxonomy of the Thalassinidea has been attempted elsewhere (Sakai, 2005). The results of that survey have been implemented in the present classification, as for as possible and desirable at the moment.

Superfamily CALLIANASSOIDEA Dana, 1852 [new sense]

Callianassidae Borradaile, 1903: 541; Pesta, 1918: 196.

Diagnosis. — Thalassinidea with a rostum of fair size, either triangular, or in the form of a spike, or else absent; linea thalassinica present. Eyestalks usually flattened and contiguous, rarely cylindrical. Maxilla 2 scaphognathite without a posterior seta. P1-2 chelate; P3 propodus oblong or broadened; P4 simple or subchelate; P5 simple, subchelate, or chelate. Abdominal somites 1-2 different in form from abdominal somites 3-5. Uropodal exopod without suture. The propyloric ossicle of the gastric mill is simple, lacking a series of transverse septa on its posterior surface.

Families included. — Callianassidae Dana, 1852, Gourretiidae Sakai, 1999, and Ctenochelidae Manning & Felder, 1991.

Remarks. — The superfamily Callianassoidea is different from the superfamily Thalassinoidea, because in Callianassoidea the P2 is chelate, and the propyloric ossicle of the gastric mill is simple, lacking a series of transverse septa on its posterior surface. By contrast, in the superfamily Thalassinoidea, including Thalassinidae Latreille, 1831, Upogebiidae Borradaile, 1903, and Laomediidae De Haan, 1849, the P2 is simple, and the propyloric ossicle is triangularly protruding downward, with its posterior surface concave and provided with a longitudinal median carina, both surfaces of which are fitted with a row of fine, transverse septa.

Based on the gastric mill character as here reported, also the families Axii-dae Huxley, 1879 and Callianideidae Kossmann, 1880, would rather have to be placed in the group with the callianassoid families than with those of the thalassinoids (Sakai, 2005). However, for the time being, this character state alone is not considered sufficient evidence for such an allocation, whence more characters will have to be taken into consideration in order to more soundly establish the true relationships, and hence the classification, within the infraorder Thalassinidea.

A new key to the 12 subfamilies and 20 genera is presented hereafter.

KEY TO SUBFAMILIES AND GENERA OF THE SUPERFAMILY CALLIANASSOIDEA

1. Carapace with dorsal oval 2
   – Carapace without dorsal oval 10
2. Abdominal somite 6 with acute lateral projections; uropodal exopod lacking lateral notch .... Callianopsinae - *Callianopsis*
   – Abdominal somite 6 without acute lateral projections; uropodal exopod bearing lateral notch 3
3. Rostral carina and cardiac prominence present .... Anacalliacinae - *Anacalliax*
   – Rostral carina and cardiac promincence absent 4
4. Male Plp2 present or absent; when present, pediform as usual .... Callianassinae - *Callianassa*
   – Male Plp2 usually present; when present, foliaceous as usual .... Callichirinae, 5
5. Abdominal somites 3-5 ornamented dorsally.. 6
   – Abdominal somites 3-5 not ornamented dorsally...................................... 7
6. Male and female Plp2 biramous .... *Callichirus*
   – Male and female Plp2 uniramous .... *Michaelcallianassa*
7. A1 peduncle longer and stouter than A2 peduncle 8
   – A1 peduncle equal to or shorter than A2 peduncle 9
8. Mxp3 ischium-merus narrow, pediform, with parallel lateral margins .... *Podocallichirus*
   – Mxp3 ischium-merus broadened, subrectangular .... *Lepidophthalmus*
9. Anterolateral spines of carapace present or absent, when present, proximally calcified .... *Neocallichirus*
   – Anterolateral spines of carapace proximally with non-calcified membrane .... *Glypturus*
10. Pleurobranchs present on P2-4 .... Pseudogourretiinae - *Pseudogourretia*
    – Pleurobranchs on P2-4 absent 11
11. Eyestalks cylindrical and set apart; A2 peduncle thick .... Calliapaguropinae - *Calliapagurops*
    – Eyestalks usually dorsoventrally flattened or subglobose, and contiguous; A2 peduncle not remarkably thick 12

12. Uropodal exopod with yellow-transparent, circular structure ....
.... Eucalliacinae - *Paraglypturus*
  – Uropodal exopod without yellow-transparent, circular structure 13
13. Mxp3 dactylus ovate 14
  – Mxp3 dactylus digitiform 15
14. Uropodal exopod with lateral notch .... *Calliax*
  – Uropodal exopod without lateral notch ..... Bathycalliacinae - *Bathycalliax*
15. P4 elongate, overreaching P2 .... Lipkecallianassinae - *Lipkecallianassa*
  – P4 not elongate, of same length as P2-3 16
16. Rostral carina present; hepatic carina and cardiac prominence present ....
.... 17
  – Rostral carina absent .... Gourretiinae - *Gourretia*
17. Major cheliped with palm subglobular and fingers elongate, pectinate........
.... Ctenochelinae - *Ctenocheles*
  – Major cheliped with palm oblong, chela of normal shape ....
.... Paracalliacinae - *Paracalliax*

FAMILY CALLIANASSIDAE DANA, 1852

Callianassidae Dana, 1852a: 12, 14; Dana, 1852b: 508; Bate, 1888: 27; Ortmann, 1891: 48; Stebbing, 1893: 183; Ortmann, 1899: 1142; Alcock, 1901: 197; Borradaile, 1903: 541; Pesta, 1918: 196; Schmitt, 1921: 114; De Man, 1928b: 18; Stevens, 1928: 318; Melin, 1939: 4; Bouvier, 1940: 100; Balss, 1957: 1581; Williams, 1965: 100; De Saint Laurent, 1973: 513; De Saint Laurent, 1979: 1395; De Saint Laurent & Le Loeuff, 1979: 46; Poore & Griffin, 1979: 254; Sakai, 1987a: 303; Sakai, 1988: 51; Poore, 1994: 101; Manning & Felder, 1991 [18 Dec.]: 766; Holthuis, 1991 [19 Dec.]: 239; Dworschak, 1992: 190; Hendrickx, 1995: 398, figs.; Sakai, 1999c: 7; Davie, 2002: 455-446; Sakai, 2002: 463; Sakai, 2004: 554.

Diagnosis. — Rostrum either more or less weakly developed, or well developed, sharp (*Lipkecallianassa*), unarmed laterally, and lacking rostral carina (exceptionally bearing a rostral carina in *Bathycalliax* and *Anacalliax*). Carapace with or without dorsal oval; linea thalassinica present. Abdominal somites 3-5 dorsolaterally with a tuft of setae. Eyestalks usually dorsoventrally flattened, or subglobose, or elongate and contiguous. A2 scaphocerite present or reduced, when present, pointed or obtuse and of small size. Maxilla 2 without posterior seta. Mxp3 pediform, subpediform, suboperculiform, or operculiform; propodus oblong or broadened, and dactylus digitiform or ovate. P1 chelate, equal, unequal, or subequal in size, and similar or dissimilar in shape; larger cheliped usually with or without proximal meral hook, palm oblong, and fingers of moderate length, not pectinate. P3 propodus often broadened in a heel shape or oblong. Plp1-2 present or absent, when present, smaller than Plp3-5. Male Plp2 with or without both appendix interna and appendix masculina, or only with appendix interna; female Plp2 with or without appendix interna. Plp3-5 with appendices internae in both sexes. Uropodal exopod usually without lateral notch (except in *Calliax*).

Type genus. — *Callianassa* Leach, 1814.

Remarks. — Under the present family concept, the Callianassidae are separated from the Callianideidae and Axiidae, because in the Callianassidae the maxilla 2 scaphognathite lacks a long posterior seta or setae, and Plp1-2 differs morphologically from Plp3-5 in size and shape, whereas in the Callianideidae and Axiidae the maxilla 2 scaphognathite bears a long posterior seta or setae, and Plp1 differs from Plp2-5. Tudge et al. (2000: 129) recently commented that: "in Ctenochelidae there is little justification for the subfamily arrangements recently proposed (by Sakai, 1999a)". However, the Ctenochelidae were not correctly defined in Manning & Felder's (1999) revision, and their concept has been uncritically followed by Tudge et al. (2000). It has long been considered that the family Ctenochelidae included *Anacalliax*, *Callianopsis*, *Ctenocheles*, *Gourretia*, and *Dawsonius*, with as a definition that the carapace lacks the dorsal oval and the uropodal exopod is simply ovate, without a secondary setal lobe (Manning & Felder, 1991: 784; Poore, 1994: 96; Tudge et al., 2000: 133). However, in contrast to the above-mentioned definiton, the carapace evidently bears a dorsal oval in *Anacalliax* and *Callianopsis*, though those genera have no secondary setal lobe on the uropodal exopod. As a consequence, they are located out of the Ctenochelidae in the present revision. In regard of cephalization, the characteristic of the dorsal oval would taxonomically be more important than that of the secondary setal lobe of the uropodal exopod. It is thus reasonable to separate *Ctenocheles* from *Anacalliax* and *Callianopsis*, and to treat the latter two genera at the same level as the subfamily Ctenochelinae, viz., as Anacalliacinae and Callianopsinae in the Callianassidae and Gourretiidae, respectively. It is to be considered that the Ctenochelinae are morphologically different from the earlier established Gourretiinae, recently raised to family level (Sakai, 2004a) as Gourretiidae, because in the Ctenochelinae a longitudinal rostral carina and a hepatic prominence are present, the scaphocerite is strong, the P3 propodus oblong, and the male Plp1 is uniramous with 3-4 segments, whereas in the Gourretiinae, a longitudinal rostral carina is absent, a hepatic prominence is absent, the scaphocerite is rudimentary, the P3 propodus broadened in a heel shape, and the male Plp1 is uniramous and distally chelate. As a consequence, Tudge et al.'s (2000) theory, resulting from a phylogenetic analysis that was not based on the actual character states to be observed, is here rejected and the Anacalliacinae and Callianopsinae are included in the families Callianassidae and Gourretiidae, respectively, with Gourretiidae and Ctenochelidae considered as separate families.

The subfamily Cheraminae Manning & Felder, 1991, is not accepted in the present work. Though the genus *Cheramus* was unequivocally characterized through the subsequent designation by Manning & Felder (1991) of a type species, i.e., *Cheramus occidentalis* Bate, 1888 [with as objective synonyms *Cheramus batei* Borradaile, 1903 and the now valid name *Callianassa profunda* Biffar, 1973], the genus *Cheramus* is, again, to be synonymized with the genus *Callianassa* Leach, 1814.

The subfamily Callichirinae is, however, accepted in the present callianassoids taxon, because in Callichirinae the Mxp3 propodus is broadened, the male Plp1-2 are blade-shaped, and the male Plp2 is often with an appendix interna and appendix masculina, whereas in the Callianassinae the Mxp3 propodus is typically oblong, the male Plp1-2 are absent or, when present, they are pediform, and the male Plp2 is without appendices interna and masculina.

Ngoc-Ho (2002: 539, 2003: 486) proposed that the genus *Calliapagurops* De Saint Laurent, 1973 should be placed in the subfamily Callichirinae Manning & Felder, 1991. However, *Calliapagurops* is to be located in the Calliapaguropinae, because it is different from other subfamilies in some characters: the eyestalks are cylindrical and set apart, and the A2 peduncles are extremely thick, whereas in the other subfamilies they are flattened, subglobose, or globose, and contiguous with each other, and the A2 peduncles are not remarkably thick.

As a result, the following 8 subfamilies, three of which new, are included in the family Callianassidae: the group in which the carapace bears a dorsal oval comprises: Callianassinae Dana, 1852; Callichirinae Manning & Felder, 1991; Eucalliacinae Manning & Felder, 1991; Calliapaguropinae Sakai, 1999c; and Anacalliacinae Manning & Felder, 1991; the second group, in which the carapace lacks a dorsal oval includes: Lipkecallianassinae n. subfam.; Bathycalliacinae Türkay & Sakai, 1999; and Paracalliacinae n. subfam.

Subfamily CALLIANASSINAE Dana, 1852

Callianassinae Dana, 1852: 12, 14; Bouvier, 1940: 100; Balss, 1957: 1582; Bǎcescu, 1967: 227; De Saint Laurent, 1973: 514; De Saint Laurent, 1979: 1395; Manning & Felder, 1991: 767; Sakai, 1999c: 10.
Cheraminae Manning & Felder, 1991: 780; Tudge et al., 2000: 136.

Definition. — Rostrum more or less weakly developed, or reduced, lacking rostral carina. Carapace with dorsal oval; cardiac prominence and cardiac sulcus usually absent. Eyestalks dorsoventrally flattened and contiguous. Abdominal somite 6 lacking lateral projections. A2 scaphocerite developed as a small process. Mxp3 ischium-merus variform, from pediform to operculiform; propodus oblong; dactylus digitiform. P1 chelate, unequal or subequal in size, and similar or dissimilar in shape; larger cheliped with or without meral hook. P2 chelate. P3 propodus broadened or oblong. Plp1 present or absent and Plp2 present or absent; Plp1-2 smaller than Plp3-5 in size and more slender in shape. Male Plp1 present or absent; when present, slender, and either biramous or uniramous. Male Plp2 present or absent; when present, slender, and either uniramous or biramous, lacking appendix interna and appendix masculina. Female Plp2 slender and biramous, lacking appendix interna. Plp3-5 biramous, foliaceous, with appendices internae in both sexes. Uropodal exopod with a secondary setal lobe.

Type genus. — *Callianassa* Leach, 1814.

Genera included. — *Callianassa* Leach, 1814.

Remarks. — Manning & Felder (1991) established a framework for the family of the American callianassids, including two subfamilies, Callianassinae Dana, 1852 and Cheraminae Manning & Felder, 1991. In their perception, the Callianassinae would consist of 10 genera, *Callianassa* Leach, 1814, *Trypaea* Dana, 1852, *Biffarius* Manning & Felder, 1991, *Neotrypaea* Manning & Felder, 1991, *Notiax* Manning & Felder, 1991, *Poti* Rodrigues & Manning, 1992, *Gilvossius* Manning & Felder, 1992, *Nihonotrypaea* Manning & Tamaki, 1998, *Necallianassa* Heard & Manning, 1998, and *Pseudobiffarius* Heard & Manning, 2000, and the Cheraminae comprise two genera, *Cheramus* Bate, 1888 and *Scallasis* Bate, 1888. In the present revision, however, it is considered that those 12 genera are to be brought into one genus, *Callianassa*, in the subfamily Callianassinae.

Poore (1994) and Tudge et al. (2000) followed Manning & Felder's (1991) revision, and Tudge et al. (2000) used a phylogenetic analysis for their revision. However, that cladistic analysis was applied to the existing taxa, without taking into account that some morphological variations are not decisive enough to be regarded as reliable character states at a generic level: series of variforms observed in the Mxp3 ischium-merus, in the rostrum, and in Plp1-2/3-5. Such variforms were, instead, evaluated at the same level as generic or higher taxonomic criteria, and no precise revision of the world-wide species assemblage was made by Tudge et al. (2000). Instead of accepting the many American genera established in the past decade, they should first have evaluated these critically.

The genus *Biffarius* Manning & Felder, 1991 is based on the type species, *B. biformis* (Biffar, 1971). Manning & Felder (1991: 769) mentioned in their remarks that: "The other genera of the Callianassinae differ from *Biffarius* as follows: *Callianassa* has a pediform Mxp3; *Neotrypaea* and *Trypaea* have the merus of Mxp3 projecting beyond its articulation with the carpus, and *Notiax* has a distinct rostral spine". However, there is no reference to other species of the Callianassinae and there is no mention made of how Mxp3 and rostral spine are different within *Biffarius*. Tudge et al. (2000: 141, 142) listed 10 species in *Biffarius*: *B. arenosus* (Poore, 1975), *B. australis* (Kensley, 1974), *B. biformis* (Biffar, 1971), *B. ceramicus* (Fulton & Grant, 1906), *B. debilis* Hernández-Aguilera, 1998, *B. delicatulus* Rodrigues & Manning, 1992,

B. diaphora (Le Loeuff & Intès, 1974), *B. fragilis* (Biffar, 1970), *B. lewtonae* (Ngoc-Ho, 1994), and *B. limosa* (Poore, 1975), and Davie (2002: 457) listed five Australian species, i.e., as *B. arenosus* (Poore, 1975), *B. ceramicus* (Fulton & Grant, 1906), *B. lewtonae* (Ngoc-Ho, 1994), *B. limosa* (Poore, 1975), and *B. poorei* (Sakai, 1999).

In *Biffarius biformis*, the Mxp3 ischium-merus is broadly rounded in shape, and the merus is slightly convex on the mesiodistal angle, so it is evidently different from that of *Callianassa subterranea*, the type species of *Callianassa*, in which it is pediform, and the merus is simply convex on the distal margin (fig. 2B). In *C. californiensis* (*Neotrypaea californiensis* of Manning & Felder, 1991), the type species of *Neotrypaea*, and *C. australiensis* (*Trypaea australiensis* of Dana, 1852), the type species of *Trypaea*, the merus is entirely projecting beyond the carpus on the mesodistal margin (fig. 1B, D). However, in *Callianassa acutirostella*, *C. amboinensis*, *C. bouvieri*, *C. filholi*, *C. japonica* (*Nihonotrypaea japonica* of Manning & Tamaki, 1998) (fig. 1F), *C. lewtonae* (fig. 1H), *C. poorei*, and *C. whitei*, it is also largely convex as in *C. californiensis*. It is thus possible to say, that the form of the Mxp3 ischium-merus is variform, so that it is difficult to separate *Biffarius* from *Callianassa* by the form of the Mxp3.

The shape of the Mxp3 ischium-merus is not correlated with the relative length of the A1-2 peduncles: in *C. californiensis*, *C. australiensis*, and *C. amboinensis* the Mxp3 ischium-merus is operculiform, and the A1 peduncle is distinctly longer than that of the A2 (fig. 1A, B, C, D), while in *C. acutirostella*, *C. arenosa* (removed to *Biffarius* by Tudge et al., 2000), *C. bouvieri*, *C. convexa*, *C. filholi*, *C. lewtonae* (removed to *Biffarius* by Tudge et al., 2000), *C. poorei*, and *C. whitei*, the Mxp3 ischium-merus is also operculiform, but the A1 peduncle is slightly longer than the A2 peduncle (fig. 1E, F, G, H).

The correlation in the relative lengths of the A1-2 peduncles is also not referred to as one of the characters of *Biffarius*. In *Biffarius biformis*, the A1 peduncle is shorter than the A2 peduncle, as in *C. ceramica* and *C. filiformis*, while in *C. acutirostella*, *C. amboinensis*, *C. bouvieri*, *Neotrypaea californiensis*, *C. lewtonae*, *C. poorei*, and *C. whitei*, the A1 peduncle is longer than that of the A2, and in *C. australiensis* the A1 peduncle is much longer than the A2 peduncle.

The correlation in the various forms of Plp1-2 is not referred to as a character of the genus *Biffarius*, either. In *C. arenosa* (*Biffarius arenosus* of Tudge et al., 2000) the male Plp1 is uniramous and biarticulate, and the male Plp2 is

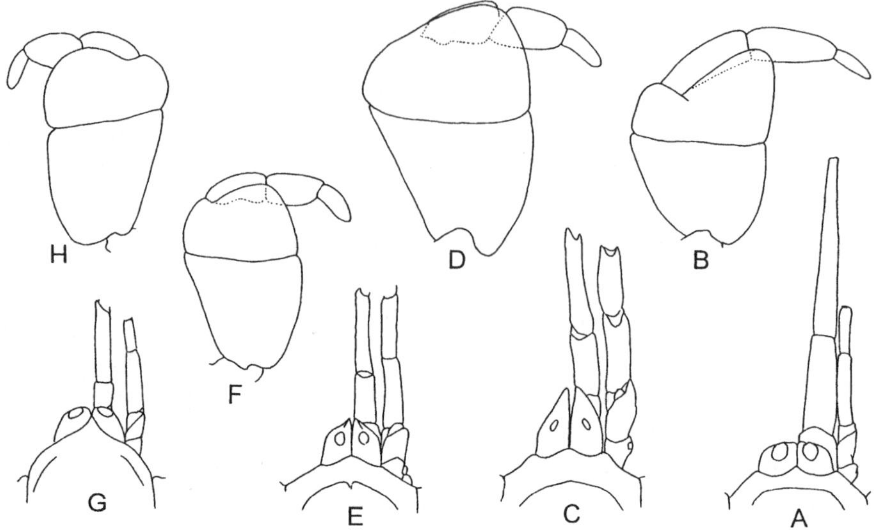

Fig. 1. The form of the A1-2 peduncles and Mxp3 in *Callianassa australiensis* Dana, 1852, *C. californiensis* Dana, 1854, *C. japonica* Ortmann, 1891, and *C. lewtonae* Ngoc-Ho, 1994, respectively. A, B, *Callianassa australiensis*, the type species of *Trypaea* by Dana (1852); C, D, *C. californiensis*, the type species of *Neotrypaea* Manning & Felder, 1991; E, F, *C. japonica*, the type species of *Nihonotrypaea* Manning & Tamaki, 1998; G, H, *C. lewtonae* Ngoc-Ho, 1994. In each case, a series could be recognized from right to left.

absent; the female Plp1 is uniramous, and the female Plp2 is biramous and filiform (Poore & Griffin, 1979, fig. 15); in *C. ceramica* (*Biffarius ceramicus* of Tudge et al., 2000) the male Plp1 is uniramous and biarticulate, and the male Plp2 is uniramous and narrow, but often absent, while the female Plp1 is uniramous and uniarticulate, and the female Plp2 is biramous and filiform (Poore & Griffin, 1979: 260, fig. 23); in *C. lewtonae* (*Biffarius lewtonae* of Tudge et al., 2000) Plp1-2 are unknown in both sexes; in *C. limosa* (*Biffarius limosa* of Tudge et al., 2000), the male Plp1 is uniramous and the male Plp2 is a minute, medially lobed, tapering papilla; the female Plp1 is uniramous, and the female Plp2 biramous and filiform (Poore, 1975: 203, fig. 5); and in *B. poorei* (*Biffarius poorei* of Tudge et al., 2000), the male Plp1 is uniramous and biarticulate, and the male Plp2 is absent; the female Plp1 is uniramous and triarticulate, and the female Plp2 is biramous (Sakai, 1999b: 374, fig. 1A). It can be safely said, that Plp1-2 are variform in *Biffarius*, because in *C. arenosa* and *C. poorei* the male Plp2 is absent, whereas in *C. ceramica* and *C. limosa* it is uniramous or merely a papilla. In *C. arenosa*, *C. lewtonae*, and *C. poorei* the A1 peduncle is longer than the A2 peduncle, while in *C. ceramica* and *C. limosa* the A1 peduncle is much shorter than the A2 peduncle.

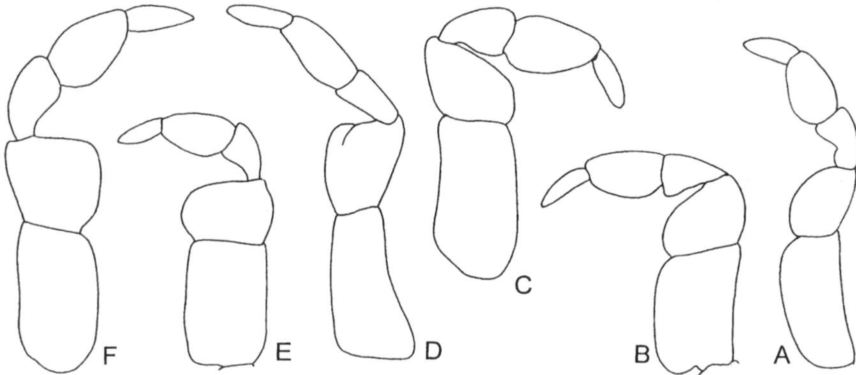

Fig. 2. The Mxp3 merus is observed to have gradually changed from the convex distal margin in *Callianassa stenomastaxa* to the concave one in *Callianassa profunda*. A, *Callianassa stenomastaxa* Sakai, 2002; B, *Callianassa subterranea* (Montagu, 1808), type species of the genus *Callianassa*; C, *Callianassa malaccaensis* Sakai, 2002; D, *Callianassa tonkinae* Grebenjuk, 1975; E, *Callianassa oblonga* Le Loeuff & Intès, 1974; F, *Callianassa profunda* Biffar, 1973.

In *Callianassa malaccaensis* the A1 peduncle is shorter than the A2 peduncle, as in *Biffarius biformis*, the type species of *Biffarius*, though the Mxp3 is pediform as in *Callianassa subterranea*; the rostrum is sharply pointed as in *Notiax brachyophthalma*, the type species of *Notiax*; and Plp3-5 are elongate and slender, bearing horn-shaped appendices internae of a form intermediate between stubby and finger-like, which are different from the stubby one in *Biffarius*, *Callianassa*, and *Notiax*. This proves that those characters are inconsistent in *Biffarius*, so it is difficult to regard *Biffarius* as a good genus in the subfamily Callianassinae.

Gilvossius setimanus (DeKay, 1844), the type species of *Gilvossius* Manning & Felder, 1992, is to be synonymized with *Callianassa* as shown by Sakai (1999c). Manning & Felder (1992: 559) mentioned that *Gilvossius* resembles *Callichirus* Stimpson, 1866 and *Lepidophthalmus* Holmes, 1904, but differs from all the other recognized genera of callianassids in the western Atlantic Ocean. However, *Gilvossius setimanus* has an elongate A1 peduncle that is much longer than the A2 peduncle as in *Callianassa australiensis*, and has the oblong Mxp3 propodus as in *Callianassa subterranea*, whereas *Callichirus* and *Lepidophthalmus* have the broadened Mxp3 propodus. *Gilvossius* is characterized by having no Plp1-2 in males, as in *Callianassa fragilis*, *C. marchali*, and *C. tyrrhena*, and no other characters can be found to separate it from other callianassids. So, *Gilvossius* is here to be considered a junior synonym of *Callianassa*. DeKay's (1844) figure of *Callianassa setimana* must be based on a female specimen, because it bears Plp1-2 (see Manning, 1987, fig. 1).

Trypaea Dana, 1852 is established by the type species, *Trypaea australiensis* Dana, 1852, but it is here treated as a junior synonym of *Callianassa*. Manning & Felder (1991) show two characteristic points in the definition of the genus *Trypaea*: (1) A1 peduncle longer and stouter than A2 peduncle; (2) Mxp3 ischium-merus operculiform; merus projecting beyond articulation with carpus. As mentioned in the remarks on the genus *Biffarius*, those characters are variform and are not sufficiently decisive to separate *Trypaea* from the other callianassids. Borradaile (1903) placed those two type species, i.e., *C. australiensis* and *C. californiensis* Dana, 1854 in the subgenus *Trypaea*. Plp1-2 are, however, different in *Trypaea australiensis* and *Neotrypaea californiensis*. In *T. australiensis*, the male Plp2 is absent; the female Plp1 is uniramous and 2-segmented, the second segment with a small lobe midway along its length (Poore & Griffin, 1979: 254); in *N. californiensis* the male Plp2 is absent; the female Plp1 is uniramous and 4-segmented (judged by the specimens present in the collections of the LACM). However, such inconsistency is also observed in the species of the genus *Biffarius*, so it is to be concluded that the genera *Trypaea* and *Neotrypaea* are not specific in the form of Plp1-2.

Pseudobiffarius caesari Heard & Manning, 2000, the type species of *Pseudobiffarius* by monotypy, is also identified here as belonging in *Callianassa*, whence *Pseudobiffarius* becomes a junior synonym of *Callianassa*. Heard & Manning (2000: 71) mentioned in their remarks that "Members of *Pseudobiffarius* can be distinguished at once from the American genera [*Biffarius, Neotrypaea, Notiax, Gilvossius, Necallianassa*, and *Poti*, all of which are here removed to *Callianassa*] with broad Mxp3, by the short, stout dorsal flagellum of the A1. They further differ from both *Neotrypaea* and *Gilvossius* in having the appendices internae of the Plp3-5 embedded in the edge of the pleopod, whereas they are projecting in *Pseudobiffarius*, as in *Biffarius, Necallianassa*, and *Notiax*. In *Notiax* male plp2 is present, whereas it is absent in *Pseudobiffarius*; members of *Notiax* also have a strong rostral spine extending almost beyond the cornea and a median distal spine on the telson. Members of *Biffarius* lack the strong ridge of teeth present on the inner margin of the Mxp3 in members of *Pseudobiffarius*". However, those characters, i.e., the A1 dorsal flagellum, Mxp3, Plp2, Plp3-5 appendices internae, and spinulation on the telson, are all variform, and it is difficult to separate *Pseudobiffarius* from other callianassids by those characters, as follows:

Callianassa amboinensis from the Indo-West Pacific region bears the broad Mxp3, the short dorsal flagellum of the A1, and the telson with a posterior median spine as in *Pseudobiffarius caesari*. In *P. caesari*, the male Plp2 is absent, as in *Callianassa acanthura, C. arenosa, C. australiensis, C. biformis, C. diaphora, C. filholi, C. fragilis, C. japonica, C. marchali, C. marginata, C. oblonga, C. sibogae, C. truncata, C. tyrrhena,* and *C. whitei.* The acute rostral spine is also found in many callianassids as in *Pseudobiffarius caesari,* i.e., in *Callianassa brachyophthalma* (transferred to *Notiax* by Manning & Felder, 1991), *C. acutirostella, C. amboinae* (the type species of *Scallasis* Bate, 1888), *C. amplimaxilla, C. anopleura, C. biformis* (transferred to *Biffarius* by Manning & Felder, 1991), *C. brevirostris, C. bouvieri, C. chakratongae, C. contipes, C. gravieri, C. gruneri, C. marginata, C. truncata, C. intermedia, C. joculatrix, C. lewtonae, C. limosa, C. lobetobensis, C. longicauda, C. malaccaensis, C. maldivensis, C. matzi, C. nieli, C. nigroculata, C. oblonga, C. orientalis* (moved from *Cheramus* in the present paper, see p. 20), *C. parvula, C. poorei, C. praedatrix, C. profunda, C. propinqua, C. propriopedis, C. pugnatrix, C. pygmaea, C. sibogae, C. spinophthalma, C. stenomastaxa,* and *C. tonkinae.* Plp3-5 have appendices internae, stubby and projecting in *Pseudobiffarius caesari* as in *Biffarius* and *Necallianassa* (transferred to *Callianassa*), but those are also found in *Callianassa subterranea,* the type species of *Callianassa, C. japonica* (*Nihonotrypaea* in Manning & Tamaki, 1998), and *C. stenomastaxa.* As mentioned above, though the rostrum is also sharply pointed in *C. anoploura* Sakai, 2002, as in *P. caesari,* Plp3-5 are slender, bearing narrow and projecting appendices internae; though the rostrum is in the form of a distinct rostral spine in *C. brachyophthalma* as in *P. caesari,* Plp3-5 are foliaceous, bearing stubby and projecting appendices internae (Manning & Felder, 1991: 773); the rostrum is acute in *C. biformis,* as in *P. caesari,* but Plp3-5 are foliaceous with stubby and projecting appendices internae (Manning & Felder, 1991: 769); and the rostrum is a sharp rostral spine in *C. berylae* (*Necallianassa* in Heard & Manning, 1998), Plp3-5 are probably foliaceous, with stubby and projecting appendices internae, though Plp3 is figured without description of whether it is foliaceous or slender (Heard & Manning, 1998, fig. 3). As a result, there are no decisive characters at all that would make it possible to separate *Pseudotrypaea* from other callianassids, so *Pseudotrypaea* is to be treated as a junior synonym of *Callianassa.*

Tudge et al. (2000: 142) listed 10 species in *Biffarius,* 46 species in *Callianassa,* one species in *Calliapagurops,* one species in *Gilvossius,* three species in *Necallianassa,* five species in *Neotrypaea,* three species in *Nihonotrypaea,* one species in *Notiax,* one species in *Poti,* and one species in *Trypaea.* However, some of those species are removed to other genera or subfamilies in the present paper. The genera of the subfamily Callianassinae as classified by Tudge et al. (2000) are listed here with removed genera and species, as follows:

(1) Genus *Biffarius* with 12 species (removed to *Callianassa*) that have already been cited in the remarks on *Biffarius*, above.

(2) Genus *Callianassa* Leach, 1814: *C. acutirostella*; *C. amboinensis*; *C. assimilis* (removed to *Glypturus*); *C. audax* (removed to *Neocallichirus*); *C. bouvieri*; *C. brevicaudata* (junior syn. of *Neocallichirus mucronatus*); *C. calmani* (removed to *Neocallichirus*); *C. candida*; *C. chilensis* (junior syn. of *Callianassa uncinata*); *C. convexa*; *C. coutierei* (removed to *Glypturus*); *C. cristata* (junior syn. of *Callianassa gravieri*); *C. filholi*; *C. gilchristi* (removed to *Podocallichirus*); *C. grandidieri* (removed to *Lepidophthalmus*); *C. gravieri*; *C. intermedia*; *C. joculatrix*; *C. kewalramanii* (junior syn. of *Podocallichirus masoomi*); *C. lignicola*; *C. lobetobensis*; *C. madagassa* (removed to *Podocallichirus*); *C. maldivensis* (junior syn. of *Callianassa bouvieri*); *C. marchali*; *C. masoomi* (removed to *Podocallichirus*); *C. mauritiana*; *C. maxima*; *C. modesta*; *C. nakasonei* (junior syn. of *Glypturus martensi*); *C. parva*; *C. parvula*; *C. pixii* (removed to *Anacalliax*); *C. pontica* (junior syn. of *Callianassa candida*); *C. pugnatrix*; *C. pygmaea*; *C. rosae* (removed to *Lepidophthalmus*); *C. rotundicaudata*; *C. subterranea*; *C. tonkinae*; *C. tyrrhena*; *C. variabilis* (syn. of *Neocallichirus indicus*); *C. vigilax* (removed to *Neocallichirus*); *C. winslowi* (removed to *Glypturus*).

(3) Genus *Calliapagurops* De Saint Laurent, 1973 (removed to the subfamily Calliapaguropinae in the family Gourretiidae): *C. charcoti*.

(4) Genus *Gilvossius* Manning & Felder, 1991 (removed to *Callianassa*): *G. setimanus*.

(5) Genus *Necallianassa* Heard & Manning, 1998 (removed to *Callianassa*): *N. acanthura* (removed to *Callianassa*); *N. berylae* (removed to *Callianassa*); *N. truncata* (removed to *Callianassa*).

(6) Genus *Neotrypaea* Manning & Felder, 1991 (removed to *Callianassa*): *N. biffari* (removed to *Callianassa*); *N. californiensis* (removed to *Callianassa*); *N. gigas* (removed to *Callianassa*); *N. rochei* (removed to *Callianassa*); *N. uncinata* (H. Milne Edwards, 1837) (removed to *Callianassa*).

(7) Genus *Nihonotrypaea* Manning & Tamaki, 1998 (removed to *Callianassa*): *N. harmandi* (Bouvier, 1901) (junior syn. of *Callianassa japonica*); *N. japonica* (removed to *Callianassa*); *N. petalura* (removed to *Callianassa*).

(8) Genus *Notiax* Manning & Felder, 1991 (removed to *Callianassa*): *N. brachyophthalma* (removed to *Callianassa*).

(9) Genus *Trypaea* Dana, 1852 (removed to *Callianassa*): *T. australiensis* (removed to *Callianassa*).

Davie (2002) followed Tudge et al. (2000) in his generic recognition of the Australian Callianassidae and Ctenochelidae, and included the genera, *Biffarius, Calliax, Cheramus, Corallianassa*, and *Trypaea* in the Callianassidae, though all of those genera except *Calliax* (removed to the subfamily Callichirinae) were synonymized with *Callianassa* by Sakai (1999c). Davie (2002: 455) mentioned that "Recently, Cheraminae and Callichirinae were placed into synonymy with Callianassinae by Sakai (1999c). It is not clear that a final consensus on subfamily definitions has emerged, and as there are relatively few genera in Australian waters, ...". The present author recognizes Callichirinae in the present paper (see the remarks on the family Callianassidae and the subfamily Callichirinae), but Cheraminae is considered a synonym of Callianassinae, because the genus *Cheramus* is herein synonymized with *Callianassa*. Therefore, there is no good reason to establish the subfamily Cheraminae for Manning & Felder's (1991) incorporated taxa.

L. B. Holthuis gave me his notion about the mess concerning the names *Cheramus occidentalis* and *Ch. orientalis* and their homonyms and replacement names, which is to be finally cleared up: "... Bate (1888) described two species, *Callianassa occidentalis* (on p. 29) and *Cheramus occidentalis* (on p. 32). Borradaile [1903: 545] evidently synonymized the two (but did not distinctly say so) and placed them in the genus *Callianassa* subgenus *Cheramus*. He furthermore showed that they were junior homonyms of *Callianassa occidentalis* Stimpson, 1856, and of each other. Because Borradaile indicated both the genera *Cheramus* and *Callianassa* as "*C*.", it is not clear for which species exactly the new name *C. Batei* was meant, probably for both. Both Biffar (1973) and Manning & Felder (1991) considered Borradaile's name a replacement name for *Cheramus occidentalis*, so that we do best to follow them. The type specimen of *Cheramus occidentalis* Bate, 1888, is now also the type specimen of *Callianassa batei* Borradaile, 1903. This specimen therefore is also the type specimen of *Callianassa profunda* Biffar, 1991 [recte: 1973], which thus now becomes the valid name of the type species of the genus *Cheramus*, as it is the valid name of the species originally described by Bate, 1888, as *Cheramus occidentalis*, which was selected for the type of that genus by Manning & Felder, 1991. No application to the nomenclature Commission is necessary. As to *Cheramus* [*orientalis*] my selection of it as the type of *Cheramus* was published one day after the selection by Manning & Felder, 1991, and thus is invalid. Biffar (1973, p. 229) pointed that *Cheramus orien-*

talis (Bate, 1888), is a junior homonym of *Callianassa orientalis* A. Milne-Edwards, 1860; but, as far as I know, no one who considered *Cheramus* and *Callianassa* synonyms has proposed a replacement name for it. Manning & Felder (1991) of course did not as they considered the two species to belong to two different genera and the homonymy is only secondary. Personally, I do not think that the type selection made for *Cheramus* by Manning & Felder is incorrect or undesirable. And therefore there is no good reason to change it." (L. B. Holthuis, in litt., 13 February 2003).

As to *Cheramus orientalis*, Holthuis (19 Dec. 1991: 239) selected it as the type species of *Cheramus* one day after the selection by Manning & Felder (18 Dec. 1991: 780) and that selection is thus invalid. Therefore, the type of *Cheramus* is to be conceived as designated by Manning & Felder (1991: 780), as *Cheramus occidentalis* Bate, 1888, with as objective synonyms *Ch. batei* Borradaile, 1903 [pre-occupied] and the now valid name *Callianassa profunda* Biffar, 1973. As the morphological characters of *C. profunda* are considered insufficient to warrant a genus separate from *Callianassa*, the genus *Cheramus* is to be synonymized, again, with the genus *Callianassa* Leach, 1814.

To arrive at this conclusion, the concept of the genus *Cheramus* was re-examined, because *Ch. orientalis*, on the one hand, and the type species of *Cheramus, C. profunda*, in the sense of both Biffar's (1973) and Manning & Felder's (1991) designation, are clearly different species.

Manning & Felder (1991: 780) mentioned on *Cheramus batei* Borradaile, 1903 (= *Callianassa profunda*) in their remarks on *Cheramus*, that: "The long slender body with a relatively small thoracic region, strong rostral spine, pediform Mxp3, and extremely elongate uropods are characteristic of members of this genus". Their remarks are self-contradictory, because they mentioned also: "Uropodal exopod elongate, length more than 2.0 times as long as width", which is applicable to *Cheramus orientalis*, but not to *C. profunda*, because in *C. profunda* the uropods are not so elongate as in *Cheramus orientalis*, and are broadened, 1.3 times as long as wide (Biffar, 1973, fig. 1b).

Further confusion arose, as Davie (2002: 459) included three Australian species, *Callianassa praedatrix* De Man, 1905, *C. propinqua* De Man, 1905, and *C. sibogae* De Man, 1905 in the genus *Cheramus*. However, those three species are not to be integrated in the genus *Cheramus* for the form of the uropodal exopod, of the male Plp2, the meral hook on the major cheliped, and the P3 propodus. The uropodal exopod is 1.5 times as long as wide, truncate distally in *C. praedatrix*, and the uropodal exopod is 1.8 times as long as wide, and truncate distally in *C. propinqua*, while it is oval rather than elongate, and

1.5 times as long as wide in *C. sibogae* (cf. Ngoc-Ho, 1994, fig. 3h). Manning & Felder (1991: 780) included *Callianassa orientalis* (Bate, 1888) (*Cheramus* in Bate, 1888), *C. oblonga* (Le Loeuff & Intès, 1974), and *C. profunda* Biffar, 1973 in the genus *Cheramus*, but the uropodal exopod is 1.8 times as long as wide in *C. orientalis* (cf. Bate, 1888, pl. 1 fig. 2), it is 2.0 times as long as wide in *C. oblonga* (fig. 4E) (Le Loeuff & Intès, 1974, fig. 9q), and the uropodal exopod is rounded, 1.3 times as long as the endopod in *C. profunda* (fig. 4F) (Biffar, 1973, fig. 1b), so that the length of the uropodal exopod is not a good character to define the genus *Cheramus*.

The form of the male Plp2 is different in *C. sibogae* and two species of *Callianassa*, *C. praedatrix* and *C. propinqua*. In *C. praedatrix* De Man, 1905, the male Plp2 is biramous; in *C. propinqua* De Man, 1905 the male Plp2 is biramous and very narrow (Ngoc-Ho, 1991: 292, fig. 4); but in *C. sibogae* De Man, 1905 the male Plp2 is absent. The character of the meral hook on the major cheliped is inconsistent in the genus *Cheramus*. In *C. praedatrix* and *C. propinqua* the major cheliped has no meral hook, while in *C. sibogae* it is armed with a single meral spine (Ngoc-Ho, 1994, fig. 3e). The P3 propodus is oblong, bearing no posterior lobe in *C. profunda* and *C. sibogae*, whereas it is short and broadened, bearing no posteroventral angle in *C. praedatrix* and *C. propinqua*.

Mxp3 ischium-merus, P3 propodus (fig. 3), Plp3-5 appendices internae, and the uropodal exopod (fig. 4) are variform in the Callianassidae. In *Callianassa orientalis* (Bate, 1888), earlier in the genus *Cheramus*, the Mxp3 ischium-merus is elongate with a truncate distal margin (De Man, 1928a, pl. 1 fig. 2a), and such a form of Mxp3 ischium-merus is also found in *Callianassa amboinae, C. gaucho, C. lobetobensis, C. malaccaensis, C. marginatus, C. nigroculata, C. oblonga, C. profunda, C. pugnatrix, C. pygmaea, C. sibogae*, and *C. tonkinae*. While it is simply elongate in *C. stenomastaxa* Sakai, 2002 (fig. 2A), *C. matzi* Sakai, 2002, *C. chakratongae* Sakai, 2002, and *C. tenuipes* Sakai, 2002 from the Andaman Sea. In *C. profunda* (fig. 3F) and *C. sibogae*, the Plp3-5 appendices internae are elongate, finger-like, and bear hooks distally (Biffar, 1973: 227), but in *C. sibogae* the Plp3-5 appendices internae are unknown. In *C. gaucho*, the type species of *Poti*, the P3 propodus is short, the posterior margin lacks a lobe, and tapers towards the dactylus; the uropodal exopod has a broadly rounded distal margin; and Plp3-5 are foliaceous, bearing stubby appendices internae (Rodrigues & Manning, 1992b). In *Callianassa marginata*, the P3 propodus is narrow and elongate, lacking a posterior lobe;

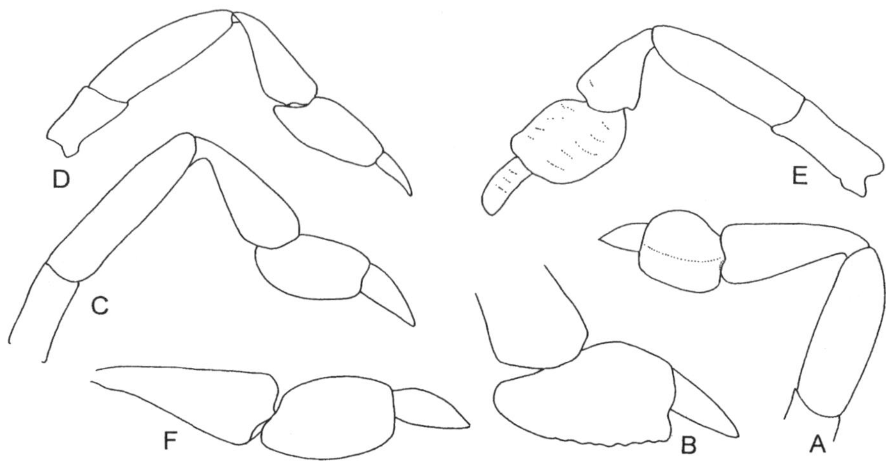

Fig. 3. The P3 propodus is observed to have gradually changed from a quadrate form in *Callianassa stenomastaxa* Sakai, 2002 to an oblong form in *Callianassa profunda* Biffar, 1973. A, *Callianassa stenomastaxa* Sakai, 2002; B, *Callianassa subterranea* (Montagu, 1808), type species of the genus *Callianassa*; C, *Callianassa malaccaensis* Sakai, 2002; D, *Callianassa tonkinae* Grebenjuk, 1975; E, *Callianassa oblonga* Le Loeuff & Intés, 1974; F, *Callianassa profunda* Biffar, 1973.

the uropodal exopod is narrow and elongate, 2.5 times as long as wide; the Plp3-5 appendices internae are rod-shaped, bearing hooks distally (Biffar, 1971: 694). In *Callianassa oblonga* the P3 propodus is square in shape (fig. 3E), the posterior margin bearing a posterior lobe; the uropodal exopod is narrow and elongate, 2.0 times as long as wide; Plp3-5 are unknown. In *Callianassa amboinae*, the type species of *Scallasis*, and in *C. pugnatrix*, *C. pygmaea*, and *C. tonkinae* (fig. 3D), the P3 propodus is oblong, the posterior margin with a small posterior lobe; the uropodal exopod is larger than the endopod, and rounded on the distal margin; Plp3-5 are narrow, bearing thick, finger-like appendices internae. In *Callianassa lobetobensis* and *C. malaccaensis* (fig. 3C) the P3 propodus is oblong, and the posterior margin bears a posterior lobe; the uropodal exopod is larger than the endopod, largely truncate on the distal margin (fig. 4C); Plp3-5 are narrow, with stubby appendices internae. In *Cheramus orientalis*, the P3 propodus is unknown, and the uropodal exopod is larger than the endopod and truncate distally; Plp3-5 are unknown. In *Callianassa nigroculata* the P3 propodus is oblong, the posterior margin bearing a small posterior lobe; the uropodal exopod is oval, 1.8 times as long as wide, and Plp3-5 are elongate, bearing thumb-like appendices internae.

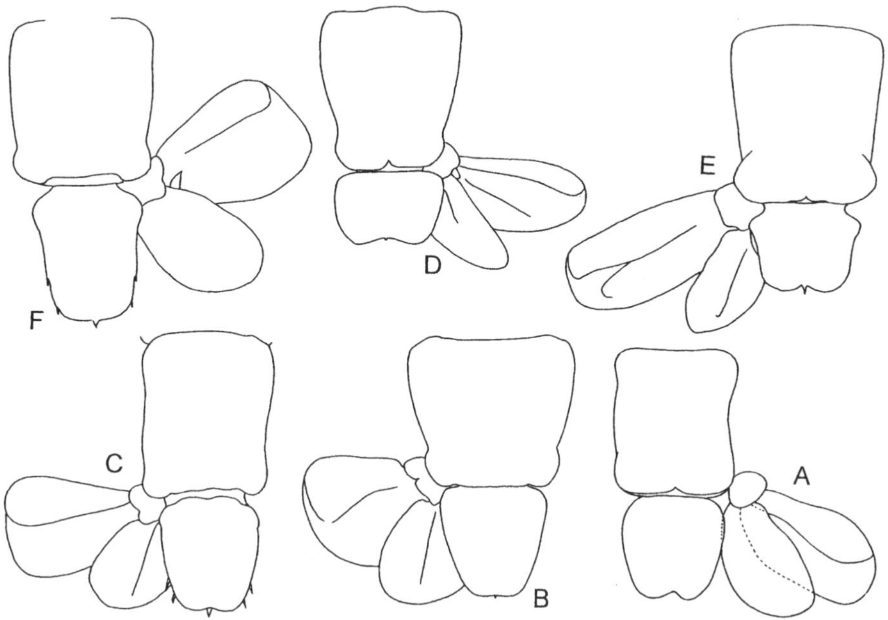

Fig. 4. The uropodal exopod has gradually changed from a square form in *Callianassa steno-mastaxa* Sakai, 2002 to an oblong form in *Callianassa oblonga* Le Loeuff & Intès, 1974. A, *Callianassa stenomastaxa* Sakai, 2002; B, *Callianassa subterranea* (Montagu, 1808), type species of the genus *Callianassa*; C, *Callianassa malaccaensis* Sakai, 2002; D, *Callianassa tonkinae* Grebenjuk, 1975; E, *Callianassa oblonga* Le Loeuff & Intès, 1974; F, *Callianassa profunda* Biffar, 1973.

Tudge et al. (2000: 145) defined *Callianassa longicauda* (Sakai, 1967), *C. rectangularis* (Ngoc-Ho, 1991), *C. spinophthalmus* (Sakai, 1970), *C. praedatrix* (De Man, 1905), and *C. propinqua* (De Man, 1905) as *Cheramus* in the sense of Manning & Felder, 1991, while the latter two were also included as *Cheramus* by Davie (2002). However, in *Callianassa longicauda* the Mxp3 ischium-merus is pediform, the distal margin bearing a median tooth; the P3 propodus is oblong, and the posterior margin lacks a posterior lobe; the uropodal exopod is 1.8 times as long as wide, and truncate distally. In *Callianassa rectangularis* the Mxp3 ischium-merus is oval, the mesiodistal angle of the meral projection over the articulation with the carpus is as in *Neotrypaea*; the P3 propodus is broadened in a heel shape; the uropodal exopod is broadened, 1.2 times as long as wide. In *Callianassa spinophthalmus* the Mxp3 ischium-merus is subsquare, the merus is slightly concave on the distal margin; the P3 propodus is triangular, the posterior margin is convex, tapering towards the dactylus; the uropodal exopod is elongated, 1.8 times as long as wide. Mxp3 ischium-merus, P3 propodus, and uropodal exopod are thus variform. It is clear that the form of the telson is not different between *Cheramus orientalis* and *Callianassa profunda*, and, therefore, the subfamily Cheraminae is to be included in the subfamily Callianassinae, as the genus *Cheramus* is invalid in the subfamily Callianassinae.

The genus *Scallasis* Bate, 1888 is based on the type species, *S. amboinae* Bate, 1888. Manning & Felder (1991) defined the species as follows "Diagnosis. Carapace with rostral spine, apparently lacking orbit. Cornea well formed, terminal, subglobular. Mxp3 with exopod, Plp elongate and slender, with finger-like appendices internae". Sakai (1999c) reported on the type specimens of *Scallasis amboinae* Bate, 1888, and, after careful examination, it is referred to as *Callianassa amboinae*.

The genus *Pestarella* Ngoc-Ho, 2003 was established by the type species, *Astacus tyrrhenus* Petagna, 1792 from the species of the genus *Callianassa*. This genus is characterized by the absence of male Plp1 and Plp2, the telson rounded in the posterior half, and the operculiform Mxp3, and includes *Callianassa candida* Olivi, 1792, *C. rotundicaudata* Stebbing, 1902, *C. convexa* De Saint Laurent & Le Loeuff, 1979, and *C. whitei* Sakai, 1999. However, Ngoc-Ho (2003) did not include *Callianassa setimanus* (DeKay, 1844), the type species of the genus *Gilvossius*. Manning & Felder (1992) established *Gilvossius* (cf. Manning & Felder, 1992: 558, fig. 1) by the type species *Gonodactylus setimanus* DeKay, 1844, but the genus *Gilvossius* is characterized by a long, stout peduncle of the antennule. *Callianassa setimana* shows the same characters as *Pestarella*: absence of Plp1-2, rounded Mxp3, and telson rounded in the posterior half. Manning & Felder (1991) established *Gilvossius* through a comparison with the species of *Callichirus* and *Lepidophthalmus*, which are considered to be different taxa in the subfamily Callichirinae, so the comparison made by Manning & Felder (1991) is neither adequate nor relevant. In consequence, *Gilvossius setimanus* is to be classified as *Callianassa setimanus*.

Obviously, the morphology of Plp1-2 is variable in the large group of species now constituting the genus *Callianassa*, and it is difficult to establish another genus, or even any other genus, with combinations of other characters in the genus *Callianassa*. As in *Pestarella*, however, it is possible to classify the species of *Callianassa* by the characters of Plp1 and Plp2, i.e., the **first** group of species is characterized as follows: Plp1-2 absent, but Mxp3 ischium-merus oblong as in *C. marchali* Le Loeuff & Intès, 1974 and *C. joculatrix* De Man, 1905; the **second** group of species characterized as follows: Plp1 absent, but Plp2 biramous as in *C. anoploura* Sakai, 2002 and *C. diaphora* Le Loeuff &

Intès, 1974; the **third** group of species as follows: Plp1 uniramous and two-segmented, and Plp2 biramous in *Callianassa subterranea* (Montagu, 1808) (type species of *Callianassa*) and also in *C. longicauda* Sakai, 1967, *C. malaccaensis* Sakai, 2002, *C. modesta* De Man, 1905, *C. parvula* Sakai, 1988, *C. praedatrix* De Man, 1905, *C. propinqua* De Man, 1905, and *C. santarita* (Thatje, 2000); the **fourth** group as follows: Plp1 uniramous and two-segmented, and Plp2 absent, in *C. acanthura* Caroli, 1946, *C. oblonga* Le Loeuff & Intès, 1974, *C. biformis* Biffar, 1971, *C. fragilis* Biffar, 1970, *C. biffari* Holthuis, 1991, *C. californiensis* Dana, 1854, *C. gigas* Dana, 1852, *C. amplimaxilla* Sakai, 2002, *C. arenosa* Poore, 1975, *C. australiensis* Dana, 1852, *C. filholi* A. Milne-Edwards, 1878, *C. japonica* Ortmann, 1891, *C. matzi* Sakai, 2002, *C. petalura* Stimpson, 1860, *C. poorei* Sakai, 1998, and *C. pugnatrix* De Man, 1905. As a result, it is difficult to define any other subdivisions of the genus than *Pestarella*, because most of the other members show complicated correlations of characters with one another. The male Plp1-2 are not related with other morphologies in the callianassid species, i.e., the form of Mxp3 and of the telson. Nevertheless, Ngoc-Ho (2003) proposed to distinguish only *Pestarella* from the remainder of the genus *Callianassa*. However, I decline to apply *Pestarella* in the callianassid taxa, because the generic factor of whether the male Plp2 is present or not, is not always compatible with other characters to define other genera, and to single out only one group of species based on this particular combination of characters thus seems arbitrary and hazardous.

The systematic position of *Callianassa celebica* De Haan, 1844 is still uncertain, because the form of Mxp3 does not match that of all the other species of the Callianassidae (cf. Sakai, 1999c: 128).

Genus **Callianassa** Leach, 1814

Cancer Montagu, 1808: 88.
Callianassa Leach, 1814: 386, 400; Leach, 1816: 116; Desmarest, 1825: 205; H. Milne Edwards, 1837a: 319; H. Milne Edwards, 1837b: 130; H. Milne Edwards, 1838: 386; Bell, 1847: 217; Nicolet, 1849: 206; Heller, 1863: 201; Stalio, 1877: 105; Boas, 1880: 84; Ortmann, 1899: 1142; Alcock, 1901: 197; Borradaile, 1903: 544; Balss, 1914: 91; Selbie, 1914: 100; Kemp, 1915: 252; Balss, 1916: 33; Pesta, 1918: 201; Schmitt, 1921: 116; De Man, 1928b: 91; Stevens, 1928: 324; Bouvier, 1940: 100; Edmondson, 1944: 44; Barnard, 1950: 505; Holthuis, 1954c: 338; Hemming, 1958: 142; Williams, 1965: 100; Băcescu, 1967: 227; Biffar, 1971a: 648, figs. 1, 2; De Saint Laurent, 1973: 514; Le Loeuff & Intès, 1974: 32; De Saint Laurent & Le Loeuff, 1979: 48; Sakai, 1987a: 303; Sakai, 1988: 57; Manning & Felder, 1991 [18 Dec.]: 767, figs. 1, 3, 4, 6, 8; Holthuis, 1991 [19 Dec.]: 239; Poore, 1994: 102; Sakai, 1999c: 11; Sakai, 2002: 488; Ngoc-Ho, 2003: 465; Sakai, 2004: 574.

Montagua, Leach, 1814: 436. (Type species: *Cancer* (*Astacus*) *subterraneus* Montagu, 1808; by monotypy. Gender feminine.)

Gebios Risso, 1822: 243. (Type species: *Gebios davianus* Risso, 1822 (= junior subjective synonym of *Cancer candidus* Olivi, 1792; by monotypy. Gender feminine.)

Gebius Agassiz, 1846: 160. (Emendation of *Gebios* Risso, 1822. Gender masculine.)

Mesostylus Bronn & Roemer, 1852: 353. (Type species: *Pagurus faujasi* Desmarest, 1822; by monotypy. Gender masculine.)

Trypaea Dana, 1852a: 14; Gurney, 1944: 83; Manning & Felder, 1991: 774, figs. 1, 3, 12; Davie, 2002: 461. (Type species: *Trypaea australiensis* Dana, 1852b; by original designation and monotypy. Gender feminine.)

Cheramus Bate, 1888: 30; Borradaile, 1903: 545 (partim); De Man, 1928b: 95; Gurney, 1944: 83; Manning & Felder, 1991 [18 Dec.]: 780 (partim) [not figs. 2, 4-6, 14 = *Callianassa*]; Holthuis, 1991 [19 Dec.]: 239; Poore, 1994: 101; Davie, 2002: 459. (Type species: *Cheramus occidentalis* Bate, 1888; by subsequent designation. Gender masculine.)

Scallasis Bate, 1888: 34; Gurney, 1944: 83; Manning & Felder, 1991: 780. (Type species: *Scallasis amboinae* Bate, 1888; by monotypy. Gender feminine.)

Trypaea s. str. Borradaile, 1903: 546; De Man, 1928b: 96; Poore, 1994: 102.

Biffarius Manning & Felder, 1991: 769, fig. 9; Poore, 1994: 102; Tudge et al., 2000: 142; Davie, 2002: 457. (Type species: *Callianassa biformis* Biffar, 1971b; by original designation.)

Neotrypaea Manning & Felder, 1991: 771, fig. 10; Poore, 1994: 102. (Type species: *Callianassa californiensis* Dana, 1854; by original designation.)

Notiax Manning & Felder, 1991: 772, figs. 6, 11; Poore, 1994: 102. (Type species: *Callianassa brachyophthalma* A. Milne-Edwards, 1870, by original designation and monotypy.)

Poti Rodrigues & Manning, 1992b: 9; Poore, 1994: 102. (Type species: *Poti gaucho* Rodrigues & Manning, 1992; by original designation and monotypy. Gender masculine.)

Gilvossius Manning & Felder, 1992: 558. (Type species: *Gonodactylus setimanus* DeKay, 1944; by original designation.)

Nihonotrypaea Manning & Tamaki, 1998: 89; Sakai, 1999c: 129; Sakai, 2001: 946. (Type species: *Callianassa japonica* Ortmann, 1891; by original designation.)

Necallianassa Heard & Manning, 1998: 883; Sakai, 1999c: 128. (Type species: *Necallianassa berylae* Heard & Manning, 1998; by original designation.)

Pseudobiffarius Heard & Manning, 2000: 70. (Type species: *Pseudobiffarius caesari* Heard & Manning, 2000; by original designation and monotypy.) [New synonymy.]

Definition. — [Revised from Sakai, 1999c.] Carapace with dorsal oval; rostral spine present or absent. Eyestalks usually dorsoventrally flattened, contiguous, cornea dorsal, subterminal, disc-shaped or flattened. A1 peduncle longer and more robust than A2 peduncle, or vice versa. Mxp3 ischium-merus variform: pediform, subpediform, suboperculiform, or operculiform; merus convex, or truncate distally, usually without a distal meral spine; propodus oblong; dactylus digitiform; exopod often absent. P1 unequal in size and dissimilar in shape; male larger cheliped with or without meral hook, carpus much expanded, high compared with merus. P3 propodus broadened or oblong. P4 subchelate or simple. P5 chelate. Male Plp1 present or absent; when present, uniramous, biarticulate or unsegmented. Male Plp2 also present or absent, and when present, uniramous, bi- or triarticulate or unsegmented, or biramous, lacking appendix interna and appendix masculina. Female Plp1 uniramous, bi- or triarticulate. Female Plp2 present or absent, when present, biramous, pediform, lacking appendix interna. Plp3-5 biramous, foliaceous, bearing appendices internae. Uropodal endopod slightly or distinctly longer than telson, and distally truncate or rounded.

Type species. — *Cancer* (*Astacus*) *subterraneus* Montagu, 1808, by monotypy. Gender of generic name, *Callianassa*, feminine.

Species included. — East Atlantic and Mediterranean species: *Callianassa acanthura* Caroli, 1946; *C. candida* (Olivi, 1792); *C. convexa* De Saint Laurent & Le Loeuff, 1979; *C. diaphora* Le Loeuff & Intès, 1974; *C. marchali* Le Loeuff & Intès, 1974; *C. oblonga* Le Loeuff & Intès, 1974; *C. subterranea* (Montagu, 1808); *C. truncata* Giard & Bonnier, 1890; *C. tyrrhena* (Petagna, 1792); *C. whitei* Sakai, 1999.

West Atlantic species: *C. berylae* (Heard & Manning, 1998) (*Necallianassa* in Heard & Manning, 1998); *C. biformis* Biffar, 1971; *C. caesari* (Heard & Manning, 2000); *C. delicatula* (Rodrigues & Manning, 1992); *C. fragilis* (Biffar, 1970); *C. gaucho* (Rodrigues & Manning, 1992); *C. marginata* Rathbun, 1901; *C. profunda* Biffar, 1973; *C. setimanus* (DeKay, 1844); *C.* sp. Rabalais, Holt & Flint, 1981.

East Pacific species: *C. biffari* Holthuis, 1991; *C. brachyophthalma* A. Milne-Edwards, 1870; *C. californiensis* Dana, 1854; *C. costaricensis* sp. nov.; *C. debilis* (Hernández-Aguilera, 1998); *C. gigas* Dana, 1852; *C. rochei* Bouvier, 1859; *C. santarita* (Thatje, 2000); *C. tabogensis* sp. nov.; *C. uncinata* H. Milne Edwards, 1837.

Indo-West Pacific species: *C. acutirostella* Sakai, 1988; *C. amboinae* (Bate, 1888); *C. amboinensis* De Man, 1888; *C. amplimaxilla* Sakai, 2002; *C. anoploura* Sakai, 2002; *C. arenosa* Poore, 1975; *C. australiensis* (Dana, 1852); *C. australis* Kensley, 1974; *C. bangensis* sp. nov.; *C. bouvieri* Nobili, 1904; *C. brachytelson* Sakai, 2002; *C. brevirostris* Sakai, 2002; *C. ceramica* Fulton & Grant, 1906; *C. chakratongae* Sakai, 2002; *C. contipes* Sakai, 2002; *C. exilimaxilla* sp. nov.; *C. filholi* A. Milne-Edwards, 1878; *C. gravieri* Nobili, 1905; *C. gruneri* Sakai, 1999; *C. intermedia* De Man, 1905; *C. japonica* Ortmann, 1891; *C. joculatrix* De Man, 1905; *C. lewtonae* Ngoc-Ho, 1994; *C. lignicola* Alcock & Anderson, 1899; *C. limosa* Poore, 1975; *C. lobetobensis* De Man, 1905; *C. longicauda* Sakai, 1967; *C. malaccaensis* Sakai, 2002; *C. matzi* Sakai, 2002; *C. mocambiquensis* Sakai, 2004; *C. modesta* De Man, 1905; *C. nieli* Sakai, 2002; *C. nigroculata* Sakai, 2002; *C. orientalis* (Bate, 1888); *C. parva* (Edmondson, 1944); *C. parvula* Sakai, 1988; *C. petalura* Stimpson,

1860; *C. persica* sp. nov.; *C. plantei* Sakai, 2004; *C. poorei* Sakai, 1999; *C. praedatrix* De Man, 1905; *C. propinqua* De Man, 1905; *C. propriopedis* Sakai, 2002; *C. pugnatrix* De Man, 1905; *C. pygmaea* De Man, 1928; *C. rotundicaudata* Stebbing, 1902; *C. sibogae* De Man, 1905; *C. spinophthalma* Sakai, 1970; *C. spinoculata* sp. nov.; *C. stenomastaxa* Sakai, 2002; *C. tenuipes* Sakai, 2004; *C. thailandica* sp. nov.; *C. thorsoni* sp. nov.; *C. tonkinae* Grebenjuk, 1975; *C.* sp. Haswell, 1882; *C.* sp. De Man, 1928; *C.* sp. Sakai, 1970; *C.* sp. 1 Sakai, 2002; *C.* sp. 2 Sakai, 2002; *C.* sp. 3 Sakai, 2002; *C.* sp. 4 Sakai, 2002.

Remarks. — Tudge et al. (2000: 143) listed *Callianassa abdominalis* White, 1847 and *C. carinaedorsis* White, 1847 but, as far as I know, these two species have never been described as Clark & Presswell (2001: 155) defined them as nomina nuda. P. Clark informed me in litt. (19 Dec. 2002) that specimens were filed as "*Callianassa* ? *abdominalis* White, 1847, reg. nos. 702a and 1843.6, Philippine Is., presented by Cuming", now labeled, however, as *Scytoleptus* sp. (Axiidae) by M. De Saint Laurent in June 1977; and as "*Callianassa* ? *carinaedorsis* White, 1847, reg. no. 1844.71, Swan River, Western Australia, purchased Dring"; now labeled as *Scytoleptus* sp. by M. De Saint Laurent in June 1977.

Davie (2002: 458) includes *Calliactites* Borradaile, 1903 (type species: *Callianassa secura* Lanchester, 1902: 555, pl. 34 fig. 2; by original designation; gender of generic name, *Calliactites*, masculine), but it should be excluded from the Callianassidae, because it is now classified as *Callianidea* in the family Callianideidae.

East Atlantic and Mediterranean species

The marine collection of Rovinj Station, Croatia, was transferred, after some vicissitudes, in 1969 to the Stazione Idrobiologica di Chioggia (SIC), a separate research centre of the University of Padova. After a careful check of the faunistic list of Marcuzzi (1972), Prof. S. Casellato, University of Padova, verified that the reported specimens of *Callianassa pestae* (= *C. candida*) from the Adriatic Sea, were collected in the vicinity of Rovinj and identified by Lutze in 1937. Marcuzzi (1972) summed up the collection of Rovinj, which had been described by Lutze (1937) while later on Vatova (1949) made its list. The present author examined the material deposited at Chioggia, Italy, and found that *Callianassa pestae* as described by Lutze (1937) includes two species, *C. candida* and *C. whitei*, and that the type specimen of *Callianassa algerica* Lutze, 1938 is confirmed to be lost.

In the present paper, the following five species of the Callianassidae: *Callianassa candida* (Olivi, 1792); *C. subterranea* (Montagu, 1808); *C. truncata* Giard & Bonnier, 1890; *C. tyrrhena* (Petagna, 1792), and *C. whitei* Sakai, 1999, are confirmed to be in the collections of the Forschungsinstitut Senckenberg, Frankfurt am Main, Germany (SMF); the Zoological Laboratory, University of Athens, Athens, Greece (ZLUA); and the Stazione Idrobiologica di Chioggia, Dipartimento di Biologia, Università degli Studi di Padova, Padova, Italy (SIC).

Callianassa acanthura Caroli, 1946

Callianassa acanthura Caroli, 1946: 66, figs. 1a, 2; Holthuis, 1953a, fig. 3; De Saint Laurent & Bozic, 1976: 21, figs. 3, 11, 19, 25, 30; Türkay, 1982: 225; d'Udekem d'Acoz, 1996: 54; Abed-Navandi & Dworschak, 1998: 605; Sakai, 1999c: 14; Ngoc-Ho, 2003: 466, fig. 8.
Callianassa (Trypaea) acanthura; Zariquiey Alvarez, 1968: 229.
Necallianassa acanthura; Heard & Manning, 1998: 884; d'Udekem d'Acoz, 1999: 155; Tudge et al., 2000: 143.

Diagnosis. — Mxp3 ischium-merus operculiform, merus rounded distally (De Saint Laurent & Bozic, 1976, fig. 3), male Plp1 two-segmented, male Plp2 absent, telson truncate with a median tooth on posterior margin, bearing a distinct pair of posteriorly directed, spine-like processes on lateral margins.

Type locality. — Bay of Naples.

Distribution. — Bay of Naples; Adriatic Sea: Kornati Archipelago and Vestar Bay near Rovinj, Croatia, 3-6 m depth; Ionian Sea; Aegean Sea.

Callianassa candida (Olivi, 1792)

Cancer candidus Olivi, 1792: 51, pl. 3 fig. 3.
Callianassa tyrrhena; Risso, 1827: 54 (partim); Forest & Guinot, 1956: 31; Forest, 1967: 6 (partim). [Not *Callianassa tyrrhena* (Petagna, 1792).]
Callianassa subterranea; Czerniavsky, 1868: 122; Giard & Bonnier, 1890: 362, figs. 1-3. [Not *Callianassa subterranea* (Montagu, 1808).]
Callianassa subterranea forma *pontica* Czerniavsky, 1884: 81 (partim). [Type locality: Black Sea.]
Callianassa pontica; Cano, 1891: 5-30; Caroli, 1940: 76; Lutze, 1941: 34; Caroli, 1946: 71; Vatova, 1949, tab. 5; Caroli, 1950: 190; Dolgopol'skaia, 1954: 179, figs. 3-4; De Saint Laurent & Bozic, 1976: 24, figs. 5, 13, 21, 32; Beaubrun, 1979: 84, figs. 58, 59, 68, 69, 70; Moncharmont, 1979: 71; García Raso, 1983: 318, fig. 3; Thessalou-Legaki & Zenetos, 1985: 311; Thessalou-Legaki, 1986: 182; Tudge et al., 2000: 143.
Callianassa (Callichirus) laticauda Pesta, 1918: 204; Bouvier, 1940: 103 (partim). [Not *Callianassa laticauda* Otto, 1821.]

Callianassa (*Callichirus*) *Pestae* De Man, 1928a: 34, pl. 9 figs. 16-16e; De Man, 1928b: 111.
[Type locality: Mediterranean.]
Callianassa pestae; Lutze, 1937: 6; Lutze, 1938: 167 (partim) (not figs. 10-21 = *Callianassa subterranea*); Vatova, 1949, tabs. 16, 18, 20, 24, 31; Marcuzzi, 1972: 193; Manning & Števčić, 1982: 295; Froglia & Grippa, 1986: 261.
Callianassa sp. Băcescu, 1949: 9, figs. 20-22.
Callianassa pestai; Holthuis, 1953a: 95, fig. 3; Dolgopol'skaia, 1969: 316, pls. 32-34; Kobyakova & Dolgopol'skaia, 1969: 286, pl. 5 fig. 1a-c; Števčić, 1971: 529; Mazmanidi et al., 1998: 129.
Callianassa algerica Lutze, 1938: 168 (partim) [not: fig. 22, eyestalks = *C. whitei*], figs. 26a, larger Mxp3 [not smaller Mxp3 = *C. subterranea*], 26b, larger Mxp3 [not smaller Mxp3 = *C. subterranea*], fig. 27. [Nomen dubium.] [Type locality: Castiglione near Algiers.]
Callianassa (*Callichirus*) *pontica*; Makarov, 1938: 73, 297, figs. 27-28; Băcescu, 1967: 231, fig. 105.
Callianassa candida; Giordani Soika, 1943: 83 (partim); Giordani Soika, 1945: 994; Mika-shavidze, 1981: 1415; Lewinsohn & Holthuis, 1986: 20; Števčić, 1990: 217; Koukouras et al., 1992: 223; Dworschak, 1992: 194 (partim); d'Udekem d'Acoz, 1996: 54; d'Udekem d'Acoz, 1999: 154; Sakai, 1999c: 14, figs. 2a-d; Tudge et al., 2000: 143; Dworschak, 2002: 63, figs. 1-2, tab. 1; d'Udekem d'Acoz, 2003, fig. on website.
Callianassa (*Callichirus*) *pestae*; Zariquiey Alvarez, 1968: 230; Kattoulas & Koukouras, 1974: 344, fig. 1.
Pestarella candida; Ngoc-Ho, 2003: 476, fig. 12.

Material examined. — SMF 27252, 2 males (Tl/Cl 49.0/11.2; 40.0/9.5, lacking P1), 1 female (42.0/10.0, lacking P1), 1 female (Tl ca. 35.0, lacking carapace), Kuvi-Bay (= Villas Rubin), area of Rovinj, Istria, Croatia, Rov 01-30d (45°03.934'N 13°39.163'E), sand and mud, 0-1 m, 23.viii.2001, excursion of Frankfurt Univ., by yabby-pump. SMF 28046, 1 male (22.0/4.8, damaged, lacking P1). Kuvi-Bay, Croatia, Rov 99-10C, (45°05.540'N 013°37.260'E), 0.3-1 m, muddy ground, 28.viii.1999, excursion of Frankfurt Univ., by yabby-pump. SMF 28047, 1 male (56.0/12.0), Limski Canal, Croatia, St. Rov 93-13, shallow water, 01.ix.1993, excursion of Frankfurt Univ., by yabby-pump. SMF 28048, 1 male (37.0/9.0), 1 female (47.0/10.5), Limski Canal, Croatia, St. Rov 93-13, shallow water, ix.1993, excursion of Frankfurt Univ., by yabby-pump. SMF 28045, 1 male (59.0/13.2, lacking major cheliped), Island Santa Katarina, Rovinj, Istria, Croatia, 2-3 m, 27.viii.1999, leg. A. D. Brandis, excursion of Frankfurt Univ., by snorkel swimming. SIC 575, 1 female (67.0/16.5), Limski Canal, 06.vi.1932, leg. et det. A. Steuer as *Callianassa pestae*. SIC 577-A, 1 male (41.0/11.0); 1 female (45.0/9.0, anterior part of carapace broken, antennal peduncles missing); SIC 577-B, 3 males (16.0/4.0-21.0/6.0), 4 females (20.0/5.0-35.0/10.0); SIC 577-C, 4 males (38.0/10.0-21.0/6.5), 4 females (19.0/5.0-38.0/11.0), Castiglione [between Koléa and Bérard], Algeria, 18.v.1937, det. J. Lutze as *Callianassa pestae*.

Diagnosis. — Mxp3 ischium-merus operculiform, merus rounded on distomesial angle (Ngoc-Ho, 2003, fig. 12F), male Plp1-2 absent, telson rounded posteriorly, lacking a median tooth on posterior margin (Sakai, 1999c, fig. 2b).

Remarks. — Lewinsohn & Holthuis (1986: 23) mentioned that "As shown by the name candida (= white or snow white), the animal is of a white colour and thus cannot be *C. tyrrhena* which is pink, while *C. pontica* (= *C. candida*) is indeed white". In the present specimens collected by the Rovinj-excursions in 1999 and 2001, it turned out that the larger specimens of *C. whitei* and *C. subterranea* are white, while *C. candida* specimens are faint-pink, and *C. tyrrhena* and *C. subterranea* (SMF 27250, Rovinj) are pink. The coloration of *C. subterranea* is variably white or pink, so it is not always valid to define the species by coloration (see for ref.: remarks on *C. subterranea*).

Type locality. — Alupka, 10 m deep and Suchumi, 2-3 m, Black Sea.

Distribution. — From Golfo de Cádiz, SW Spain, eastern Atlantic to Alupka, Black Sea, up to 10 m depth.

[E. Atlantic — Golfo de Cádiz, El Chato, SW Spain (García Raso, 1985b); Algeciras (Marcuzzi, 1972; García Raso, 1983). Mediterranean — Morocco: Strait of Gibraltar, Tanger (Sakai, 1999c). Algeria: Castiglione (De Saint Laurent & Bozic, 1976). Tunisia: Salammbô; Les Bibans (De Saint Laurent & Bozic, 1976); N. Punic Port (Sakai, 1999c). France: Vieux Port, Villefranche sur Mer. France: Côte d'Azur - Iles d'Hyères; Ile de Port-Cros, 1.8 m (Sakai, 1999c). Italy: Porto Cesareo, Golfo di Taranto (De Saint Laurent & Bozic, 1976); Liguria, Portofino (Sakai, 1999c). Tyrrhenian Sea — Italy: Gulf of Naples (De Man, 1928a; Sakai, 1999c); Sicily, Stromboli (Sakai, 1999c); Moncharmont, 1979); Genova. Malta: M'Xlakk Bay, 3 m rock (Sakai, 1999c). Cyprus: Bogaz, S. of Karpas peninsula, N. Cyprus (Lewinsohn & Holthuis, 1986). Adriatic Sea (Manning & Števčić, 1982) — Italy: Punta Sabbioni, Venice lagoon (Sakai, 1999c). Italy: Trieste, Aurisina; Zaule; Lido di Staranzano; Lagoon of Grado; Lesina (Sakai, 1999c). Slovenia, Strunjan (Sakai, 1999c); Piran Gulf (Manning & Števčić, 1982; Sakai, 1999c). Croatia, Istria, Kuvi-Bay (Dworschak, 1992; Sakai, 1999c). E. Mediterranean (De Saint Laurent & Bozic, 1976): Ionian Sea (De Saint Laurent & Bozic, 1976; Thessalou-Legaki & Zenetos, 1985; d'Udekem d'Acoz, 1996). Greece: Northern Sporades, Peristera, Vasilikos Ormoz (39°11.500'N 023°58.350'E), 0.2 m (Sakai, 1999c); Aegean Sea (Koukouras et al., 1992; d'Udekem d'Acoz, 1996); N. Euboikos Gulf (Kattoulas & Koukouras, 1974); Saronikos in Kefalonia Island, S. Euboikos Gulf, and Mytlene Island (Thessalou-Legaki, 1986); Rhodes Island (Lewinsohn, 1976). Cyprus (Lewinsohn & Holthuis, 1986). Black Sea (Dolgopol'skaia, 1954; Băcescu, 1967; Mikashavidze, 1981); — Alupka, 10 m deep and Suchumi, Black Sea, 2-3 m (Czerniavsky, 1884).]

Habitat. — In coarse sand or mud under stones in the intertidal and shallow subtidal, and in sandy silt or mud in 7-9 m.

Callianassa convexa De Saint Laurent & Le Loeuff, 1979

Callianassa convexa De Saint Laurent & Le Loeuff, 1979: 53, fig. 10a-e; Sakai, 1999c: 17; Tudge et al., 2000: 143.

Remarks. — Mxp3 ischium-merus subsquare, straight distally and rounded on distomedial angle, male Plp1-2 absent, telson rounded in posterior half (De Saint Laurent & Le Loeuff, 1979: 53, fig. 10d, e).

Type locality. — South of Cape Bald, Gambia, 18 m.

Distribution. — Gambia; Mauritania.

Callianassa diaphora Le Loeuff & Intès, 1974

Callianassa diaphora Le Loeuff & Intès, 1974: 32, fig. 7a-v; De Saint Laurent & Le Loeuff,
    1979: 49, fig. 9a, b, e, g; Sakai, 1999c: 18; Sakai, 2004: 581-585, figs. 13-14.
Callianassa guineensis; Longhurst, 1958: 31 (partim).
Biffarius diaphora; Tudge et al., 2000: 143.

Remarks. — Mxp3 ischium-merus subpediform with an oblique distal margin, male Plp1 absent, male Plp2 biramous, telson trapezoid in posterior half (Le Loeuff & Intès, 1974: 32, fig. 7k, q).

Type locality. — Grand Lahou, Ivory Coast, 5°07'N 5°04.5'W, 22 m.

Distribution. — Siera Leone to Ivory Coast, 10-60 m.

Callianassa marchali Le Loeuff & Intès, 1974

Callianassa marchali Le Loeuff & Intès, 1974: 35, fig. 8a-r; De Saint Laurent & Le Loeuff,
    1979: 51, fig. 8c, d, f, h; Sakai, 1999c: 18; Tudge et al., 2000: 143.
Callianassa guineensis; Longhurst, 1958: 31 (partim).

Remarks. — Mxp3 ischium-merus subpediform with an oblique distal margin, male Plp1-2 absent, telson convex medially on posterior margin (Le Loeuff & Intès, 1974: 32, fig. 8j, p).

Type locality. — Ivory Coast, between Vridi and Jacqueville (5°09.5'N 4°09'W), 70 m.

Distribution. — Pointe-Noire, Congo; Ivory Coast; Sierra Leone; Senegal; 70-250 m.

Callianassa oblonga Le Loeuff & Intès, 1974

Callianassa oblonga Le Loeuff & Intès, 1974: 38, fig. 9a-r; De Saint Laurent & Le Loeuff,
    1979: 55; Sakai, 1999c: 18.
Cheramus oblongus; Tudge et al., 2000: 145.

Diagnosis. — Mxp3 ischium-merus subsquare, merus straight on distal margin, male Plp1 uniramous, two-segmented, male Plp2 absent, telson concave medially on posterior margin, bearing a median spine (Le Loeuff & Intès, 1974: 32, fig. 9k, q).

Type locality. — Ivory Coast, Grand Bassam (4°59'N 3°48'W), 200 m.

Distribution. — Known only from the type locality.

Callianassa subterranea (Montagu, 1808)
(figs. 2B, 3B, 4B, 5-7)

Cancer (*Astacus*) *subterraneus* Montagu, 1808: 89, pl. 3 figs. 1-2.

Callianassa subterranea; Leach, 1815: 343; Leach, 1816, pl. 32; Desmarest, 1825, tab. 36, fig. 2; H. Milne Edwards, 1837a: 309 (partim?); H. Milne Edwards, 1837b: 130 (partim?); Thompson, 1844: 268; Bell, 1846/53: 219, fig.; White, 1847: 70; A. Milne-Edwards, 1870: 80, 101; Stalio, 1877: 106; Neumann, 1878: 34; Czerniavsky, 1884: 76, 80; Adensamer, 1898: 620; Lagerberg, 1908: 53; Blohm, 1913: 17 (partim); Selbie, 1914: 101; Athanasso-poulos, 1917: 32; Gustafson, 1934: 14; Nobre, 1936: 121, pl. 40 fig. 101; Lutze, 1938: 170, figs. 28-51; Makarov, 1938: 62, 63 (partim); Poulsen, 1940: 229, figs. 10-12; Lutze, 1941: 34; Gordon, 1957: 240; Holthuis & Gottlieb, 1958: 115; Holme, 1961: 397; Bourdon, 1965: 1; Holme, 1966: 401; Allen, 1967: 18, 57; Christiansen, 1972: 41, fig. 49; Pastore, 1972: 107; De Saint Laurent & Bozic, 1976: 17, figs. 1, 9, 17, 28; Moncharmont, 1979: 71; Števčić, 1979: 282, fig. 1; Domenech et al., 1981: 150; Adema et al., 1982: 23, fig. 6a-c; Christiansen & Greve, 1982: 213; Thessalou-Legaki, 1985: 457; Thessalou-Legaki, 1986: 182; Garcia So-cias & Massuti Jaume, 1987: 75; Witbaard & Duineveld, 1989: 209-219, fig. 1; Števčić, 1990: 217; Dworschak, 1992: 203; Hayward et al., 1995: 432, fig. 8-52; Stamhuis et al., 1997: 155; Nickell et al., 1998: 733; Simboura et al., 1998: 300; Števčić, 1998: 652; Sakai, 1999c: 14 (key), 19; d'Udekem d'Acoz, 1999: 155; Hughes et al., 2000: 189; Tudge et al., 2000: 143; Ngoc-Ho, 2003: 468, figs. 9, 10.

Callianassa (*Cheramus*) *subterranea*; Borradaile, 1903: 545; De Man, 1928a: 6, pl. 1 fig. 1-1h; De Man, 1928b: 27, 91, 92, 94, 97; Makarov, 1938: 63, fig. 21; Bouvier, 1940: 101 (partim), fig. 67; Gordon, 1957: 249; O'Céidigh, 1962: 163.

Callianassa (*Cheramus*) *subterranea* var. *minor*; Pesta, 1918: 205.

Cheramus subterraneus; Colosi, 1923: 6.

Callianassa pestae; Lutze, 1938, figs. 10-21 (= *C. subterranea*).

Callianassa algerica Lutze, 1938: 168 (partim), figs. 22-26 [not: fig. 22, eyestalks = *C. whitei*); 26a, smaller Mxp3 = *C. subterranea* [not larger Mxp3 = *C. candida*]; 26b, smaller Mxp3 = *C. subterranea* [not larger Mxp3 = *C. candida*], [not fig. 27 = *C. candida*]; [type locality: Castiglione near Algiers, 36°37'N 02°40'E]; Lutze, 1941: 34. [Nomen dubium.]

Callianassa helgolandica Lutze, 1938: 174, figs. 52-61; Lutze, 1941: 34. [Type locality: Hel-goland.]

Callianassa stebbingi; Lutze, 1938, figs. 10-21 [= *Callianassa subterranea*].

Callianassa hertlingi Lutze, 1941: 34 [a junior synonym of *Callianassa helgolandica* Lutze, 1938].

Callianassa tyrrhena, [not *Callianassa tyrrhena* Petagna]; Holthuis & Gottlieb, 1958: 62 (par-tim), fig. 13 (= *C. subterranea*); Zariquiey Alvarez, 1968: 230.

Callianassa (*Callianassa*) *subterranea*; Zariquiey Alvarez, 1968: 229; Lagardère, 1973: 88, 94.

Not *Callianassa subterranea*; H. Milne Edwards, 1837b: pl. 48 fig. 3-3e; Heller, 1863: 202, pl. 6 fig. 9-11; Ortmann, 1891: 55, pl. 1 fig. 10 [= *C. tyrrhena* (Petagna, 1792)]; Czerniavsky, 1868: 122; Giard & Bonnier, 1890: 362, figs. 1-3 [= *C. pontica* Czerniavsky, 1884 = *C. can-dida* (Olivi, 1792)]; Adensamer, 1898: 620 [= *Gourretia denticulata* (Lutze, 1937)].

Not *Callianassa*; Kinahan, 1859: 266.

Not *Callianassa subterranea* forma *pontica* Czerniavsky, 1884: 81 [nomen dubium] [= *C. pon-tica* Czerniavsky, 1884 = *C. candida* (Olivi, 1792)].

Not *Callianassa subterranea* var. *minor* Gourret, 1887: 1034; Gourret, 1888: 96, pl. 8 figs. 1-15
    [= *Gourretia denticulata* (Lutze, 1937)].
Not *Callianassa subterranea* var. *japonica* Ortmann, 1891: 56 [= *C. japonica*].

Material examined. — SMF 25800, 1 male (Tl/Cl 30.0/7.3), Canary Islands, Atlantic Ocean,
Va St. VI, 6r 6KG, 75-270 m depth, leg. R/V "Valdivia", 13.viii.1975; SMF 30081, 1 male
(18.0/4.5), Cimotoe sea mount, off Sicily, 36°59.90'N 12°30.82'E, Mediterranean Sea 211 m,
leg. MIPA-Mehrmast-I Expedition, R/V "Poseidon", Pos. 172/4-726KG, 07.v.1990; SMF
27250, 1 male (27.0/6.5 with *Bopyrus* sp. in right branchial chamber), 2 ovig. females (20.5/4.8;
18.5/4.2), 1 female (25.0/5.0), by Sotto Castello, Limski-Canal, Istria, Croatia, 45°07.863'N
13°39.389'E, muddy bottom, ca. 30 m, leg excursion of Frankfurt Univ., R/V "Burin", Rov 01-
03, by ring dredge, 14.viii.2001; SMF 27250 (cont.), 3 males (19.0/4.1-11.0/2.8), 1 female
(19.0/4.5), 1 damaged, 1 juv., front of Sotto Castello, Limski-Canal, Croatia, 45°07.863'N
013°39.389'E, muddy bottom, 30 m, leg. excursion of Frankfurt Univ., R/V "Burin", Rov 01-03,
by ring dredge, 14.viii.2001; SMF 28041, 7 males (16.0/4.0-11.0/2.5), 1 ovig. female (20.0/4.5),
3 females (20.0/4.5-14.0/2.8), front of Sotto Castello, Limski-Canal, Croatia, 45°08.020'N
013°39.030'E, 29 m, leg. excursion of Frankfurt Univ., R/V "Burin", St. Rov 99-21, by ring
dredge, 01.ix.1999; SMF 28042, 1 male (15.0/3.5), Valdibora Bay, Istria, Croatia, 45°05.250'N
013°38.000'E, 25.9 m depth, leg. excursion of Frankfurt Univ., R/V "Burin", St. Rov 99-12, by
Van Veen Grab, 30.viii.1999; SMF 28043, 2 males (10.0/2.4-9.5/2.4), 1 female (9.5/2.5), Limski-Canal, Rovinj, Istria, Croatia, St. Rov 97-06, 45°07.842'N 013°38.980'E, muddy bottom, leg.
excursion of Frankfurt Univ., R/V "Burin", by ring dredge, 22.viii.1997; SMF 28107, 1 female
(35.0/8.0), detached larger cheliped, German Bight, St. P12, Ku, 53°59.08'N 7°51.85'E-
53°58.98N 7°50.45'E, 30-40 m depth, leg. J. Doerjes, R/V "Senckenberg", by beam-trawl,
07.vi.1975; SMF 28108, 1 female (40.0/10.0), detached larger and smaller chelipeds, and 1 male
larger cheliped, German Bight, P 12-4, St. P12, Ku, 53°59.08'N 7°51.85'E-53°58.98'N
7°50.45'E, 30-40 m, leg. J. Doerjes, R/V "Astarte", by beam-trawl 23.vii.1980; ZLUA BE 17a, 1
male (21.0/16.0), 2 females (19.0/6.0, 20.0/6.0), 21.xi.1991, Greece, leg. M. Thessalou-Legaki;
ZLUA 17b, 1 male (24.0/6.0), Greece, leg. M. Thessalou-Legaki, 21.xi.1991.

Diagnosis. — Mxp3 ischium-merus subpediform, merus obliquely declined
on distal margin, male Plp1 uniramous, two-segmented, male Plp2 biramous,
sometimes absent in small individuals (Ngoc-Ho, 2003: 472), telson truncate
with a median spine on posterior margin.

Remarks. — This is the first time that a female specimen of *C. subterranea*
from the Canary Islands has become available for study (75-270 m). This female (fig. 7) shows specific characters almost as found in Mediterranean
specimens (see figs. 6-7), but it also shows morphological variation in the larger cheliped (fig. 7D).

Adensamer (1898) described one female of *C. subterranea* from the Aegean
Sea, 92 m deep, which was identified by Pesta (1918: 205), with a question
mark, as *Callianassa* (*Cheramus*) *subterranea* var. *minor* Gourret, 1887. Sakai
(1999c: 19) identified this Aegean species, collected by Adensamer (1898),
and referred to by Pesta (1918) as *Callianassa subterranea* var. *minor*, as a

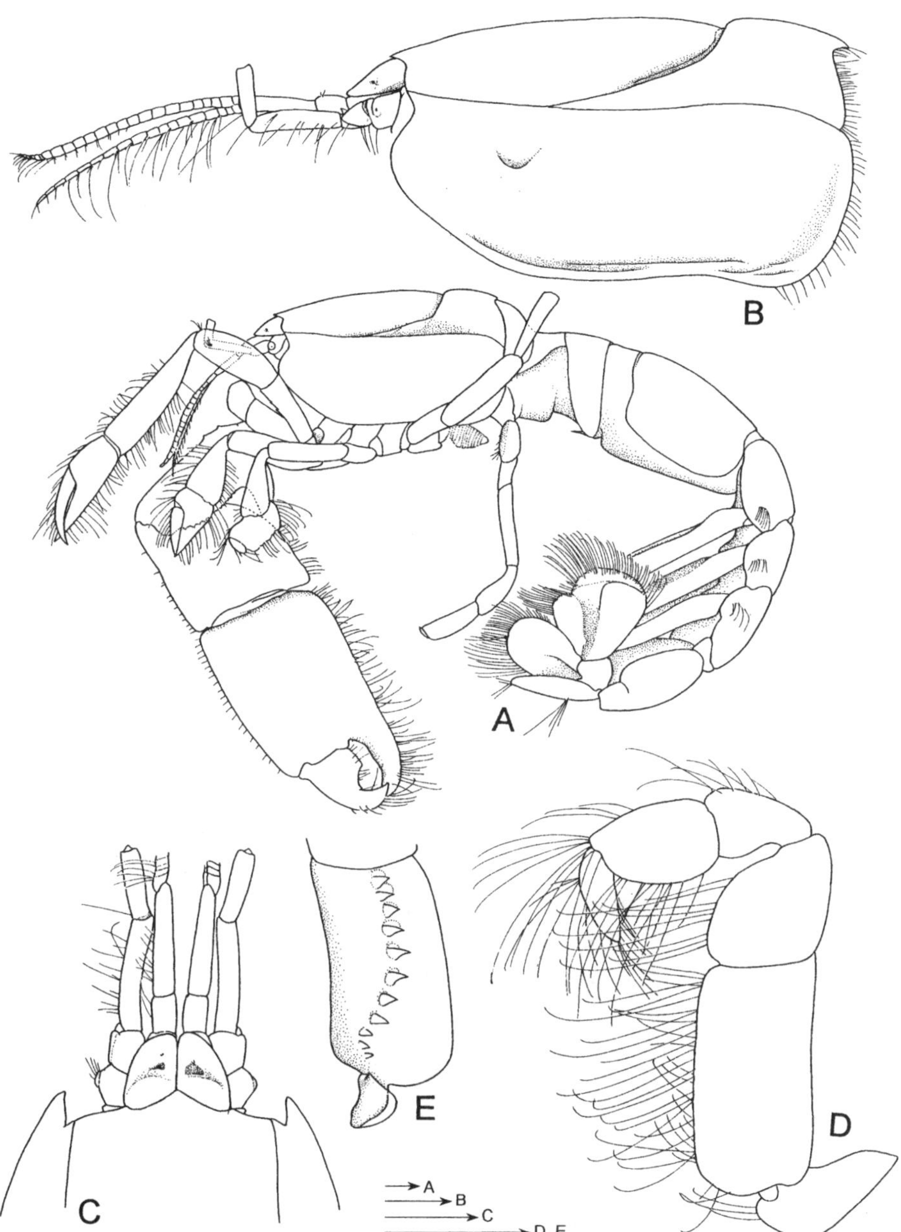

Fig. 5. *Callianassa subterranea* (Montagu, 1808). SMF 25800, male, Sotto Castello, Limski-Canal, Istria, Croatia, Rov 01-03 (45°07.863'N 13°39.389'E), muddy bottom, ca. 30 m. A, whole body; B, carapace, lateral view; C, anterior part of carapace, dorsal view; D, Mxp3, lateral view; E, ischium of Mxp3, mesial view. Scales 1 mm.

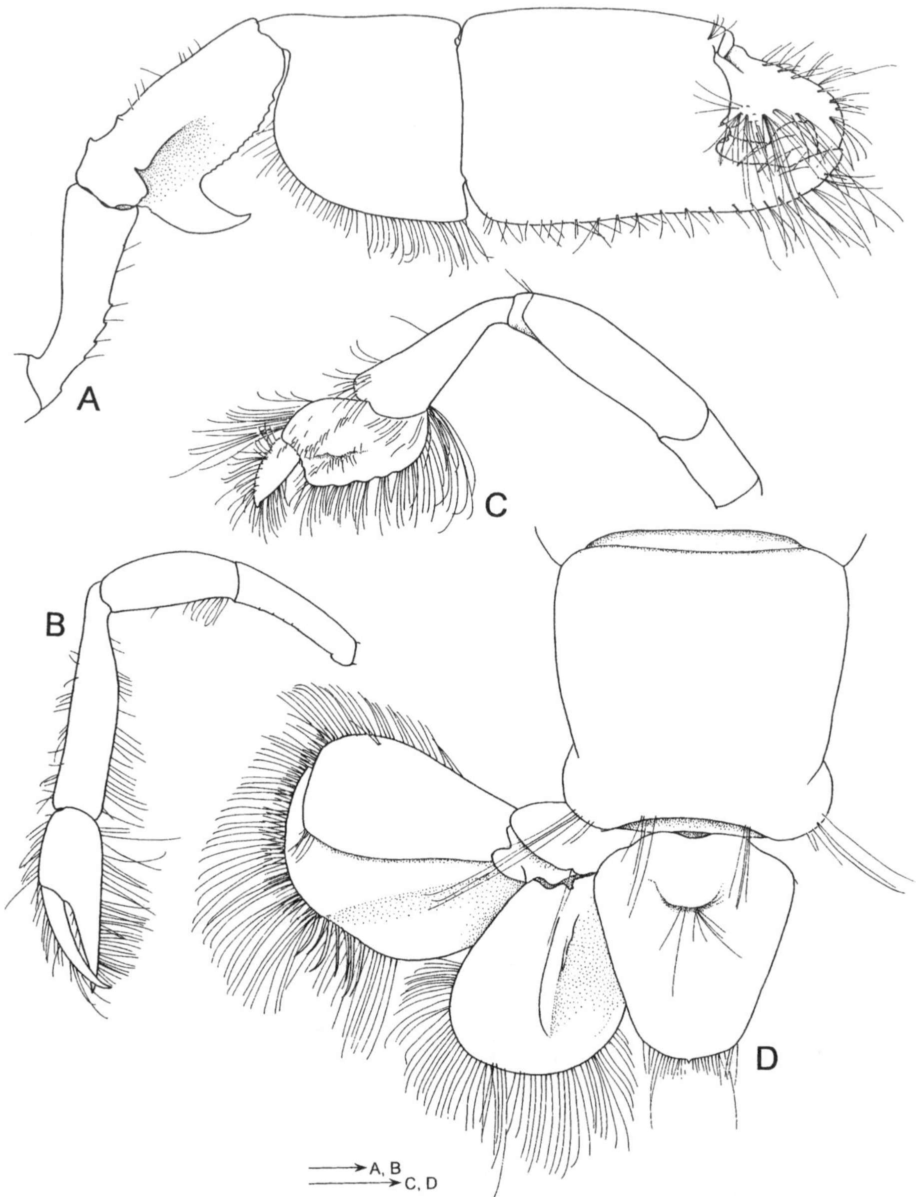

Fig. 6. *Callianassa subterranea* (Montagu, 1808). SMF 25800, male, Sotto Castello, Limski-Canal, Istria, Croatia, Rov 01-03 (45°07.863'N 13°39.389'E), muddy bottom, ca. 30 m. A, larger cheliped, lateral view; B, smaller cheliped, lateral view; C, pereiopod 3, lateral view; D, abdominal somite 6 and tail fan, dorsal view. Scales 1 mm.

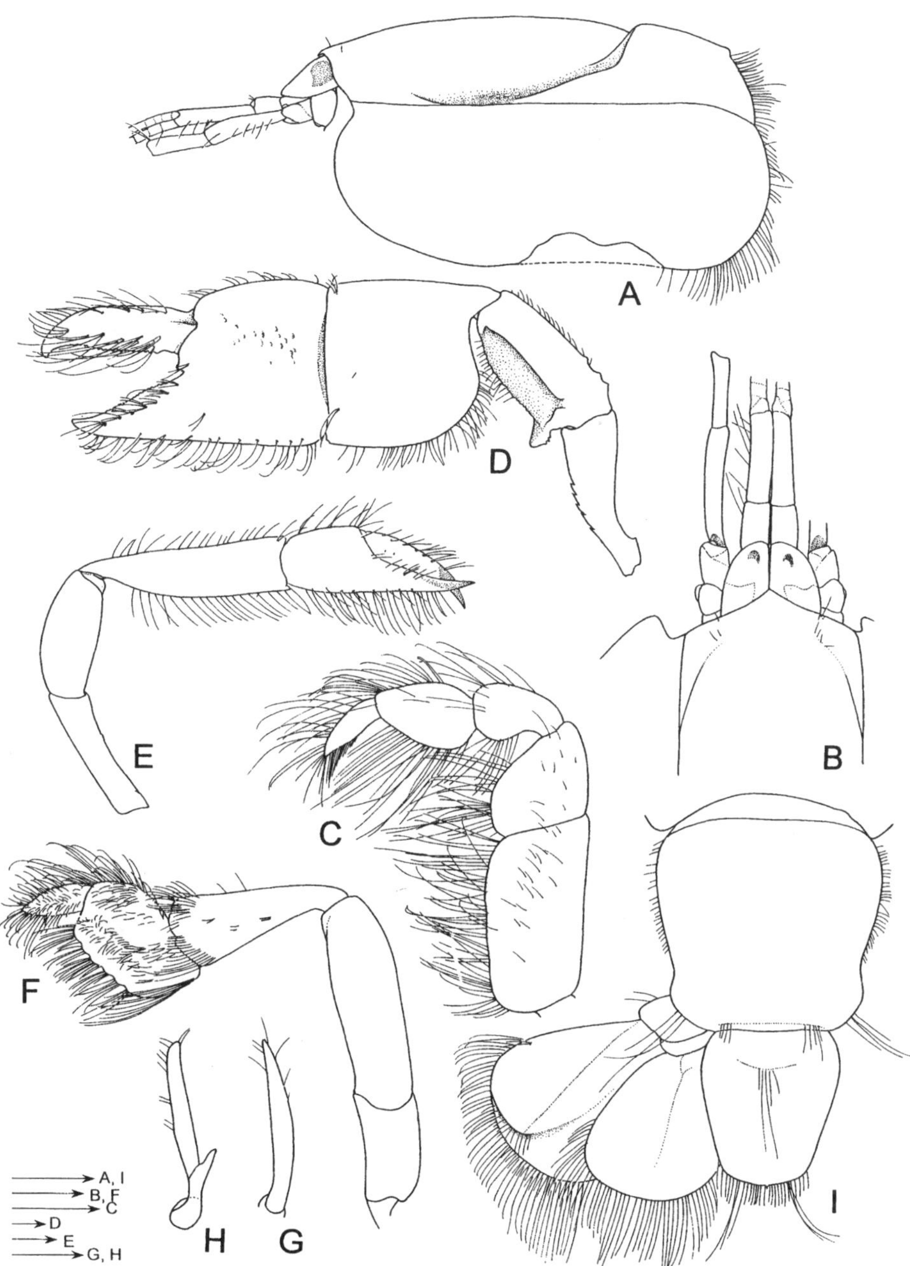

Fig. 7. *Callianassa subterranea* (Montagu, 1808). SMF 27250, male (Tl/Cl 30.0/7.3), Canary Islands, Atlantic Ocean, leg. R/V "Valdivia". A, carapace, lateral view; B, anterior part of carapace, dorsal view; C, Mxp3, lateral view; D, larger cheliped, lateral view; E, smaller cheliped, lateral view; F, pereiopod 3, lateral view; G, male Plp1; H, male Plp2; I, abdominal somite 6 and tail fan, dorsal view. Scales 1 mm.

Gourretia (= *Gourretia denticulata* (Lutze, 1937)). However, it is most probable that the species collected by Adensamer (1898) from the depth is *C. subterranea*, because Pesta (1918) compared Adensamer's specimen with his own specimen collected from 30 m depth on the coral bottom of the bay of Marseille, and the other possible species, *C. tyrrhena*, is found only in the lower intertidal and shallow subtidal zones of the Adriatic Sea, as mentioned by Dworschak (2000a: 155-156).

Bell (1846: 221) mentioned that: "its claim to be considered as an Irish species is thus stated by Mr. Thompson: "March 25th, 1839. On examining the contents of the stomach of several individuals of the *Platessa Pola* which were taken off Newcastle (County Down), two of the larger arms of this species, so peculiar in form, and still retaining their beautiful pink colour, were detected."..." In the male specimen from Limski-Canal, Istria, Croatia (SMF 27250), the chelipeds are salmon pink, abdominal somites 1-2 are yellow, and the rest of the abdominal somites are faintly salmon pink. From this fact, it is possible to consider that the species described by Bell is not *C. tyrrhena*, but *C. subterranea* instead.

Ngoc-Ho (2003) says about the type-species, *Callianassa subterranea* (Montagu, 1808) (her p. 472, column 1, line 13): "Male Plp2 small, sometimes absent", while about the other *Callianassa* species she mentions in, e.g., *C. acanthurus* Caroli, 1946 and *C. truncata* Giard & Bonnier, 1890: "Male Plp2 is absent". However, in *C. subterranea* it was observed in the larger male specimen from SMF 21879 (German Bight, North Sea, 54°39.90'N 6°0.00'E, 42.5 m, R/V "Senckenberg", 24.v.1987), that the male Plp2 is biramous, as shown by Manning & Felder (1991: 774, fig. 8). It is thus considered that the male Plp2 of *Callianassa subterranea*, the type species of *Callianassa* is biramous in adult males. Type locality. — Devon, England.

Distribution. — East Atlantic from Scandinavia to Canary Islands and the eastern Mediterranean, usually subtidal, 10-270 m depth.

[E. Atlantic Ocean. — Norway: near Bergen (Christiansen, 2000) to Skagerrak (Christiansen, 1972), 5-100 m depth. Sweden: Gullmar fjord; Trillingarna; Bonden; Sotefjord; Kristineberg (Gustafson, 1934); Kattegat (Lagerberg, 1908; Sakai, 1999c); Dalsvik-Tova, Gullmaren, 30-40 m depth, mud (Sakai, 1999c); Bläckhall, Rännan, 40 m depth, sandy mud, mud with shells (Sakai, 1999c); Gullmaren, Kölvik, 30 m (Sakai, 1999c); Utanför Blåbärsholmen, Kristineberg (Sakai, 1999c). — Denmark: Skagen-Nidingarne, 35 fathoms [c. 52 m] (Sakai, 1999c); Wäderöarne, 60 fathoms (110 m) depth, mud (Sakai, 1999c). Ireland: Ireland to Devon coasts (Thompson, 1844). — England: Plymouth (De Saint Laurent & Bozic, 1976). — S. Scandinavia (Gustafson, 1934; Poulsen, 1940; Christiansen & Greve, 1982; Christiansen & Stene, 1998). — S. W. Scotland (Allen, 1967; Nickell et al., 1998). — Southern North Sea (De Saint Laurent & Bozic, 1976); in areas with fine or muddy sediments; Oyster Ground and the Outer Silver Pit, Markham's Hole; Helgoland Channel in sandy mud; Botney Cut, S. of 53°20'N; Southern Bight;

Brown Ridge; 52°19.59'N 03°10.40'E (Adema et al., 1982). E. of Homboröy, Homborsund, 58°15'N c. 08°35'E, 60-100 m (Christiansen & Greve, 1982). — North Sea: 53°41.45'N 03°55'W (De Saint Laurent & Bozic, 1976). — German Bight: 54°00.000'N 06°00.000'E; 55°00.02'N 6°20.02'E-55°00.00'N 6°21.47'E, 46 m; 54°00.100'N 07°45.000'E, 34.7 m; 55°00.02'N 5°00.10'E, 42 m; White Bank: 55°18'N 06°07'E, 47 m (Sakai, 1999c); S. Elich Doggerbank: 55°15.28'N 04°30.10'E-55°14.54'N 04°29.88'E, 48.5-48.5 m; North Western: 54°01.00'N 7°45.03'E, 34.1 m; 54°00.97'N 7°44.99'E, 34.1 m; Helgoland (54°57'N 7°54.0'E), (Sakai, 1999c); Oyster ground (Dworschak, 1992). N. W. Atlantic Ocean. — France: Brest (De Saint Laurent & Bozic, 1976); English Channel (Gordon, 1957; Holme, 1961, 1966); Bay of Biscay, 96-400 m (Lagardère, 1973; De Saint Laurent & Bozic, 1976). — Portugal (De Saint Laurent & Bozic, 1976). — Spain: Región Cantábrica, Gijón, Palma de Mallorca, Rosas (Domenech et al., 1981); Canary Islands, Atlantic Ocean, 5-270 m. Mediterranean Sea: W. Mediterranean. — Spain: Balearic Islands. — France: Banyuls (Števčić, 1990). — Italy: Campania, Naples (Sakai, 1999c); Naples (Moncharmont, 1979). — Adriatic Sea: Kvarner region including Rijeka Bay, 50-93 m (Števčić, 1979); Croatia, Limski-Canal, near Rovinj, Istria (Sakai, 1999c). — Aegean Sea (Thessalou-Legaki, 1986, 1987; Dworschak, 1992); Saronikos Gulf (Athanassopoulos, 1917); N. Euboikos Gulf; Rhodos; S. Euboikos Bay (Simboura et al., 1998); N. Euboikos Gulf and Rhodes (Thessalou-Legaki, 1986); Pagassitikos Gulf (Thessalou-Legaki, 1978, 1979). — Israel (Holthuis & Gottlieb, 1958). — North African coast (Lutze, 1938, 1941, as *C. algerica*).]

Callianassa truncata Giard & Bonnier, 1890

Callianassa truncata Giard & Bonnier, 1890: 362, figs. 2, 4; Caroli, 1940: 73; Lutze, 1941: 34; Reverberi, 1942a: 89; Reverberi, 1942b: 225; Caroli, 1946: 66, figs. 1b, 3; Holthuis, 1953a: 91, fig. 4; Lagardère, 1966: 195, pls. 2-5; Faure, 1970: 751; Lagardère, 1973: 94; De Saint Laurent & Bozic, 1976: 19, figs. 2, 10, 18, 29; Beaubrun, 1979: 84, figs. 56, 57, 62, 63, 64; Zavodnik, 1979: 313; Kocatas, 1981: 161-162; Mikashavidze, 1981: 1415; Thessalou-Legaki & Zenetos, 1985: 309; Thessalou-Legaki, 1986: 183; Petrescu & Balasescu, 1995: 99; d'Udekem d'Acoz, 1996: 54; Ziebis et al., 1996a: 619; Ziebis et al., 1996b: 227; Abed-Navandi & Dworschak, 1997: 565, figs. 1-9; Sakai, 1999c: 20; Ngoc-Ho, 2003: 473, fig. 11; d'Udekem d'Acoz, 2003, fig. on website.
Callianassa (Trypaea) truncata; Borradaile, 1903: 546; Bouvier, 1940: 102, fig. 68; Zariquiey Alvarez, 1950: 81, fig. 1, pl. 2 figs. 1-6, pl. 3 fig. 2; Dolgopol'skaia, 1954: 186, figs. 5-9; Zariquiey Alvarez, 1968: 229.
Callianassa italica Parisi, 1915: 64, figs. 1, 2. [Type locality: Italy.]
Callianassa (Trypaea) italica De Man, 1928a: 11, figs. 5-5h; De Man, 1928b: 27, 101.
?*Callianassa truncata*; Dolgopol'skaia, 1954: 186, figs. 5, 6; Dolgopol'skaia, 1969: 316, pls. 35-38; Kobyakova & Dolgopol'skaia, 1969: 286.
Callianassa (Trypaeta) truncata; Băcescu, 1967: 229, fig. 104A, D, d, E (B, C, after Giard & Bonnier, 1890) [misspelling].
Necallianassa truncata; d'Udekem d'Acoz, 1999: 156; Tudge et al., 2000: 143.

Diagnosis. — Mxp3 ischium-merus operculiform, merus rounded on distal margin, male Plp1 uniramous, two-segmented, male Plp2 absent, telson truncate with a median spine on posterior margin (Ngoc-Ho, 2003, fig. 11D, J, L).

Type locality. — Gulf of Naples, Italy.

Distribution. — From Atlantic Ocean: Bay of Biscay, France, 44-57 m depth to Romanian coast, south-eastern Black Sea, 15-40 m depth.

[Atlantic Ocean. — Bay of Biscay, France, 44-57 m depth (Lagardère, 1973; De Saint Laurent & Bozic, 1976). — East Atlantic. Morocco: Temara (De Saint Laurent & Bozic, 1976). — Mediterranean. Melilla, Morocco; Tyrrhenian Sea: Isola del Giglio; Naples (De Saint Laurent & Bozic, 1976). Adriatic Sea, 3-10 m depth (Dworschak, 1997: 566). Ionian Sea (Thessalou-Legaki & Zenetos, 1985); South Patras Gulf, Kefalonia Island, S. Euboikos, Saronikos Gulf, W. Peloponnessos, and Kefalonia. Aegean Sea, 5 m depth (Thessalou-Legaki, 1986); Turkey (Kocatas, 1981). Crete and Israel (De Saint Laurent & Bozic, 1976). Romanian coast, south-eastern Black Sea, 15-40 m depth (Mikashavidze, 1981).]

Callianassa tyrrhena (Petagna, 1792)

Astacus tyrrhenus Petagna, 1792: 418, pl. 5 fig. 3.

Alpheus Tyrhenus Risso, 1816: 94 (only reference), not pl. 2 fig. 2.

Callianassa laticauda Otto, 1821: 11; Otto, 1828: 345, pl. 21 fig. 3; Heller, 1863: 203; Stalio, 1877: 107; Caroli, 1940: 73; Lutze, 1941: 34; Caroli, 1946: 71; Zariquiey Alvarez, 1946: 106; Caroli, 1950: 189; Holthuis, 1953a: 91; Forest & Gantès, 1960: 348. [Type locality: Nice, S. France, Mediterranean.]

Gebios Davyanus Risso, 1822: 243; Risso, 1827: 52. [Type locality: Nice, Mediterranean.]

Callianassa tyrrhena; Risso, 1827: 54 (partim); Holthuis, 1947: 320, fig. 1; Holthuis, 1950: 113 (partim), fig. 40; Holthuis, 1953a: 91, 93, fig 1; Picard, 1957: 48; Forest & Guinot, 1958: 10; Holthuis & Gottlieb, 1958: 62 (partim), not fig. 13 (= *C. subterranea*); Bourdon, 1965: 1; Forest, 1967: 6 (partim); Glaçon, 1971: 1; Števčić, 1971: 529; Pastore, 1972: 107, 111; Neves, 1974: 13, fig. 4; Pastore, 1976: 107; De Saint Laurent & Bozic, 1976: 22, figs. 4, 12, 20, 31; Holthuis, 1977: 57; Zarkanellas & Bogdanos, 1977: 155; Beaubrun, 1979: 90, figs. 55, 60, 61, 66, 67; De Saint Laurent & Le Loeuff, 1979: 53; Domenech et al., 1981: 149; Adema et al., 1982: 26, fig. 7a-c; Manning & Števčić, 1982: 295; García Raso, 1983: 323, fig. 2; Thessalou-Legaki, 1986: 182; d'Udekem d'Acoz, 1986: 101; Türkay et al., 1987: 92; d'Udekem d'Acoz, 1989: 176; Števčić, 1990: 217; Thessalou-Legaki, 1990: 659; Holthuis, 1991: 252, 264, fig. 457; Pérez Sánchez & Moreno Batet, 1991: 1; Dworschak, 1992: 206; Koukouras et al., 1992: 223; Gruner, 1993: 997; Mayoral et al., 1994: 236; Hayward et al., 1995: 432, fig. 8-52; d'Udekem d'Acoz, 1996: 53; Hadjichristophorou et al., 1997: 22; Thessalou-Legaki et al., 1997: 439; Thessalou-Legaki & Kiortsis, 1997: 127; Sakai, 1999c: 21, fig. 3a-c; Thessalou-Legaki et al., 1999: 635; d'Udekem d'Acoz, 1999: 155; Dworschak, 2000b: 155, figs.; Dworschak et al., 2000: 301; Tudge et al., 2000: 143; Vanhaelen, 2001: 147; d'Udekem d'Acoz, 2003, fig. on website.

Callianassa subterranea; H. Milne Edwards, 1837a: 309 (partim?); H. Milne Edwards, 1837b: 130 (partim?), pl. 48 fig. 3-3e; Heller, 1863: 202, pl. 6 figs. 9-11; Ortmann, 1891: 55, pl. 1 fig. 10 (partim?).

Gebios davianii; Risso, 1844: 94.

?*Callianassa subterranea*; Bell, 1847: 219 (partim).

Callianassa Laticauda; Hope, 1851: 14.

Callianassa Candida; Hope, 1851: 14.

Gebia Daviana; Hope, 1851: 15.

Callianassa (Callichirus) laticauda; Stalio, 1877: 664; Carus, 1885: 489; Giard & Bonnier, 1890: 366; Borradaile, 1903: 547; Pesta, 1912: 105; Pesta, 1918: 204; De Man, 1928a: 33, pl. 8 fig. 15-15d; De Man, 1928b: 28, 91, 92, 111; Bouvier, 1940: 102, fig. 69 (partim); O'Céidigh, 1962: 164. [Not *Callianassa laticauda* Otto, 1821.]

Callianassa subterranea laticauda var.; Czerniavsky, 1884: 76, 80.

Callianassa sp. Stebbing, 1893: 184.

Callianassa (Callichirus) Stebbingi Borradaile, 1903: 547. [Type locality: Jersey, N. E. Atlantic.]

Callianassa Stebbingi; Selbie, 1914: 100, pl. 14 figs. 8-10.

Callianassa stebbingi; Caroli, 1921: 245 (not *C. subterranea* after Thessalou-Legaki, 1990); Schellenberg, 1928: 78, figs. 59, 60; Steinitz, 1933: 147; Bodenheimer, 1937: 281; Lutze, 1937: 6, fig.; Lutze, 1938: 165, figs. 1-9; Gottlieb, 1953: 441; Števčić, 1969b: 347.

Astacus tyrrhenus; Monod, 1931: 123.

Callianassa (Cheramus) subterranea; Bouvier, 1940: 102 (partim).

Callianassa (Callichirus) tyrrhena; Zariquiey Alvarez, 1968: 230; Andaloro et al., 1979: 86; Moncharmont, 1979: 72.

Pestarella tyrrhena, Ngoc-Ho, 2003: 479, figs. 13, 14.

Not *Alpheus Tyrhenus* Risso, 1816: 94 (only references), not pl. 2 fig. 2 (= *Pontonia pinnophylax* (Otto, 1821)).

Material examined. — SMF 27253, 1 male (Tl/Cl 48.0/11.0), 1 ovig. female (34.0/8.0, lacking chelipeds), 1 female (49.0/11.2, damaged, lacking minor cheliped), Kuvi-Bay, Croatia, 45°03.934'N 13°39.163'E, by yabby-pump, Rov 01-30d, sand and muddy bottom, 1-2 m, leg. excursion of Frankfurt Univ., 23.viii.2001; SMF 27254, 4 males (36/5.5-19.0/4.0), 1 female (32.0/8.5), Sotto Castello, Limski-Canal, Croatia, 45°03.934'N 13°39.163'E, by yabby-pump, Rov 01-30d, sand and muddy bottom, 1-2 m, leg excursion of Frankfurt Univ., 23.viii.2001; SMF 28044, 1 female (31.9/9.4), Kuvi-Bay, Croatia, 45°05.540'N 13°37.260'E, by yabby-pump, St. Rov 99-10C, muddy bottom, leg. excursion of Frankfurt Univ., 28.viii.1999; SMF 28049, 1 male (50.0/11.5, lacking minor cheliped) , 2 females (41.0/9.0, lacking minor cheliped; 36.0/8.5, lacking P1), Kuvi-Bay, Croatia, 45°05.540'N 13°37.260'E, yabby-pump, Rov 99-10C, muddy bottom, excursion of Frankfurt Univ., 28.viii.1999; SMF 28050, 2 males (50.0/12.5-45.0/10.0), 2 females (54.0/13.5-28.0/6.2), Kuvi-Bay, Croatia, 45°03.940'N 13°39.160'E, by yabby-pump, Rov 97-10c, sand, leg. excursion of Frankfurt Univ., 26.viii.1997; SMF 28051, 3 males (40.0/9.0-26.0/6.5), 1 female (31.0/7.3), Kuvi-Bay, Croatia, 45°03.940'N 13°39.160'E, by yabby-pump, Rov 97-10c, sand, leg. excursion of Frankfurt Univ., 26.viii.1997; SMF28052, 1 male (28.0/6.0), Kuvi-Bay, Croatia, by yabby-pump, Rov 93-03e, sand and mud bottom, 1 m depth, leg. excursion of Frankfurt Univ., 24.viii.1993.

Diagnosis. — Mxp3 ischium-merus subpediform, merus broadly rounded on distomesial angle (Ngoc-Ho, 2003, fig. 13G), male Plp1-2 absent, telson rounded in distal half (Sakai, 1999c, fig. 3b).

Type locality. — Naples, Italy, Mediterranean.

Distribution. — From Isle of Man to Canary Islands, and Zarzis, Tunisia, intertidal to Tantura, Israel, eastern Atlantic Ocean, 5-20 m depth in the Mediterranean Sea.

[East Atlantic: Ireland Ballynakill; Isle of Man (Selbie, 1914). — North Sea: Kattegat. Atlantic Ocean. Jersey, Channel Islands (Borradaile, 1903); Netherlands, Dutch waters: Eierland Ground; Texel Ground; Texel coast; lightship Noord-Hinder (51°32.15'N 02°40.10'E) (Holthuis, 1950); 30 miles off IJmuiden (Holthuis & Heerebout, 1976); 52°20'N 04°00'E; 52°10'N 02°15'E; 52°11'N 04°21'E, near Scheveningen; Southern Bight of North Sea (52°19.08'N 03°13.55'E; 52°16.29'N 03°32.14'E) (Adema et al., 1982). Belgian Coast, Koksijde (Vanhaelen, 2001); off Zeebrugge, southwestern North Sea (51°32.56'N 02°40.28'E-51°34.12'N 02°41.26'E), 37-42.5 m (Sakai, 1999c). Atlantic coast of France: Iles Chausey, S. W. side of Grand Ile (Sakai, 1999c); eastern English Channel (Glaçon, 1971); S. W. English Channel (Bourdon, 1965; d'Udekem d'Acoz, 1986); English Channel, St. Malo (Dworschak, 1992); Coutainville, Roscoff (De Saint Laurent & Bozic, 1976); Concarneau, rocky shore (De Saint Laurent & Bozic, 1976; Sakai, 1999c); Bay of Biscay (De Saint Laurent & Bozic, 1976). Portugal coast (Neves, 1974); S. W. Spain: (García Raso, 1983); Ría de Villaviciosa, Asturias (Domenech et al., 1981; Mayoral et al., 1994); Mauritanie (De Saint Laurent & Le Loeuff, 1979); Canary Islands (Pérez Sánchez & Moreno Batet, 1991); Morocco (Forest & Gantès, 1960; De Saint Laurent & Bozic, 1976); Temara (De Saint Laurent & Bozic, 1976); Rabat (Beaubrun, 1979).

Mediterranean Sea: Alboran Sea (García Raso, 1983). Spain: Cadaqués (Zariquiey Alvarez, 1968; Sakai, 1999c); Cataluña, near Cadaqués, Port Lligat, 0.1-0.5 m, shallow water, muddy sand (Sakai, 1999c); France: Marseille (Bouvier, 1940; Sakai, 1999c); Nice (Otto, 1821; De Saint Laurent & Bozic, 1976); Italy: Emilia Romagna, Cattolica (Sakai, 1999c); Napoli (Petagna, 1792; Otto, 1821; Ortmann, 1891; De Saint Laurent & Bozic, 1976; Sakai, 1999c); Aliki Thalassos, Rimini; Banco Mulla di Muggia near Grado, and Lido di Staranzano near Monfalcone (Dworschak, 1992); Golfe de Tarente, Porto Cesareo (Forest, 1967; De Saint Laurent & Bozic, 1976); Napoli (Moncharmont, 1979); Sicily (Dworschak, 1992). Eastern Mediterranean (De Saint Laurent & Bozic, 1976). Adriatic Sea (Stalio, 1877; Heller, 1863; Števčić, 1990): Slovenia: Piran Gulf (Manning & Števčić, 1982); Strunjan near Piran (Dworschak, 1992). Croatia: Rovinj, Zaule, Aurisina, Lido di Stranzano, Lagoon of Grado, and Punta Sabbioni (Dworschak, 1992); Kuvi-Bay, Croatia, 0.3-1 m, muddy and sandy bottom (Sakai, 1999c). Ionian Sea: (Pastore, 1976; d'Udekem d'Acoz, 1996). Aegean Sea: Saronikos Gulf (Picard, 1957; Zarkanellas & Bogdanos, 1977); Pagassitikos Gulf (Bogdanos & Satsmadjis, 1983); S. Euboikos Gulf; Amvrakikos Gulf, Mytilene and Naxos Islands (Thessalou-Legaki, 1986; Koukouras et al., 1992; d'Udekem d'Acoz, 1996); Peristera (Türkay et al., 1987; Sakai, 1999c); Thalassos Limenas (Dworschak, 1992); Karpathos (Sakai, 1999c); Ornoma Peristeri (39°10.000'N 23°58.000'E), Peristera, Northern Sporades, littoral (Sakai, 1999c); Dodecanesos, Southland (Sakai, 1999c). Crete: Bay of Sauda (Sakai, 1999c); Cyprus (Hadjichristophorou et al., 1997). South-central Mediterranean (Forest & Guinot, 1956). Levantine basin (= Israel, Jordan, Lebanon) (Forest & Guinot, 1958; De Saint Laurent & Bozic, 1976; Kocatas, 1981). Israel: Tantura (De Saint Laurent & Bozic, 1976); Tunisia: Zarzis (Dworschak, 1992).]

Habitat. — Burrows in muddy sand, 5-20 m or deeper. All coasts of British Isles, not uncommon, elsewhere from southern Norway to Mediterranean (Hayward et al., 1995), intertidal.

Callianassa whitei Sakai, 1999

Callianassa Davyana White, 1847: 70 (not *Gebios Davyana* Risso, 1822); De Man, 1928a: 5, 37.
Callianassa (*Callichirus*) *stebbingi*; Pesta, 1918: 201, fig. 63.

Callianassa algerica Lutze, 1938: 168 (partim), fig. 22 (eyestalks = *C. whitei*) [not: figs. 26a (smaller Mxp3, not larger Mxp3), 26b (smaller Mxp3, not larger Mxp3), fig. 27]. [Nomen dubium.]
Callianassa candida; Dworschak, 1992: 194 (partim).
Callianassa whitei Sakai, 1999c: 14 (key), 23, fig. 4a-d; Dworschak, 2002: 63.
Pestarella whitei, Ngoc-Ho, 2003: 484, fig. 15.

Material examined. — SMF27251, 1 male (Tl/Cl 78.0/14.5), isle Santa Katarina, Istria, Croatia, 45°04.651'N 13°37.891'E, S. E.-coast, 0-1 m, St. Rov 01-02, stony coast, leg. excursion of Frankfurt Univ., 13.viii.2001, M. Türkay by snorkel swimming; SIC 578, 1 female (64.0/16.0) (det. J. Lutze as *Callianassa pestae*).

Diagnosis. — Mxp3 ischium-merus operculiform, merus rounded on distomesial angle (Ngoc-Ho, 2003, fig. 15E), male Plp1-2 absent, telson rounded in distal half (Sakai, 1999c, fig. 4b).

Remarks. — Lutze (1938: 162) described a new species, *Callianassa algerica* from Castiglione near Algiers (= Alger), northern Africa. This species is similar to *C. whitei* Sakai, 1999 from Rovinj, Croatia in the shape of the eyestalks, bearing a distolateral projection (Lutze, 1938, fig. 22), but still different, because in *C. whitei* the A1 peduncles are longer than those of the A2, while in *C. algerica* the A1 peduncles are almost as long as the A2 peduncles (Lutze, 1938, fig. 23); the larger Mxp3 (Lutze, 1938, fig. 26a, b) are as in *C. candida*, and the smaller Mxp3 (Lutze, 1938, fig. 26a, b) as in *C. subterranea*. The type specimen of *C. algerica* turned out to be lost, after examining the collections of the Stazione Idrobiologica di Chioggia (SIC), so the present author treats *C. algerica* as a nomen dubium.

This species occurs in the Mediterranean in coarse sand or mud, under stones, from the intertidal to the shallow subtidal (Dworschak, 2002: 64). On Santa Katarina island, Rovinj, the material was caught from the shallow subtidal zone.

Type locality. — Mediterranean.

Distribution. — Mediterranean; rarely found in the Adriatic Sea, Croatia.

[Mediterranean (De Man, 1928a; Sakai, 1999c). Croatia: Montauro, Rovinj (Dworschak, 1992; sand under stone, Sakai, 1999c); Kap Monsena, S. Val Salina, near Rovinj (Sakai, 1999c); S Salu, Rovinj (Dworschak, 1992; intertidal in mud, under stones, Sakai, 1999c); Gustinja, Bale Bay, Istria (0-4 m, sea grass meadow, Sakai, 1999c); Kuvi-Bay, Croatia (Dworschak, 2002).]

West Atlantic species

Callianassa berylae (Heard & Manning, 1998)

Necallianassa berylae Heard & Manning, 1998: 884, figs. 1-3; Tudge et al., 2000: 143.
Callianassa berylae; Sakai, 1999c: 129.

Diagnosis. — Mxp3 ischium-merus operculiform, merus broadly rounded on distomesial angle (Heard & Manning, 1998, fig. 2f), male Plp1 uniramous, two-segmented, male Plp2 uncertain but probably absent, telson truncate with a median spine on posterior margin, bearing two distinct pairs of posteriorly directed, spine-like processes on lateral margins (Heard & Manning, 1998: 888).

Type locality. — South Carolina, 32°00'57"N 79°31'03"W, 43 m, sand.

Distribution. — South Carolina (32°00'57"N 79°31'03"W, 43 m; 32°01'04"N 79°31'05"W, 35 m, sand); Georgia (31°33'38"N 79°39'01"W, 75 m; 31°33'36"N 79°40'21"W, 65 m; 31°33'30"N 79°41'38"W, 53 m).

Callianassa biformis Biffar, 1971

Callianassa biformis Biffar, 1971b: 225, fig. 1; Rabalais et al., 1981: 101; Manning, 1987: 397; Squires, 1990: 349, figs. 184-186; Sakai, 1999c: 26.
Biffarius biformis; Tudge et al., 2000: 143.

Diagnosis. — Mxp3 ischium-merus operculiform, merus rounded on distomesial angle, male Plp1 uniramous, two-segmented, male Plp2 absent, telson truncate, without a median spine on posterior margin (Holthuis, 1991, fig. 443).

Type locality. — Mouth of Doboy Sound, south end of Sapelo Island, McIntosh County, Georgia, sandy lower intertidal.

Distribution. — Bass River, Yarmouth, Massachusetts to Florida, western Atlantic; Gulf of Mexico, sandy lower intertidal.

Callianassa caesari (Heard & Manning, 2000)

Pseudobiffarius caesari Heard & Manning, 2000: 71, figs. 1, 2, 3a-k, n-o, 4, 5a.

Diagnosis. — Mxp3 ischium-merus operculiform, merus rounded on distomesial angle, male Plp1 uniramous, two-segmented, male Plp2 absent, telson truncate with a median spine on posterior margin (Heard & Manning, 2000, figs. 3i, k, 4g).

Remarks. — As cited in the remarks on the genus *Callianassa*, *Pseudobiffarius* is considered a synonym of *Callianassa*.

Type locality. — Buccoo Reef (11°11'N 60°49'W), Tobago.

Distribution. — Northwest corner of Man o'War Bay (11°19'N 60°34'W), Lover's Beach, Tobago, ca. 2 m; Pirate's Cove, east side of Man o'War Bay; Buccoo Reef (11°11'N 60°49'W); Coral Gardens, Buccoo Reef; Pigeon Point (11°10'N 60°51'W); Lowlands Lagoon (= Petit Trou; 13°50'N 61°05'W, 1-3 m).

Callianassa delicatula (Rodrigues & Manning, 1992)

Biffarius delicatulus Rodrigues & Manning, 1992a: 324, fig. 1; Tudge et al., 2000: 143.
Callianassa delicatula; Sakai, 1999c: 27.

Diagnosis. — Mxp3 ischium-merus operculiform, merus rounded on distomesial angle, male Plp1 uniramous, two-segmented, male Plp2 uniramous, two-segmented, telson truncate without a median spine on the posterior margin (Rodrigues & Manning, 1992a, fig. 1j, r, s, w).

Type locality. — Praia do Araça, São Sebastião, Brazil.

Distribution. — São Sebastião, State of São Paulo, Brazil.

Callianassa fragilis Biffar, 1970

Callianassa fragilis Biffar, 1970: 45, fig. 3; Biffar, 1971a: 667, figs. 7, 8; Manning, 1987: 397; Sakai, 1999c: 27.
Biffarius fragilis; Manning & Felder, 1991: 769; Tudge et al., 2000: 143.

Diagnosis. — Mxp3 ischium-merus subsquare, merus straight distally with a rounded distomesial angle, male Plp1 uniramous, two-segmented, male Plp2 absent, telson truncate and slightly concave medially on posterior margin, lacking a posteromedian spine (Biffar, 1971a: 671, figs. 7g, 8b, f).

Type locality. — Puerto Rico, Punta Arenas, sandy flat.

Distribution. — Miami, southeastern Florida; Puerto Rico; Antigua; Venezuela, sandy flat.

Callianassa gaucho (Rodrigues & Manning, 1992)

Poti gaucho Rodrigues & Manning, 1992b: 9, fig. 1; Tudge et al., 2000: 145.
Callianassa gaucho; Sakai, 1999c: 28, fig. 3d-e.

Diagnosis. — Mxp3 ischium-merus subsquare, merus straight distally with an obtuse distomesial angle, male Plp1 uniramous, unsegmented, male Plp2 biramous, telson slightly convex on posterior margin, lacking a posteromedian spine (Rodrigues & Manning, 1992b: 9, fig. 1p, q, j, u).

Type locality. — Off Chui, near the border between Brazil and Uruguay (33°43'S 51°13'W), 150 m.

Distribution. — Off Chui, between Brazil and Uruguay (33°43'S 51°13'W), 150 m.

Callianassa marginata Rathbun, 1901

Callianassa marginata Rathbun, 1901: 92, fig. 15; Schmitt, 1935b: 4; Biffar, 1971a: 654, 689, figs. 15, 16; Coelho & Ramos, 1973: 161; Rabalais et al., 1981: 100; Manning, 1987: 397; Sakai, 1999c: 28.
Callianassa (Callichirus) marginata; Borradaile, 1903: 547; Bouvier, 1925: 472; De Man, 1928b: 29, 94, 113; Schmitt, 1935a: 195, fig. 56.
?*Callichirus marginatus*; Bouvier, 1905: 804.
Cheramus marginatus; Manning & Felder, 1991: 780, fig. 14; Blanco Rambla et al., 1994: 18, figs. 2-3; Tudge et al., 2000: 145.

Diagnosis. — Mxp3 ischium-merus subsquare, merus straight distally with obtuse distomesial angle, male Plp1 uniramous, two-segmented, male Plp2 absent, telson truncate and slightly concave medially on posterior margin, bearing a distinct posteromedian spine (Manning & Felder, 1991: 780, fig. 14b, g, e).

Type locality. — Puerto Rico, Mayaguez Harbor, 315 m.

Distribution. — Northeastern Gulf of Mexico; Arrowsmith Bank, Barbados to Puerto Rico; Colombia in Caribbean Sea, 55-640 m.

Callianassa profunda Biffar, 1973

Callianassa occidentalis Bate, 1888: 29, pl. 2 fig. 2k; Borradaile, 1903: 548; Balss, 1925: 212; De Man, 1928b: 115; Schmitt, 1935b: 3; Biffar, 1971a: 649. [Type locality: 18°29'3N 63°24.6'W, off Sombrero Is., West Indies, 686-724 m.] [Not *Callianassa occidentalis* Stimpson, 1856.]

Cheramus occidentalis Bate, 1888: 32, pl. 2 fig. 1. [Type locality: 18°29'3N 63°24.6'W, off
   Sombrero Is., West Indies, 686-724 m.] [Not *Callianassa occidentalis* Stimpson, 1856 (=
   subjective junior synonym of *Callianassa californiensis* Dana, 1852).]
Callianassa (Cheramus) Batei Borradaile, 1903: 546; De Man, 1928a: 10, pl. 1 fig. 3; De Man,
   1928b: 26, 98. [New name for *Callianassa occidentalis* Bate, 1888.]
Callianassa batei; Schmitt, 1935b: 5; Biffar, 1971a: 649, 654; Manning, 1987: 398.
Callianassa profunda Biffar, 1973: 225, figs. 1-2 [new name for *Callianassa batei* Borradaile,
   1903]; Sakai, 1999c: 28.
Cheramus batei; Manning & Felder, 1991: 780.
Cheramus profundus; Tudge et al., 2000: 145.

Diagnosis. — Mxp3 ischium-merus subsquare, merus straight distally with
an obtuse distomesial angle, male Plp1 uniramous, two-segmented, male Plp2
absent, telson truncate and slightly concave medially on posterior margin,
bearing a distinct posteromedian spine (Manning & Felder, 1991: 780, fig.
14b, g, e).

Remarks. — *Callianassa profunda* Biffar, 1973 was a new name for
Cheramus batei Borradaile, 1903, when Manning & Felder (1991: 780) selected *Cheramus occidentalis* Bate, 1888 as the type species of the genus
Cheramus. However, *Cheramus batei* is a subjective junior homonym of a fossil species, *Callianassa batei* Woodward, 1868. Therefore, Biffar (1973) proposed the replacement name *Callianassa profunda*, which now is the valid
name for the species.

Type locality. — 18°29'3N 63°24.6'W, off Sombrero Is., West Indies, 686-
724 m.

Distribution. — Off Sombrero Is., West Indies, 686-724 m.

Callianassa setimanus (DeKay, 1844)

Gonodactylus setimanus DeKay, 1844: 34, pl. 8 fig. 23; Manning, 1987: 386.
Callianassa Stimpsoni Smith, 1873: 549, pl. 2 fig. 8; Kingsley, 1878: 327; Borradaile, 1903:
   548; Hay & Shore, 1917: 406, pl. 29 fig. 5. [Type locality: "Our species ranges from the
   coast of the Southern [United] States north to Long Island Sound" (Smith, 1873).] [Junior
   homonym of *Callianassa stimpsoni* Gabb, 1864, a fossil species.]
Callianassa atlantica Rathbun, 1926: 107; Schmitt, 1935b: 4; Biffar, 1971a: 654; Rabalais et al.,
   1981: 101, fig. 2; Manning, 1987: 397; Squires, 1990: 334, figs. 181-183 [new name for *C.*
   stimpsoni].
Callianassa (Callichirus) atlantica; De Man, 1928a: 37, pl. 9 fig. 17-17d; De Man, 1928b: 28,
   94, 112; Williams, 1965: 102, fig. 79.
Callianassa stimpsoni; Manning, 1987: 397.
Gilvossius setimanus; Manning & Felder, 1992: 558, fig. 1; Tudge et al., 2000: 143.
Callianassa setimana; Sakai, 1999c: 28.

Diagnosis. — Mxp3 ischium-merus broadened and subsquare, merus rounded on distomesial angle, male Plp1 uniramous, two-segmented, male Plp2 biramous, telson rounded in posterior half, lacking a distinct posteromedian spine (Manning & Felder, 1992, fig. 1c, g).

Type locality. — New York.

Distribution. — Bass River, Nova Scotia to South Carolina; Franklin County, Florida, 38 m; northwestern Gulf of Mexico, 134 m; Colombia.

Callianassa sp. Rabalais, Holt & Flint, 1981

Callianassa sp. Rabalais, Holt & Flint, 1981: 106, fig. 4; Manning, 1987: 397; Sakai, 1999c: 30.

Diagnosis. — Mxp3 ischium-merus subpediform, male Plp1-2 unknown, telson convex and medially concave in one-third the length of the posterior margin, bearing a posteromedian spine (Rabalais et al., 1981, fig. 4B, E).

Distribution. — N. W. Gulf of Mexico.

East Pacific species

Callianassa biffari Holthuis, 1991
(fig. 8)

Callianassa affinis Holmes, 1900: 162, pl. 2 figs. 29-30; Rathbun, 1904: 154; Schmitt, 1921: 119, fig. 81; Stevens, 1928: 341, fig. 18; Haig & Abbott, 1980: 580, pl. 166 fig. 24.3 [a junior primary homonym of *Callianassa affinis* A. Milne-Edwards, 1861, for a fossil species].
Callianassa (*Callichirus*) *affinis*; Borradaile, 1903: 547.
Callianassa (*Trypaea*) *affinis*; De Man, 1928b: 27, 101.
Neotrypaea affinis; Manning & Felder, 1991: 771.
Callianassa biffari Holthuis, 1991: 242, fig. 243 (partim, *C. tabogensis* sp. nov.) [new name for *Callianassa affinis* Holmes, 1900].
Neotrypaea biffari; Tudge et al., 2000: 143.

Material examined. — Panama: ZMUC CRU-3765, 1 male (Tl/Cl 17.0/3.0), 1 female (23.0/4.3.0), S. W. point of Rey Island, Arch. de las Perlas, dredging at depths from about 8 to 15 fms (15-27 m), Th. Mortensen's Pacific Expedition, leg. Th. Mortensen, 26.i.1916; ZMUC CRU-3766, 1 female (15/3.2), S. W. point of Rey Island, Arch. de las Perlas, dredging at depths from about 8-15 fms (15-27 m), Th. Mortensen's Pacific Expedition, leg. Th. Mortensen, 26.i.1916; ZMUC CRU-3767, 1 female (19.0/3.8), Contadora Is., Arch. de las Perlas, dredging at 1-7 fms (2-13 m) depth, Th. Mortensen's Pacific Expedition, leg. Th. Mortensen, 28.i.1916; ZMUC CRU-3768, 2 females (15.0/3.2-21.0/4.0), S. of Rey Is., Arch. de las Perlas, dredging, Th. Mortensen's Pacific Expedition, leg. Th. Mortensen, 27.i.1916. California: ZMUC CRU-

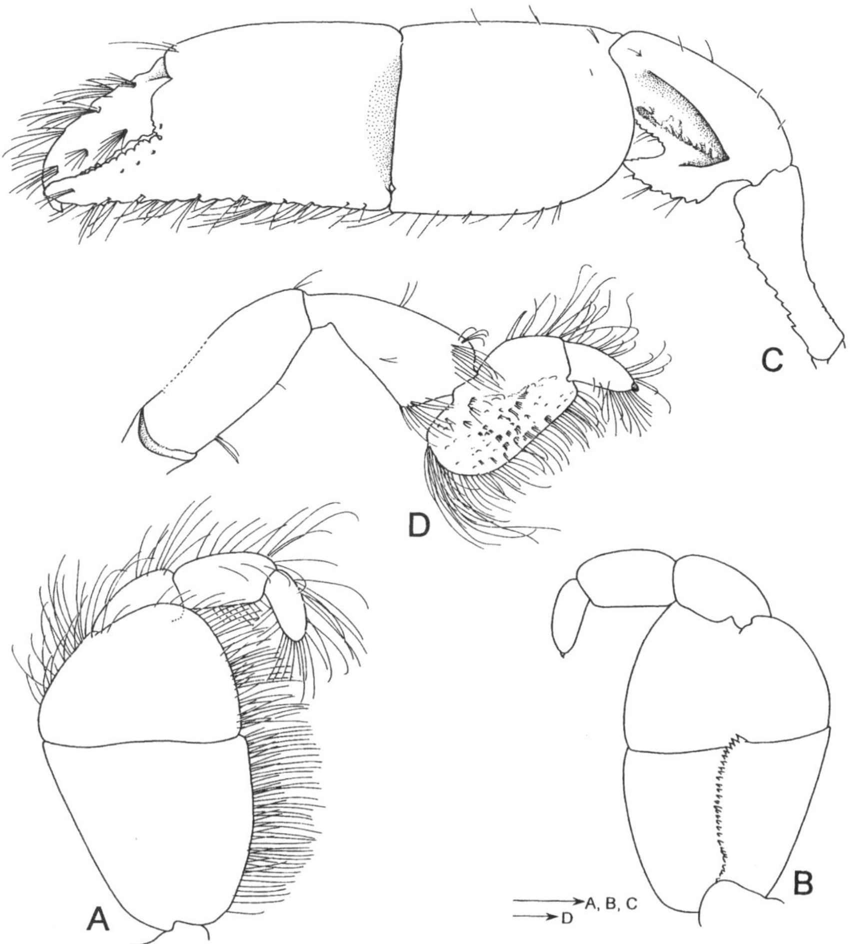

Fig. 8. *Callianassa biffari* Holthuis, 1991. LACM, 1 male (Tl/Cl 45.0/8.2), South Lunada Bay, Palos Verdes Peninsula, 17.iii.1968, leg. M. Wicksten. A, Mxp3, lateral view; B, Mxp3, mesial view; C, larger cheliped, lateral view; D, pereiopod 3. Scales 1 mm.

3769, 1 male (34.0/8.5), shore collections, La Jolla, California, Th. Mortensen's Pacific Expedition, leg. Th. Mortensen, 27.viii.1915; ZMUC CRU-3770, 1 male (56.0/11.5), shore, La Jolla, California, Th. Mortensen's Pacific Expedition, leg. Th. Mortensen, 25.viii.1915; LACM, 1 male (44.0/10.4), Flat Rock Point, Los Angeles County, tide pools in sandstone, leg. M. Wicksten, xii.1968; LACM, 1 male (45.0/8.2), South Lunada Bay, Palos Verdes Peninsula, leg. M. Wicksten, 17.iii.1968.

Diagnosis. — Rostrum a low, blunt angle, frontal margin of carapace with a pair of small lateral processes. A1 peduncle almost as long as A2 peduncle. Mxp3 ischium-merus operculiform, merus rounded on distomesial angle. Palms of larger cheliped in males and females with ridge on ventral margin.

Male Plp1 uniramous and two-segmented, male Plp2 absent. Telson convex on posterior margin, lacking a median spine.

Remarks. — This species is recorded for the first time from Panama. Holthuis (1991: 243) mentioned that *C. biffari* is characterized by the telson being armed with two very small denticles at each posterolateral angle. However, this character was not found in the material examined, though he also showed a telson lacking such lateral denticles as in *C. biffari*, whence the actual position and shape of those denticles are well documented.

Type locality. — Point Loma, California.

Distribution. — Santa Monica Bay, Los Angeles, San Diego and San Clemente Is., U.S.A.; Ensenada and San Quintin Bay, Baja California, Mexico; Panama. On beaches.

Callianassa brachyophthalma A. Milne-Edwards, 1870

Callianassa brachyophthalma A. Milne-Edwards, 1870: 85; De Man, 1928b: 115; Sakai, 1999c: 31.
Callianassa (Trypaea) brachyophthalma; Borradaile, 1903: 546; De Man, 1928b: 27, 94, 115, 134; Holthuis, 1952: 92, fig. 19; Ferrari, 1981: 16, fig. 2.
Notiax brachyophthalma; Manning & Felder, 1991: 772, figs. 6, 11; Tudge et al., 2000: 143.

Diagnosis. — Mxp3 ischium-merus operculiform, merus rounded on distomesial angle, male Plp1 uniramous, two-segmented, male Plp2 uniramous, unsegmented, telson truncate on posterior margin, bearing a median spine (Ferrari, 1981, figs. 2-7, 16, 17, 18).

Type locality. — Chiloé Is., Chile.

Distribution. — Chile.

Callianassa californiensis Dana, 1854

Callianassa californiensis Dana, 1854: 175; Stimpson, 1857b: 4, 89, pl. 21 fig. 4; Stimpson, 1860: 24; A. Milne-Edwards, 1870: 82, 101; Lockington, 1878: 301; Bouvier, 1895: 8; Holmes, 1900: 159, pl. 2 fig. 27; Rathbun, 1904: 154; Hilton, 1916: 63; Schmitt, 1921: 116, 117, fig. 78; Stevens, 1928: 325, 333, figs. 10-13, 16-17, 55-71; MacGinitie, 1934: 166-176, pls. 5-6; MacGinitie, 1935: 709, fig. 14; Haig & Abbott, 1980: 579, pl. 166 fig. 24.2; Hart, 1982: 58, fig. 15; Holthuis, 1991: 244, 264, figs. 445, 446; Dworschak, 1992: 192, fig. 2a, c, e; Sakai, 1999c: 32.
Callianassa occidentalis Stimpson, 1856: 88; Manning, 1987: 399.
Callianassa (Trypaea) californiensis; Borradaile, 1903: 546; De Man, 1928b: 27, 105.
Neotrypaea californiensis; Manning & Felder, 1991: 771, fig. 10; Feldman et al., 2000: 141; Tudge et al., 2000: 143.

Not *Callianassa* (*Trypaea*) *californiensis*; Parisi, 1917: 23 (= *Callianassa japonica*).

Material examined. — ZMUC CRU-3771, 1 male (Tl/Cl 46/10.0), 1 female (48/10.0), False Bay, La Jolla, California, shore collecting, Th. Mortensen's Pacific Expedition, leg. Th. Mortensen, 16.ix.1915.

Diagnosis. — Mxp3 ischium-merus operculiform, merus rounded on distomesial angle, male Plp1 uniramous, two-segmented, male Plp2 absent, telson truncate on posterior margin, bearing a median spine.

Type locality. — California.

Distribution. — Mutiny Bay, Alaska; Vancouver Is. British Columbia to mouth of Tia Juana River, San Diego, California; and Bahia de San Quintin, Mexico. Inhabits sandy sediments in the upper tidal zone (Dworschak, 1992: 196).

Callianassa costaricensis sp. nov.
(figs. 9-10)

Material. — SMF 25816, holotype, 1 male, Tl/Cl 38.0/3.0, Golfo de Nicoya, Bahia Herradura, Costa Rica, Pacific side, 09°38'N 84°41'W, 45 m, coll. R/V "Victor Hensen", GN-50 KG, Box-corer-1, 05.ii.1994.

Diagnosis. — Rostrum shows a low triangular shape. Eyestalks triangular, slightly apart from ventral margin of rostrum, tips obtuse apically, merged with one another on median line, slightly longer than distal end of antennular basal article; cornea small, located medially, pigmented black in alcohol specimen. A1 peduncle obviously longer than that of A2. Mxp3 ischium-merus subsquare, divergent distally, merus rounded on distomesial angle. Male Plp1 uniramous, two-segmened, male Plp2 absent. Telson subsquare, as long as broad, reduced in breadth posteriorly, the posterior margin truncate and not armed with a median spine. Uropodal endopod oval in form, exopod broadly rounded distally, with secondary setal lobe in its anterior half.

Description of male holotype. — Carapace smooth, rostrum showing a low, triangular shape; frontal margin with a pair of low anterolateral processes; dorsal oval conspicuous; cervical groove located in about one-fifth length of carapace. Linea thalassinica extends entire length. Eyestalks triangular, longer than broad, slightly apart from ventral margin of rostrum, tips obtuse apically, merged with one another on median line, slightly longer than distal end of an-

TABLE II

Branchial formula of *Callianassa costaricensis* sp. nov.

	Maxillipeds			Pereiopods				
	1	2	3	1	2	3	4	5
Exopods	1	1	–	–	–	–	–	
Epipods	1	–	–	–	–	–	–	–
Podobranchs	–	r	–	–	–	–	–	–
Arthrobranchs	–	–	2	2	2	2	2	–
Pleurobranchs	–	–	–	–	–	–	–	–

(r = rudimentary)

tennular basal article; cornea small, located medially, pigmented black in alcohol specimen. A1 peduncle (fig. 9A-B) conspicuously longer than that of A2, terminal segment 2.8 times as long as penultimate; flagella short, and about half length of peduncle. A2 scaphocerite vestigial; terminal article distinctly shorter than penultimate segment; antennal flagellum about three times as long as antennal peduncle.

Mxp3 (fig. 9C, D) without exopod; ischium-merus plate of endopod subsquare, slightly increasing in breadth distally, 2.0 times as long as broad; ischium rectangular, 1.3 times as long as broad, crista dentata with sparse row of denticles; merus rectangular, 0.7 times as broad as long, distal margin broadly convex, distolateral margin largely rounded; distal three segments of carpus, propodus, and dactylus pediform, carpus triangular, 1.5 times as long as broad; propodus rectangular, 1.3 times as long as broad; dactylus digitiform, 2.0 times as long as broad, oval at tip. Branchial formula as shown in table II.

P1 unequal in size and dissimilar in shape. Larger cheliped (fig. 9E, F) massive; ischium slender, dorsal margin slightly sinuous and unarmed, ventral margin bearing a row of three indistinct denticles; merus almost as long as ischium, 2.2 times as long as high, dorsal margin slightly arcuate and smooth, ventral margin bearing sharp triangular proximal lobe, and exterior surface inclined in ventral half under indistinct median line. Carpus broadened with rounded posterior margin, 0.9 times as long as high, and 0.9 length of merus. Chela heavy, 2.0 times as long as carpus; palm 1.3 times as long as carpus, about 1.2 times as long as high; dorsal margins smooth, ventral margin carinate in proximal half, bearing a row of distal setae along its ventromesial margin, and distal gap armed with a single small swelling at its ventral corner, below it with a broadly rounded hollow leading to prehensile margin of fixed finger; fixed finger three-fourths length of palm, prehensile margin smooth;

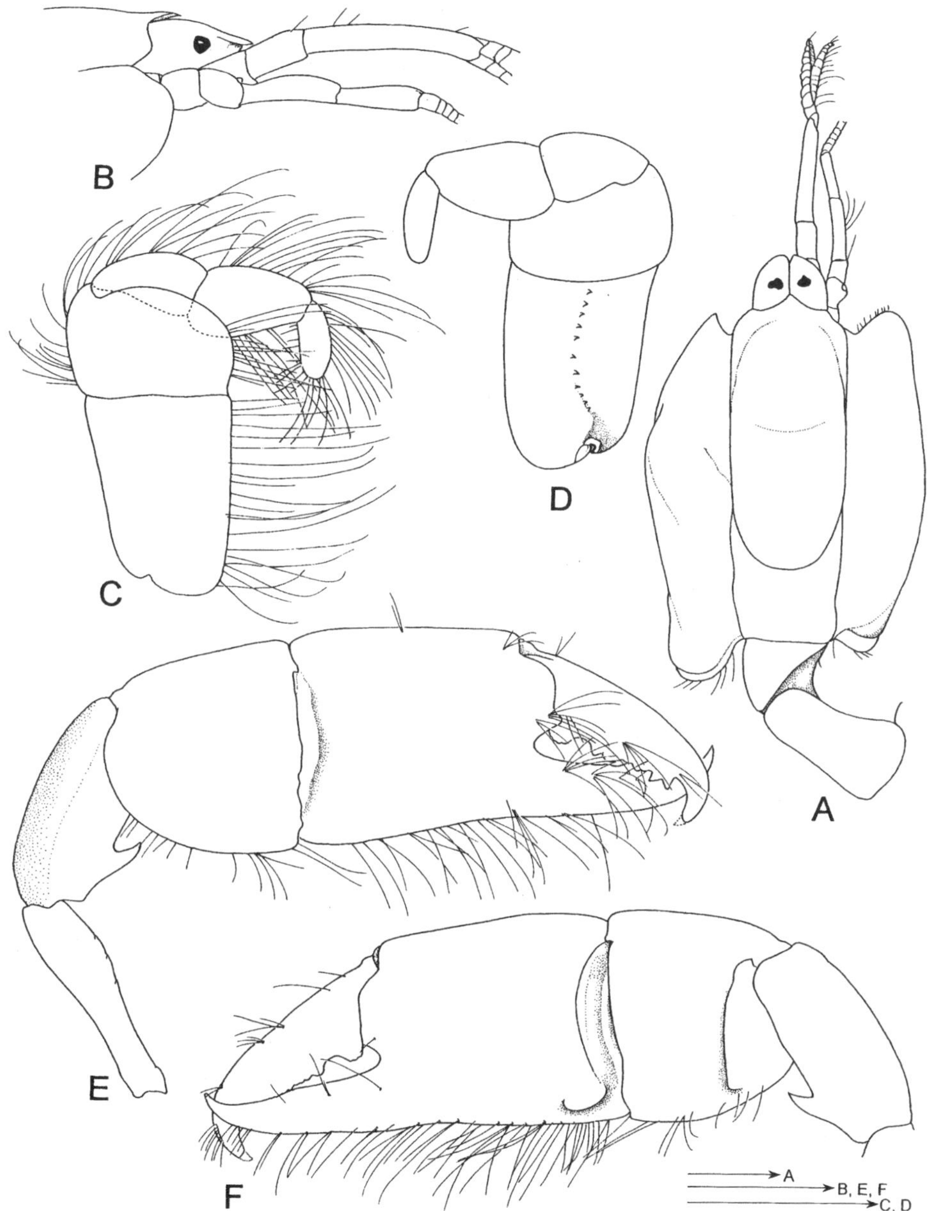

Fig. 9. *Callianassa costaricensis* sp. nov. A, Carapace, dorsal view; B, anterior part of carapace, lateral view; C, Mxp3, lateral view; D, Mxp3, mesial view; E, male larger cheliped, lateral view; F, same, mesial view. A-F, SMF 25816, holotype male (Tl/Cl 38.0/3.0 mm), Golfo de Nicoya, Bahia Herradura, Costa Rica, Pacific side (09°32'38"N 84°32'41"W), 45 m depth, coll. R/V "Victor Hensen". Scales 1 mm.

dactylus sharply incurved distally downward, prehensile margin armed with denticles in irregular shape. Smaller cheliped (fig. 10A) slender and less massive than larger cheliped; ischium slender, dorsal and ventral margins unarmed; merus slender, of a narrow spindle-shape, about as long as ischium, dorsal and ventral margins slightly arched, respectively, and unarmed; carpus elongate and triangular, 3 times as long as high, proximoventral margin descending forwards in straight line. Chela as long as carpus, slightly increasing in breadth distally along its ventral margin; palm subsquare, about 1.7 times as long as high, distal margin unarmed, and protruded in its ventral half leading to fixed finger; fixed finger 0.6 times as long as palm, prehensile margin unarmed; dactylus slender, slightly shorter than palm, slightly longer than fixed finger, prehensile margin unarmed.

P2 chelate, ischium about as long as broad; merus 2.5 times as long as high, and 2.8 times as long as ischium; carpus 1.8 times as long as high, and 0.6 times as long as merus, chela about as long as carpus, and dactylus about 2.0 times as long as palm.

P3 (fig. 10B) ischium rectangular, 1.8 times as long as broad; merus 2.8 times as long as high and 2.2 times as long as ischium; carpus triangular in form, 2.0 times as long as broad, and 0.8 times as long as merus; propodus kidney-shaped, as long as high, posteroventral angle squarely protruded, and ventral margin entirely truncate; dactylus triangular in shape, slightly shorter than palm.

P4 (fig. 10C) simple, ischium 2.8 times as long as broad; merus 2.8 times as long as high, and 0.8 times as long as ischium; carpus 2.8 times as long as high, and about 0.7 times as long as merus; propodus rectangular, slightly more than 2.0 times as long as high, and 0.8 times as long as carpus, ventrodistal corner densely setose and not protruded; dactylus about half length of propodus.

P5 chelate (fig. 10D); propodus protruded ventrodistally, forming a chela with dactylus, ventral surface with dense setation; dactylus hooked towards external side of fixed finger, tip deflected.

Abdominal somites smooth, glabrous dorsally; pleura 2-5 each with a tuft of setae laterally; abdominal somite 6 (fig. 10H) about as long as broad, smooth on lateral margin.

Telson (fig. 10H) subsquare, about as long as broad, lateral margins divergent proximally and then convergent to distolateral angles in distal fifth; posterior margin almost in straight line, and without setosity and a median spine;

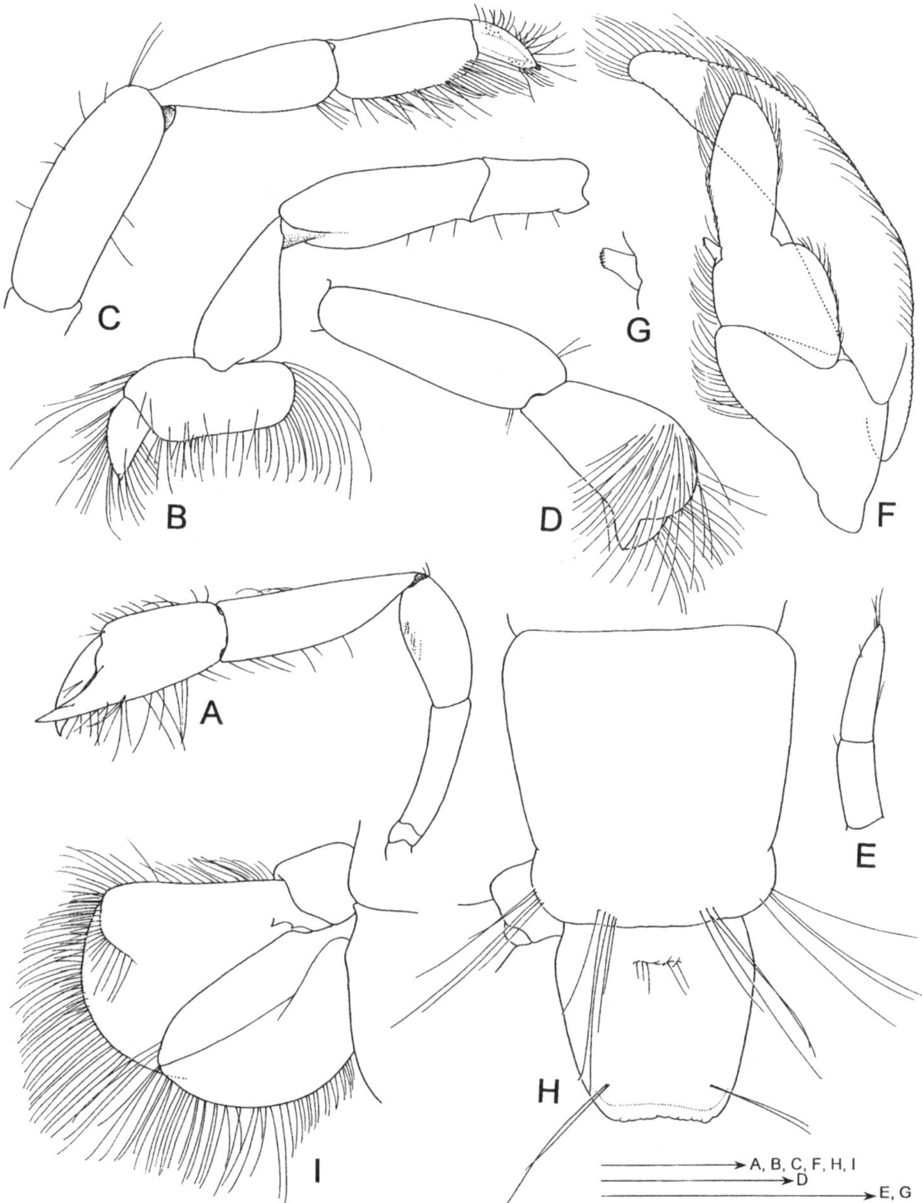

Fig. 10. *Callianassa costaricensis* sp. nov. A, smaller cheliped, lateral view; B, pereiopod 3, lateral view; C, pereiopod 4, lateral view; D, carpus, propodus, and dactylus of pereiopod 5, lateral view; E, male Plp1; F, Plp3, posterior view; G, appendix interna of Plp3; H, abdominal somite 6 and telson, dorsal view; I, uropods on left side. A-I, SMF 25816, holotype male (Tl/Cl 38.0/3.0 mm), Golfo de Nicoya, Bahia Herradura, Costa Rica, Pacific side (09°32'38"N 84°32'41"W), 45 m, coll. R/V "Victor Hensen". Scales 1 mm.

dorsal surface with transverse row of setae posteromedially. Uropodal endopod (fig. 10I) oval in form; anterior margin slightly convex, extending to entirely rounded posterior margin over posterolateral angle; dorsal surface with longitudinal median carina. Uropodal exopod larger than endopod, and slightly longer than broad, broadly rounded distally, dorsal surface armed with transverse anterodistal ridge with setosity at anterolateral angle.

Plp1 (fig. 10E) uniramous, 2-segmented. Plp2 absent. Plp3 (fig. 10F) to Plp5 biramous, slender, foliaceous, each bearing a small, stubby appendix interna (fig. 10G) projecting medially on mesial margin of endopod.

Etymology. — Named after the type locality. The specific name is an adjective agreeing in gender with the (feminine) generic name.

Remarks. — This specimen was sent to me for study through the courtesy of Dr. M. Apel, of the Museum Wiesbaden, Germany. Ten *Callianassa* species are known from the eastern Pacific: *Callianassa californiensis* Dana, 1854, *C. costaricensis* sp. nov., *C. gigas* Dana, 1852, *C. biffari* Holthuis, 1991, *C. brachyophthalma* A. Milne-Edwards, 1870, *C. debilis* (Hernández-Aguilera, 1998), *C. rochei* Bouvier, 1895, *C. santarita* (Thatje, 2000), *C. tabogensis* sp. nov., and *C. uncinata* H. Milne Edwards, 1837. The present new species, *C. costaricensis* sp. nov., is morphologically similar to *C. biffari* by the A1 peduncle distinctly longer than that of the A2, and by the eyestalks overreaching the distal margin of the A1 proximal segment, but different from *C. biffari* by the form of Mxp3, P1 palm, P3 propodus, and telson.

In *C. biffari*, the Mxp3 ischium-merus (fig. 8A, B) is oval, 1.5 times as long as broad, and entirely convex on its lateral margin, crista dentata with a row of many denticles; merus roundly convex around mesiodistal margin; P1 palm of larger cheliped denticulate on ventromesial margin in both sexes; P3 propodus (fig. 8D) kidney-shaped, as long as high, posteroventral angle squarely protruded, and lateral margin straight and parallel to dorsal margin; the telson is broadly rounded.

In *C. costaricensis*, the Mxp3 ischium-merus (fig. 9A, B) is subquadrate, 2.0 times as long as broad, and almost straight on its exterior margin; crista dentata with a sparse row of denticles; merus broadly convex on mesiodistal margin; P1 palm of larger side smoothly carinate in proximal half, bearing a row of distinct setae on ventromesial margin; P3 propodus kidney-shaped, 1.2 times as long as high, posteroventral angle squarely protruded and lateral margin inclined towards posterior angle; the telson is distally truncate.

Type locality. — Golfo de Nicoya, Bahia Herradura, Costa Rica, GN-50 KG (09°32'38"N 84°32'41"W), 45 m.

Distribution. — Known only from the type locality.

Callianassa debilis (Hernández-Aguilera, 1998)

Biffarius debilis Hernández-Aguilera, 1998, 303, fig. 1; Tudge et al., 2000: 143.

Diagnosis. — Mxp3 ischium-merus subsquare, merus rounded on distomesial angle, male Plp1 uniramous, two-segmented, male Plp2 absent, telson concave on posterior margin, lacking a median spine (Hernández-Aguilera, 1998, fig. 1h, i, l).

Type locality. — Revillagigedo Archipelago: 18°20'52"N 114°43'43"W, Azufre Bay, Clarión Island.

Distribution. — Revillagigedo Archipelago, Pacific side of Mexico.

Callianassa gigas Dana, 1852

Callianassa gigas Dana, 1852a: 19; Dana, 1852b: 512; Dana, 1855, pl. 32 fig. 3; Stimpson, 1857b: 489, pl. 21 fig. 3; A. Milne-Edwards, 1870: 81, 101; Lockington, 1878: 302; Holmes, 1900: 162; Rathbun, 1904: 154; Schmitt, 1921: 116, 119, fig. 80; Stevens, 1928: 325, figs. 6-9, 14-15, 38-54; MacGinitie, 1935: 712; Haig & Abbott, 1980: 579; Hart, 1982: 56, fig. 14; Holthuis, 1991: 245, 264, figs. 447, 448; Dworschak, 1992: 194, fig. 2b, d, f; Sakai, 1999c: 32.
Callianassa longimana Stimpson, 1857a: 86; Stimpson, 1857b: 490, pl. 21 fig. 5; A. Milne-Edwards, 1870: 83, 101; Lockington, 1878: 302; Neumann, 1878: 34; Holmes, 1900: 161, pl. 2 fig. 28; Rathbun, 1904: 154; Hilton, 1916: 63, fig. 14; Schmitt, 1921: 116, 117, fig. 79. [Type locality: Puget Sound, Washington State, U.S.A.]
Callianassa (Trypaea) gigas; Borradaile, 1903: 546; De Man, 1928b: 27, 101 ,134, 180, 181.
Callianassa (Trypaea) longimana; Borradaile, 1903: 546; De Man, 1928b: 27, 102, 106, 134.
Neotrypaea gigas; Tudge et al., 2000: 143.

Diagnosis. — Mxp3 ischium-merus subsquare, merus rounded on distomesial angle, male Plp1 uniramous, two-segmented, male Plp2 absent, telson concave on posterior margin, bearing a median spine (Holthuis, 1991, fig. 447).

Type locality. — Puget Sound.

Distribution. — Vancouver Island, British Columbia; Puget Sound; Elkhorn Slough, Monterey Bay, San Juan Island, Poulsbo, Allyn, Washington, San Francisco, California to San Quentin Bay and Gulf of Farallones. Lower tidal in muddy sediments.

Callianassa rochei Bouvier, 1895

Callianassa Rochei Bouvier, 1895: 7.
Callianassa (Trypaea) Rochei; De Man, 1928a: 17, fig. 8-8d; De Man, 1928b: 28, 104.
Callianassa rochei; Sakai, 1999c: 33.
Neotrypaea rochei; Tudge et al., 2000: 143.

Diagnosis. — Mxp3 ischium-merus suboperculiform, merus rounded on distomesial angle, male Plp1-2 undescribed, telson undescribed.

Type locality. — Baja California.

Distribution. — Baja California.

Callianassa santarita (Thatje, 2000)

Notiax santarita Thatje, 2000: 289, figs. 1-6, tab. 1.

Diagnosis. — Mxp3 ischium-merus operculiform, male Plp1 uniramous, two-segmented, male Plp2 absent, telson straight with a median spine (Thatje, 2000: 295, 297, figs. 3B, 6C).

Remarks. — This species is included in *Callianassa*, because the carapace has a dorsal oval; the A1 peduncle is longer than the A2 peduncle; and the Mxp3 propodus is broadened. Fig. 6D in Thatje (2000) is referred to as the left pleopod 2, however, this is in error as shown by the editional board of Crustaceana in the erratum, 2000: 1018, because the description says that Plp2 is absent (Thatje, 2000: 295).

Type locality. — Seno Ponsonby, 55°12.0'S 68°87.03'W, Chile, sublittoral, 65 m, mud to sandy-mud.

Distribution. — Seno Ponsonby, Chile.

Callianassa tabogensis sp. nov.
(figs. 11-12)

Callianassa biffari Holthuis, 1991: 242 (partim).

Material examined. — ZMUC CRU-3772, holotype, male (Tl/Cl 21.0/4.1), shore on a small island off the northern coast of Taboga, Bahia de Panama, Panama, Th. Mortensen's Pacific Expedition, leg. Th. Mortensen, 03.iii.1916.

TABLE III

Branchial formula of *Callianassa tabogensis* sp. nov.

	Maxillipeds			Pereiopods				
	1	2	3	1	2	3	4	5
Exopods	1	1	–	–	–	–	–	
Epipods	1	–	–	–	–	–	–	–
Podobranchs	–	r	–	–	–	–	–	–
Arthrobranchs	–	–	2	2	2	2	2	–
Pleurobranchs	–	–	–	–	–	–	–	–

(r = rudimentary)

Diagnosis. — Rostrum triangular; front margin of carapace with pair of obscure anterolateral protuberances. Eyestalks triangular and convergent distally, reaching about to distal end of antennular basal segment. Antennular peduncle longer than antennal one. Mxp3 ischium-merus oval; carpus, propodus, and dactylus pediform. Male P1 unequal and dissimilar; larger cheliped with ischium narrow, and armed with a few denticles on ventral margin; merus with a sharply pointed proximal lobe on ventral margin; dactylus distinctly incurved distally, bearing a thick, truncate median tooth on prehensile margin. Male Plp1 uniramous, male Plp2 absent. Telson subsquare, lateral margins with two spines at rounded posterior corner, posterior margin convex and slightly concave medially, with a median spine.

Description of male holotype (fig. 11A). — Rostrum (fig. 11B-C) triangular and pointed at tip. Frontal margin of carapace with a pair of obscure anterolateral projections; dorsal oval conspicuous; cervical groove located in posterior fourth of carapace; linea thalassinica entire. Eyestalks (fig. 11B) triangular, convergent distally, reaching about to distal end of antennular basal seg-ment, dorsal surface convex; cornea rounded, located medially, pigmented brown in alcohol specimen. Antennular peduncle distinctly longer than antennal one, terminal segment about 2.0 times as long as penultimate. A2 terminal segment about as long as penultimate; scaphocerite oval and vestigial; antennal flagellum about 5 times as long as antennular flagella. Mxp3 ischium-merus (fig. 11D) oval; ischium divergent distally, about as long as broad; crista dentata (fig. 11E) with a sparse row of denticles; merus subtriangular, 0.7 times as long as broad, largely protruded and rounded on mesiodistal margin. Carpus, propodus, and dactylus pediform; carpus triangular, 2.0 times as long as broad; propodus subquadrate, longer than broad; dactylus digitiform, 0.8 times as long as propodus; no exopod present. Branchial formula as shown in table III.

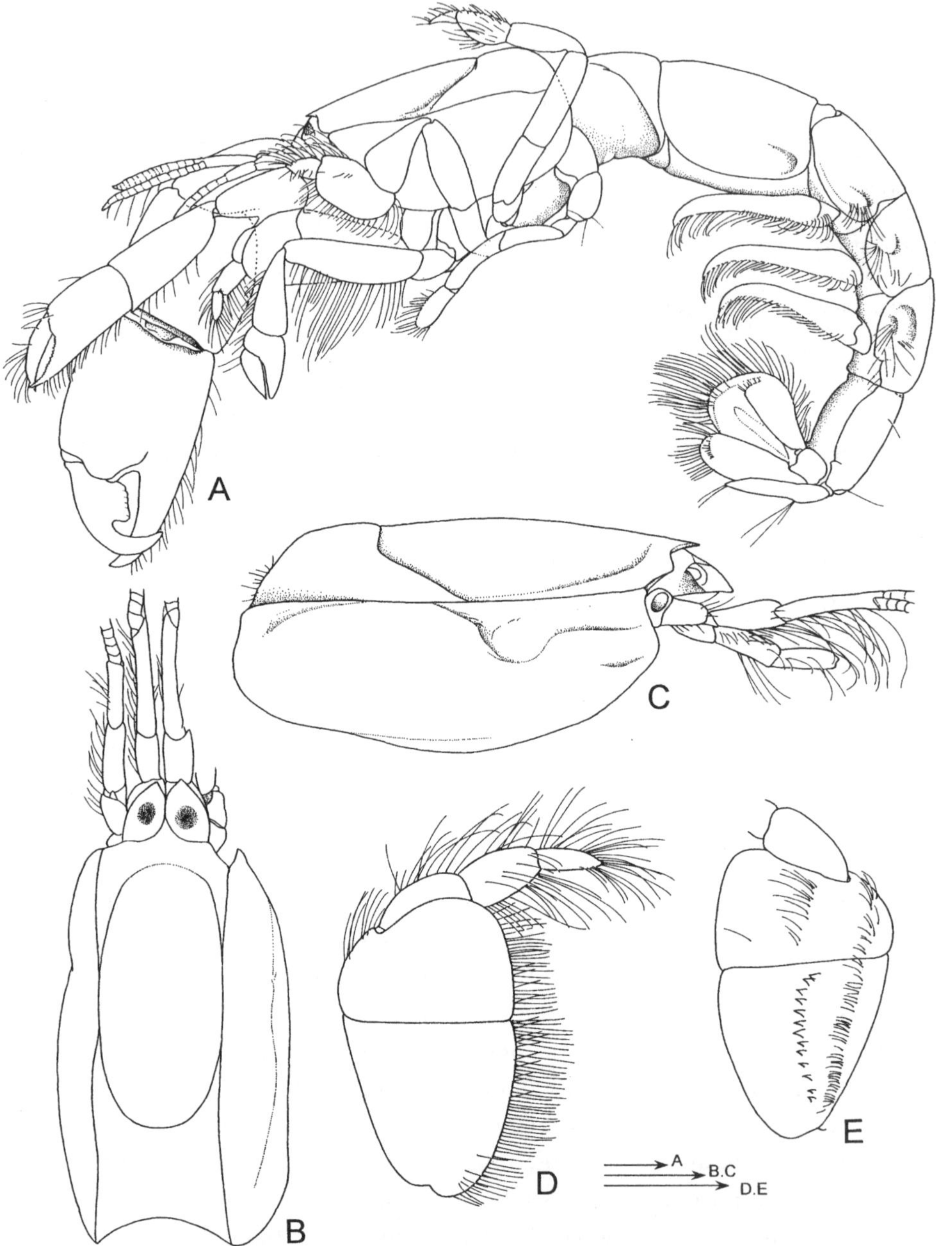

Fig. 11. *Callianassa tabogensis* sp. nov. A, whole body, lateral view; B, carapace with eyes, and antennal and antennular peduncles, dorsal view; C, same, lateral view; D, Mxp3, lateral view; E, same, mesial view. A-E, ZMUC 211, holotype male, Taboga, Bahia de Panama, Panama, shore on a small island off the northern coast. Scales 1 mm.

P1 unequal and dissimilar. Larger cheliped (fig. 12A) massive; ischium slender; dorsal margin entirely concave and unarmed, ventral margin with a row of five denticles medially. Merus about as long as ischium, about 2.0 times as long as high, dorsal margin slightly arcuate and roughly crenulate in its proximal half; ventral margin provided with sharply pointed proximal lobe, exterior surface convex in upper half, and concave around basal part of proxi mal lobe; proximal lobe denticulate on ventral margin. Carpus broadened, about as high as long, about as long as merus, and largely rounded on posteroventral margin. Chela heavy, 1.9 times as long as carpus; palm 1.2 times as long as carpus, about 1.2 times as long as high, dorsal and ventral margins smooth, distal gap convex with a denticle above broad concavity at the base of fixed finger, fixed finger 0.6 times as long as palm, prehensile margin smooth; dactylus distinctly incurved downward distally, prehensile margin with a thick, truncate median tooth. Smaller cheliped (fig. 12B) slender and less massive than larger cheliped; ischium narrow, dorsal and ventral margins unarmed; merus spindle-shaped, about as long as ischium, ventral margin with a small median tooth; carpus elongate, 2.1 times as long as high, largely divergent on proximoventral margin. Chela 1.2 times as long as carpus; palm subsquare, about 1.4 times as long as high; distal gape with a small triangular tooth at its dorsal corner; fixed finger 0.7 times as long as palm, prehensile margin finely denticulate; dactylus slender and incurved distally, slightly shorter than palm, about as long as fixed finger; prehensile margin unarmed. P2 (fig. 11A) chelate; ischium subsquare, about as long as broad; merus 3.5 times as long as high and 4.5 times as long as ischium; carpus 0.7 times as long as merus; chela about as long as carpus, dactylus 1.8 times as long as palm. P3 (fig. 12C) ischium 1.8 times as long as broad; merus more than 2.0 times as long as ischium and 2.8 times as long as high; carpus triangular, 0.8 times as long as merus; propodus bean-shaped, about as long as high, posteroventral margin roundly protruded, lateral surface setose; dactylus 0.7 times as long as palm, pointed at tip. P4 (fig. 11A) simple; ischium 2.5 times as long as high, merus 1.5 times as long as ischium; carpus 0.7 times as long as merus; propodus rectangular, 1.2 times as long as carpus, lateral surface setose, ventrodistal corner not protruded; dactylus half the length of propodus and setose on external surface. P5 subchelate; merus 3.5 times as long as ischium, carpus 0.8 times as long as merus, propodus slightly shorter than carpus, forming a short fixed finger ventrodistally; dactylus hooked towards external side of fixed finger, tip incurved.

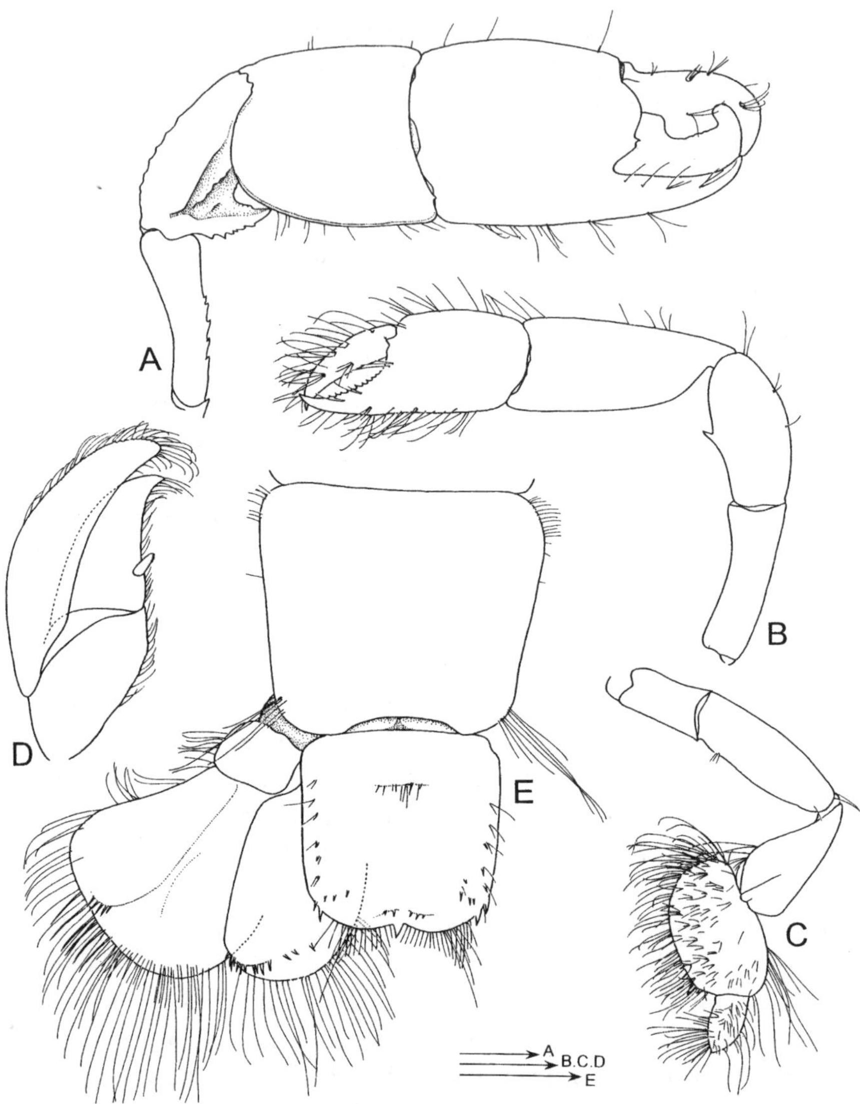

Fig. 12. *Callianassa tabogensis* sp. nov. A, larger cheliped, lateral view; B, smaller cheliped, lateral view; C, pereiopod 3; D, Plp3; E, abdominal somite 6 and telson, with uropod on left side, dorsal view. A-E, ZMUC 211, holotype male, Taboga, Bahia de Panama, Panama, shore on a small island off the northern coast. Scales 1 mm.

Abdominal somites smooth, glabrous dorsally; pleura 2-5 each with a tuft of setae laterally; abdominal somite 6 (fig. 12E) smooth, 0.8 times as long as broad, slightly convergent laterally. Plp1 uniramous, 2-segmented. Plp2 absent. Plp3 (fig. 12D) to Plp5 biramous, narrowly foliaceous, each bearing a small triangular appendix interna on the mesial margin of the endopod. Telson square, almost as long as broad, almost parallel laterally, lateral margins posteriorly armed with two sharp spines at rounded posterior corner; posterior margin convex and slightly concave medially, with a median spine; dorsal surface medially armed with a transverse row of setae in proximal third, and with 5-6 tufts of setosity and spinules, arranged longitudinally along each lateral margin; a row of a few spinules around each posterolateral angle and median posterior margin. Uropodal endopod rounded distally, longer than telson; anterior and posterior margins parallel and forming straight lines, dorsal surface with a row of spinules near the posterior margin. Uropodal exopod truncate distally, larger than endopod, about as long as broad; dorsal surface elevated, in anterior half bordered by longitudinal carina, bearing 1-3 denticles distally.

Etymology. — The species, *Callianassa tabogensis*, is named after the type locality, Taboga, Bahia de Panama, Panama. The specific name is an adjective agreeing in gender with the (feminine) generic name.

Remarks. — This species is closely similar to *Callianassa biffari* Holthuis, 1991 from California. *C. tabogensis* differs in having a triangular rostrum with a pointed tip, eyestalks that reach to about the distal end of the antennular basal segment; in the male larger cheliped the palm is smooth on the ventral margin, in the smaller cheliped the merus has a median spine on the ventral margin, the telson is armed, and the dorsal surface has 5-6 tufts of setae and spinules along each lateral margin. In contrast, in *C. biffari* the rostrum is blunt distally, the eyestalks overreach the antennular basal segment; in the male larger cheliped the palm is denticulate on the ventral margin, in the smaller cheliped the merus is unarmed on the ventral margin, and the telson is unarmed, but with two very small denticles at the posterolateral angles, and without a median spine on the posterior margin; the dorsal surface has three tufts of setae along each lateral margin. In *C. biffari* the palm of the larger cheliped is entirely denticulate, with a row of setae on the ventral margin; in *C. costaricensis* the palm is smoothly carinate in the proximal half of the ventral margin, bearing a row of distinct setae along its ventromesial edge, while in *C. tabogensis* the palm is smoothly carinate entirely, bearing a row of fine setae along its ventromesial margin.

Type locality. — Taboga, Bahia de Panama, Panama.

Distribution. — Known only from the type locality.

Callianassa uncinata H. Milne Edwards, 1837

Callianassa uncinata H. Milne Edwards, 1837a: 310, pl. 25 fig. 1; Nicolet, 1849: 208; Guérin-
   Méneville, 1857: 43; A. Milne-Edwards, 1870: 83, 101; Czerniavsky, 1884: 76; Sakai,
   1999c: 33, fig. 3f.
Callianassa chilensis A. Milne-Edwards, 1860: 302, pl. 16 fig. 2a, 2A; A. Milne-Edwards,
   1870: 84, 101; Tudge et al., 2000: 143. [Type locality: Chile.]
Callianassa (Trypaea) uncinata; Borradaile, 1903: 546; Schmitt, 1921: 119, fig. 80; Stevens,
   1928: 325, figs. 6-9, 14-15, 38-54; MacGinitie, 1935: 712.
Callianassa (Trypaea) chilensis; Borradaile, 1903: 546; De Man, 1928a: 15, fig. 7-7c; De Man,
   1928b: 27, 94, 103.
Neotrypaea uncinata; Manning & Felder, 1991: 771; Tudge et al., 2000: 143.

Diagnosis. — Mxp3 ischium-merus suboperculiform, merus rounded on distomesial angle, male Plp1 uniramous, two-segmented, male Plp2 absent, telson
concave on posterior margin, bearing a median spine (Holthuis, 1991, fig.
447).

Type locality. — Coast of Chile.

Distribution. — Island of Quehuy, off Chile; Puerto Montt, province of
Llanquihue, southern Chile; Tumbes, near Lago Llanquihue; Talcahuano,
Chile; Capon, Peru. Tide mark.

Indo-West Pacific species

Callianassa acutirostella Sakai, 1988
(figs. 13-14)

Callianassa acutirostella Sakai, 1988: 57, fig. 2; Sakai, 1999c: 37; Tudge et al., 2000: 143; Davie, 2002: 458.

Material examined. — ZMUC CRU-3773, 2 ovig. females (Tl/Cl 21.0/4.6-23.0/4.8), 2 females (18.0/4.2-21.0/4.5), 10°43'S 139°17'E, Arafura Sea, coral-sand and gravel, 54 m depth,
"Galathea" Exped. 1950-1952, Station 502, leg. R/V "Galathea" 27.ix.1951; ZMUC CRU-3774,
1 female (18.0/4.1), 10°37'S 139°19'E Arafura Sea, coral-sand and gravel, 57 m depth, "Galathea" Exped. 1950-1952, Station 503, leg. R/V "Galathea", 27.ix.1951.

Diagnosis. — Mxp3 ischium-merus suboperculiform, merus rounded on distomesial margin, male Plp1-2 unknown, telson almost straight on posterior
margin, bearing a median spine.

Supplementary description of female (fig. 13A). — Rostrum acute (fig.
13B, C); frontal margin of carapace with triangular anterolateral process. Eyes
extending anteriorly slightly beyond antennular basal segment. Antennular pe-

duncle distinctly longer than antennal peduncle, distal segment 3 times as long as penultimate segment. Scaphocerite small, rod-like. Cervical groove located in posterior fifth of carapace. Mxp3 (fig. 13D) ischium-merus subpediform, 1.8 times as long as wide; merus broad, half as long as ischium, mesiodistal margin entirely convex; crista dentata with row of 18 denticles. Carpus, propodus, and dactylus narrow.

P1 unequal and dissimilar. Larger cheliped (fig. 14A) massive; ischium slender, dorsal margin almost straight, unarmed, ventral margin armed with row of sharp denticles; merus spindle-shaped, as long as ischium, about 1.8 times as long as high, dorsal margin entirely arcuate and smooth, ventral margin arched, with distinct medial tooth. Carpus triangular, divergent on entire ventral margin, 1.5 times as long as high, 1.1 times as long as merus. Chela solid, robust, 1.8 times as long as carpus; palm slightly shorter than carpus, about as long as high, dorsal and ventral margins smooth, distal gap protruding on lower half; fixed finger shorter than palm, proximal two-thirds of prehensile margin armed with minute denticles, distal third concave, smooth; dactylus slender, incurved, overreaching fixed finger, prehensile margin unarmed. Smaller cheliped (fig. 14B) slender, less massive than larger; ischium narrow, dorsal margin unarmed, ventral margin armed with row of sharp denticles; merus spindle-shaped, about as long as ischium, dorsal margin smooth, entirely arched, entire ventral margin slightly convex and armed with small median tooth; carpus triangular, 1.8 times as long as high, entire ventral margin divergent. Chela 1.8 times as long as carpus; palm subsquare, about as long as high; distal gape convex on lower half; fixed finger as long as palm, prehensile margin in proximal three-fifths minutely denticulate, distal two-fifths concave, unarmed; dactylus slender and incurved, 1.5 times as long as palm, conspicuously overreaching fixed finger, prehensile margin unarmed. P2 chelate; merus broad, 2.8 times as long as ischium, flexor margin with closely set setae; carpus 0.6 times as long as merus, setose on dorsal and flexor margins; chela 1.2 times as long as carpus, margins setose; both fingers 1.8 times as long as palm, corneous on prehensile margins and tips. P3 (fig. 14C) simple; ischium rectangular, 1.8 times as long as broad; merus 2.0 times as long as ischium, carpus triangular, 0.8 times as long as merus; propodus rounded, 0.5 times as long as carpus, as long as high, ventral margin entirely rounded; dactylus digitiform. P4 subchelate, ischium rectangular, 2.2 times as long as broad; merus 3.0 times as long as ischium; carpus 0.7 times as long as merus; propodus about as long as carpus, ventrodistal angle not protruded; dactylus slightly less than 0.5

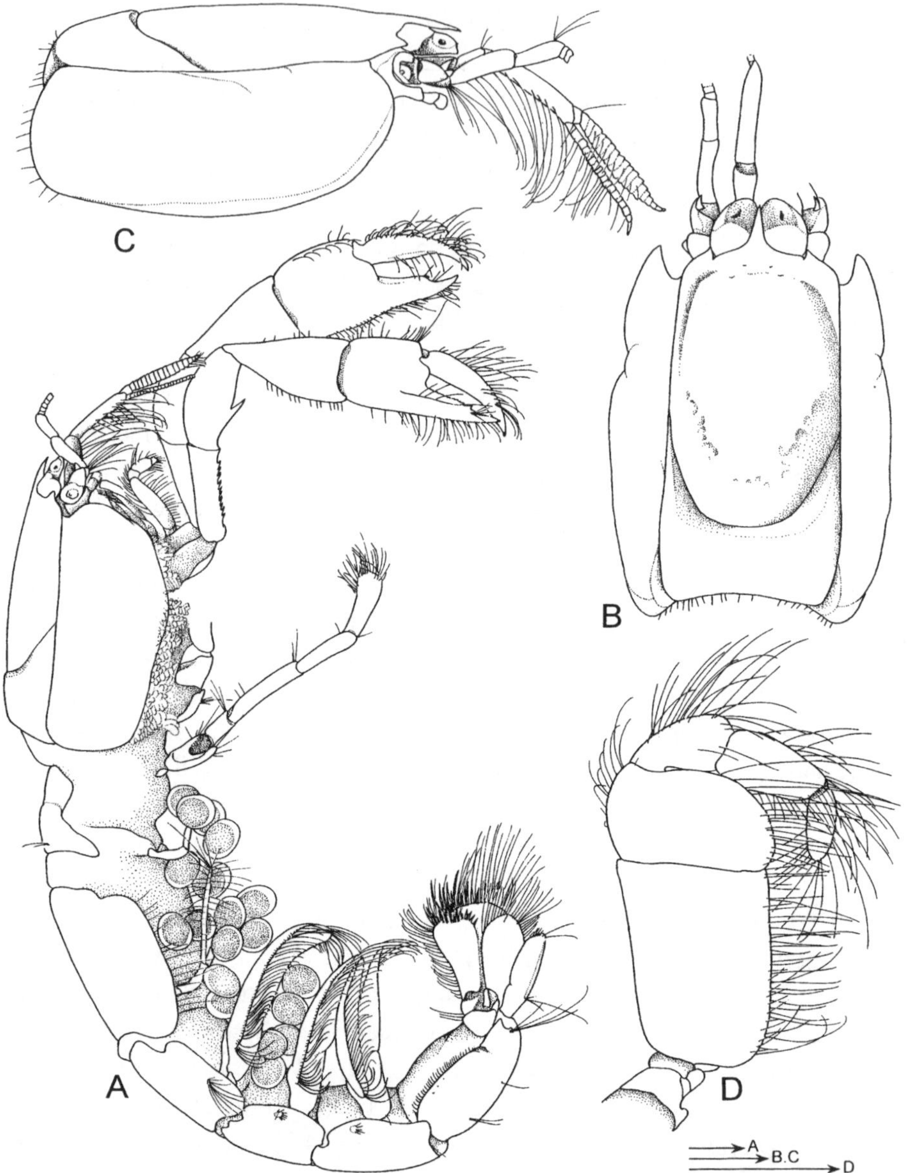

Fig. 13. *Callianassa acutirostella* Sakai, 1988. A, female, body, lateral view; B, carapace, dorsal view; C, same, lateral view; D, maxilliped 3, lateral view. A-D, ZMUC 61, ovig. female from Arafura Sea. Scales 1 mm.

length of propodus. P5 chelate; ischium 1.8 times as long as broad; merus 3.0 times as long as ischium; carpus 0.7 times as long as merus; chela slightly

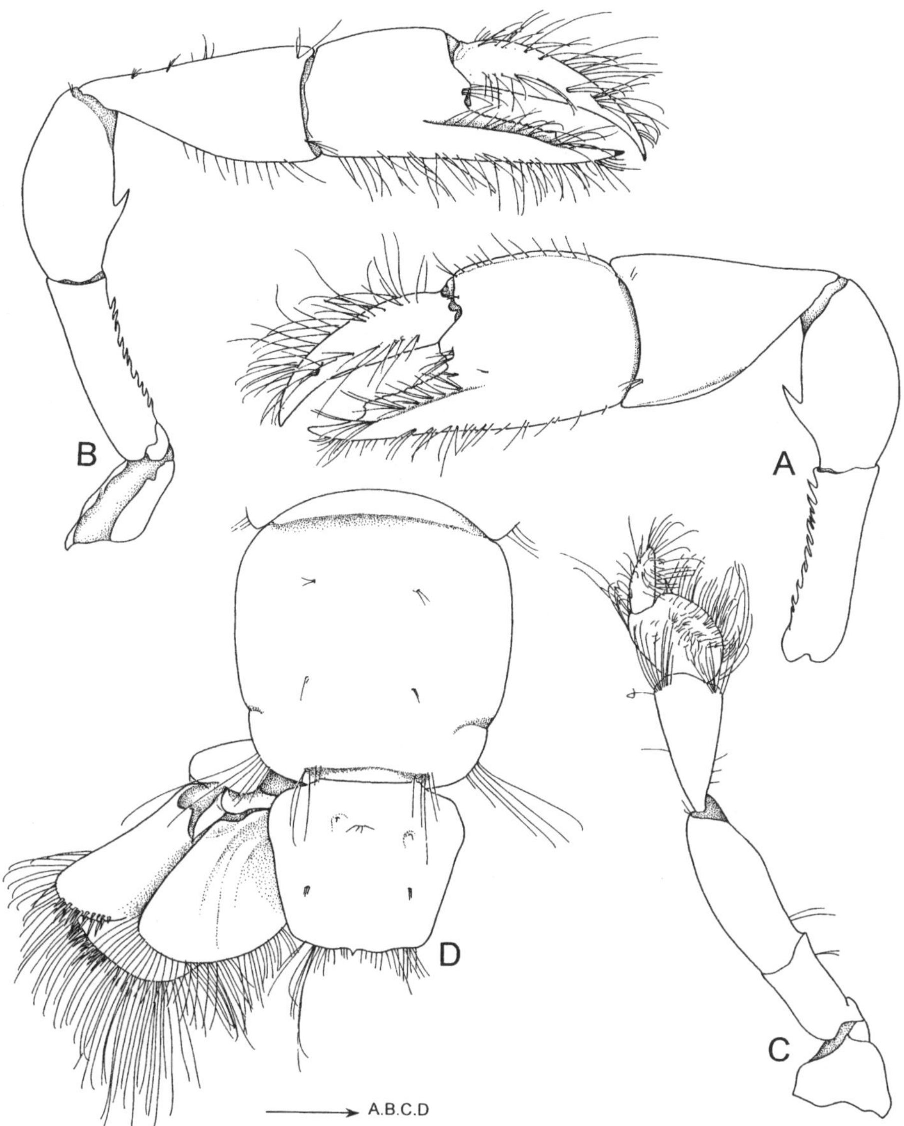

Fig. 14. *Callianassa acutirostella* Sakai, 1988. A, female larger cheliped, lateral view; B, female smaller cheliped, lateral view; C, pereiopod 3, lateral view. D, abdominal somite 6 and telson, with uropod on left side, dorsal view. A-D, ZMUC 61, ovig. female from Arafura Sea. Scale 1 mm.

longer than carpus, forming broad fixed finger ventrodistally, ventral surface with dense setation; dactylus hooked towards fixed finger, tip deflected.

Telson (fig. 14D) trapezoid, slightly broader than long, lateral margins convex proximally, convergent posteriorly; posterior margin almost straight, armed with distinct median spine; medial dorsal surface with transverse row of setae. Uropodal endopod overreaching telson. Uropodal exopod larger than endopod, truncate on distal margin.

Female Plp1 uniramous, 3-segmented; Plp2 biramous; exopod shorter than endopod; endopod 2-segmented. Plp3-5 biramous, broadly foliaceous, each endopod with small appendix interna near tip, on mesial margin.

Remarks. — The female holotype from Western Australia is an imcomplete specimen, lacking pereiopods 1 and 3. The female specimens from the Arafura Sea are entire and in good condition, which makes it possible to expand the known description. Male specimens remain unknown.

Type locality. — Western Australia, North West Shelf (19°05.1'S 118°53.7'E), 82 m depth.

Distribution. — North West Shelf of Western Australia and Arafura Sea.

Callianassa amboinae (Bate, 1888)
(figs. 15-16)

Scallasis amboinae Bate, 1888: 34, pl. 2 figs. 3, 4; Manning & Felder, 1991: 780 (list); Tudge et al., 2000: 145.
Callianassa (Scallasis) amboinae; Borradaile, 1903: 547; De Man, 1928a: 93.
Callianassa (Scallasis) Amboinae; De Man, 1928b: 30.
Callianassa amboinae; Sakai, 1999c: 37, fig. 5d-f.
Callianassa caledonica Ngoc-Ho, 1991: 285, fig. 2. [Type locality: East Lagoon, New Caledonia, 21 m.]

Material examined. — ZMUC CRU-3775, 1 ovig. female, (Tl/Cl 17.0/3.8), with larger cheliped on right side; 1 ovig. female (15.0/3.3), larger cheliped on left side, 13°13'N 100°34'E, Gulf of Thailand, mud with a little fine sandy mud, 20 m depth, "Galathea" Exped. 1950-1952, Station 394, leg. R/V "Galathea", 11.vi.1951; ZMUC CRU-3776, 2 ovig. females (Tl 15-16, lacking chelipeds and P3), 1 female (Tl 16.0 lacking chelipeds and P3), 21°20'S 165°24'E, off Burail Bay, Fiji Is., New Caledonia, 200 m depth, "Dana" II (1928-1930), Station 3612, leg. R/V "Dana", 27.xi.1928; ZMUC CRU-3777, 1 female (Tl 12.0, with both chelipeds), 1 detached larger cheliped, 10°43'S 139°17'E, Arafura Sea, coral-sand and gravel, 54 m depth, "Galathea" Exped. 1950-1952, Station 501, leg. R/V "Galathea", 27.ix.1951; ZMUC CRU-3778, 1 female (12.0/3.2, lacking chelipeds), 20°51'N 87°58'E, Bay of Bengal, mud, 50 m depth, "Galathea" Exped. 1950-1952, Station 305, leg. R/V "Galathea", 26.iv.1951; ZMUC CRU-3779, 3 males (Tl 16-12, lacking larger chelipeds), 5 ovig. females (Tl 16.0-12.0), 7 females (Tl 11.0-16.0). [Jour. 1.4.67] West Malay Peninsula, 9°44'N 98°22'E, 14 m depth, Thai/Danish Exped. 1966, Station 1163, 01.iii.1966.

Diagnosis. — Mxp3 ischium-merus subpediform, merus obliquely truncate with a rounded distomesial angle, male Plp1 uniramous, two-segmented, male Plp2 absent, telson concave on posterior margin, lacking a median spine.

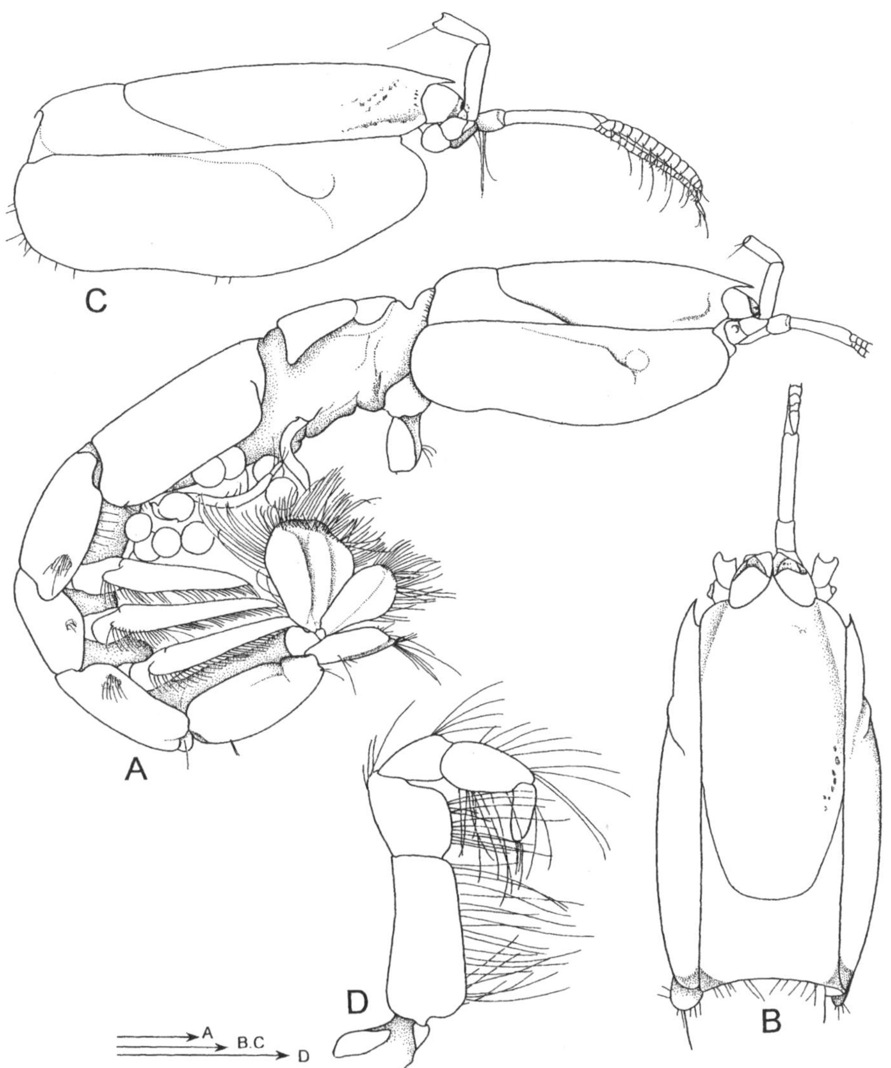

Fig. 15. *Callianassa amboinae* (Bate, 1888). A, female, body in lateral view; B, carapace, dorsal view; C, same, lateral view; D, Mxp3, lateral view. A-D, ZMUC 10b, ovig. female. Scales 1 mm.

Remarks. — The series of specimens examined here is identified as *C. amboinae* (Bate, 1888) by the form of the acute, triangular rostrum (fig. 15A-C), the rectangular Mxp3 ischium-merus (fig. 15D), the oblong P3 propodus (fig. 16C), the tail fan (fig. 16D), and the elongate abdominal somite 6, though the P3 propodus was unknown in the male holotype of *C. amboinae* (cf. Sakai, 1999c: 37, fig. 5). The shape of the female chelipeds of *C. amboinae* is here described for the first time (fig. 16A, B).

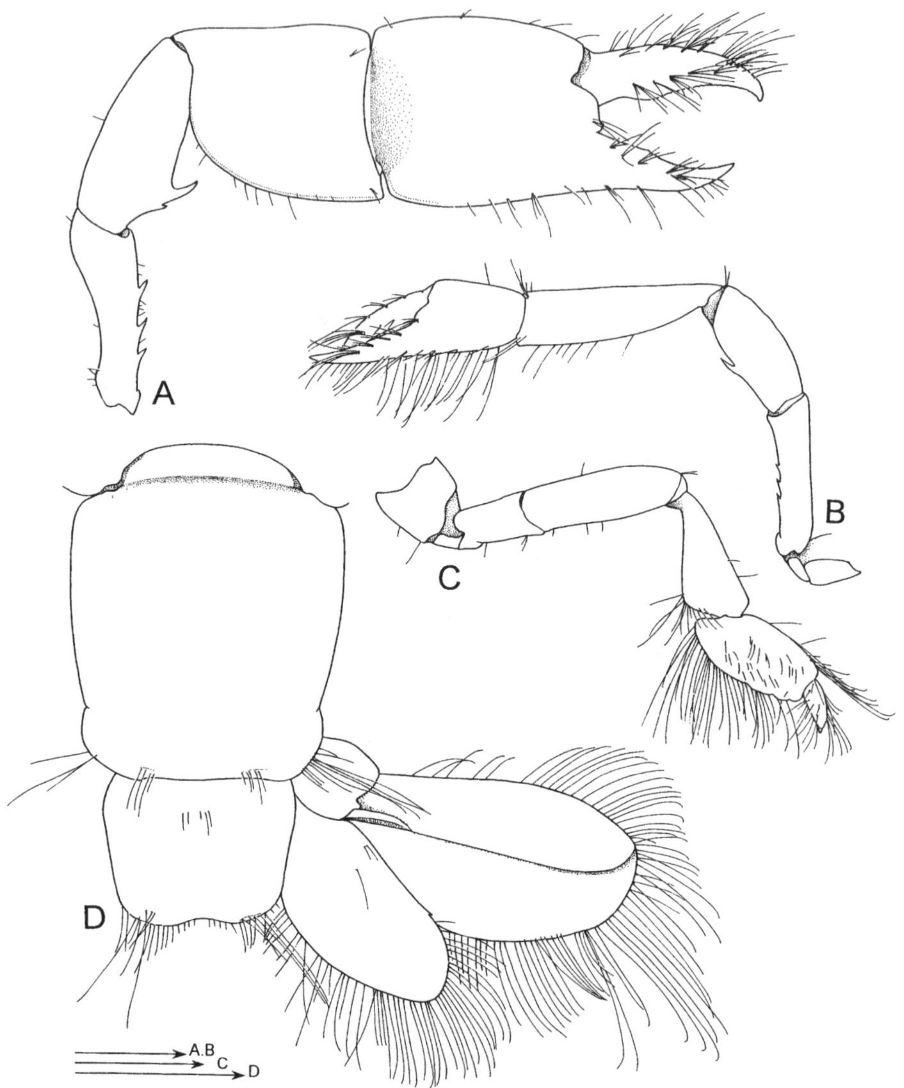

Fig. 16. *Callianassa amboinae* (Bate, 1888). A, female larger cheliped, lateral view; B, female smaller cheliped, lateral view; C, pereiopod 3, lateral view. D, abdominal somite 6 and telson, with uropod on right side, dorsal view. A-D, ZMUC 10b, ovig. female. Scales 1 mm.

In one of the present female specimens (fig. 16; ZMUC CRU-3775, 1 ovig. female, Tl 17.0, Cl 3.8) the telson is unarmed on the posterior margin, whereas in the other female specimens (ZMUC CRU-3776, 2 ovig. females, Tl 15-16), it is armed with a median spine on the posterior margin, as in *C. pygmaea* (cf. De Man, 1928b, pl. 16 fig. 24b).

The present species is very similar to *C. contipes* Sakai, 2002 from the Andaman Sea, being closely similar in the appearance of the uropodal endopod and exopod, P3 propodus, Mxp3, and tail fan. However, *C. contipes* Sakai, 2002 is slightly different from *C. amboinae*, having an elongate sixth abdominal somite, 1.3 times as long as broad, and having the anterior margin of the uropodal endopod unarmed, while in *C. amboinae* abdominal somite 6 is slightly longer than broad (fig. 16D), and the uropodal endopod has a spine on the anterior margin.

Type locality. — Ambon (= Amboina), Indonesia.

Distribution. — Bay of Nhatrang, Vietnam (Sakai, 1997); Ambon (= Amboina), Indonesia (Bate, 1888; De Man, 1928); East Lagoon, New Caledonia (Ngoc-Ho, 1991).

Callianassa amboinensis De Man, 1888

Callianassa amboinensis De Man, 1888: 480, pl. 20 fig. 4; Zehntner, 1894: 194; Borradaile, 1903: 545; Holthuis, 1958: 35; Poore & Griffin, 1979: 248, fig. 14; Sakai, 1984: 96, figs. 1, 2; Sakai, 1988: 53, 57, fig. 1; Ngoc-Ho, 1991: 283, fig. 1; Tudge et al., 2000: 143; Davie, 2002: 458.
Callianassa (Trypaea) amboinensis; De Man, 1928a: 27, 93, 107, 165, pl. 18 fig. 28-28c.
Callianassa (Calliactites) amboinensis; Borradaile, 1903: 545.
Callianassa ngochoae Sakai, 1999c: 49. [Misidentification for *Callianassa amboinensis*; Ngoc-Ho, 1991: 283.]

Material examined. — ZMUC CRU-3780, 1 female (Tl/Cl 18.0/3.9, P2-4 absent, detached P1 present), 12°27'N 124°03'E, San Beruardino Strait, Philippines, 91-183 m depth, bottom temp. 61°F (16.1°C), 03.viii.1911; ZMUC CRU-3781, 1 female (19.0/3.2), Java Sea, 05°47'S 106°07'E, 49 m depth, hard bottom, Sigsbee trawl, sponges, Danish Exped. to Kei Islands, 1922, Station 107, leg. Th. Mortensen, 05.viii.1922; ZMUC CRU-3782, 1 ovig. female (Tl/Cl 31.0/6.2), 5°47'S 106°14'E, Java Sea, stones, numerous sponges, 35 m depth, Danish Exped. to Kei Islands, 1922, Station 68, leg. Th. Mortensen, 27.vii.1922; ZMUC CRU-3783 1 ovig. female (19.0/4.4), ca. 200 fms (366 m) depth, off Tombeau Bay, Mauritius, hard bottom with sand, Sigsbee-trawl, Th. Mortensen's Pacific Expedition, Station 34, leg. Th. Mortensen, 26.ix.1929.

Diagnosis. — Mxp3 ischium-merus rounded in shape, merus rounded on distomesial margin, male Plp1 uniramous, unsegmented, male Plp2 uniramous, two-segmented (Ngoc-Ho, 1991: 283, fig. 1), telson truncate on posterior margin, bearing a median spine.

Remarks. — No male specimen of this species has been examined in the present study. The female material from the Philippines (ZMUC CRU-3780) was captured at 91-183 m, bottom temperature 16.1°C. Although these specimens were collected from an unusual habitat, they are identified as the present species because the chelipeds are the same as those from Heron Island, Australia, which Sakai (1984, fig. 2A, B) found to be identical to the type specimens; the carpus is distinctly wider than long, and the fixed finger of the cheliped is concave distally on the prehensile margin.

The present author misidentified Ngoc-Ho's species of *C. amboinensis* from New Caledonia as a new species, *C. ngochoae*, but the differences in the length of the A1 peduncle and in the form of the P1 merus and of the telson, are to be considered within the range of variation.

Type locality. — Ambon (= Amboina), Indonesia.

Distribution. — Philippines, 91-183 m depth. Amboina, off Lirung, Salibabu Island; Java, reef to 18 m, Indonesia. Northern Territory, Table Head, Port Essington; Dampier Island, Dampier Archipelago, northern Western Australia, reef to 18 m; Heron Island, Queensland, Australia. New Caledonia, 18-80 m off Tombeau Bay, Mauritius. Elat, Israel, Gulf of Aqaba.

Callianassa amplimaxilla Sakai, 2002

Callianassa amplimaxilla Sakai, 2002: 501, figs. 22A-D, 23A-D.

Diagnosis. — Mxp3 ischium-merus suboperculiform, merus rounded on distomesial margin, male Plp1 uniramous, two-segmented, male Plp2 absent, telson almost straight on posterior margin, lacking a median spine.

Type locality. — Andaman Sea, 9°00.009'N 98°02.962'E, 40.0 m.

Distribution. — Andaman Sea, 7°35.995'N 98°25.119'E; 7°59.839'N 98°13.625'E; 9°00.009'N 98°02.962'E; 9°14.939'N 97°54.212'E, 40.0-58.0 m; muddy sand, sandy mud.

Callianassa anoploura Sakai, 2002

Callianassa anoploura Sakai, 2002: 490, fig. 17A-H.

Diagnosis. — Mxp3 ischium-merus suboperculiform, merus rounded on distomesial margin, male Plp1 absent, male Plp2 small, biramous, telson truncate on posterior margin, bearing a median spine.

Type locality. — Andaman Sea, 9°30.351'N 97°57.168'E, 60.7 m, sandy mud, fine sand and shell fragments.

Distribution. — Andaman Sea, 7°29.786'N 98°17.348'E to 9°30.351'N 97°57.168'E, 60.7-71.8 m, sandy mud, fine sand with shell fragments.

Callianassa arenosa Poore, 1975

Callianassa arenosa Poore, 1975: 197-201, figs. 1-2; Poore & Griffin, 1979: 250, figs. 15-18; Sakai, 1988: 57; Sakai, 1999c: 39.
Biffarius arenosus; Tudge et al., 2000: 142; Davie, 2002: 457.

Diagnosis. — Mxp3 ischium-merus suboperculiform, merus rounded on distomesial margin, male Plp1 uniramous, two-segmented, male Plp2 absent, telson almost straight on posterior margin, lacking a median spine (Poore & Griffin, 1979, figs. 15k, 16b,18b).

Type locality. — Port Phillip Bay, Victoria, Australia.

Distribution. — Moreton Bay, Queensland; New South Wales; Victoria; Tasmania; intertidal to shallow water.

Callianassa australiensis (Dana, 1852)

Trypaea australiensis Dana, 1852b: 513; Dana, 1855, pl. 32 fig. 4; Fulton & Grant, 1906: 14; Manning & Felder, 1991: 774, figs. 1, 3, 12; Tudge et al., 2000: 143; Davie, 2002: 461.
Trypaea porcellana Kinahan, 1856: 130, pl. 4 fig. 2 [type locality: Port Philip, Victoria, Australia].
Callianassa (Trypaea) porcellana; Borradaile, 1903: 546.
Callianassa (Trypaea) australiensis; Borradaile, 1903: 546; De Man, 1928b: 27, 93, 104, 134; Stephenson et al., 1931: 56; Dakin & Colefax, 1940: 182-184, figs. 270, 271; Gurney, 1944: 83, figs. 8, 9; Dakin et al., 1952: 199, pl. 44; Hailstone & Stephenson, 1961: 259-285, figs. 1-15, pls. 1-3; Hailstone, 1962: 29-31, 2 figs.; McNeill, 1968: 26; Healy & Yaldwyn, 1970, pl. 30.
Callianassa australiensis; Poore & Griffin, 1979: 250, figs. 18-20; Sakai, 1988: 57; Holthuis, 1991: 241, 264, figs. 441, 442; Sakai, 1999c: 39.

Material examined. — ZMUC CRU-3784, 23 males (Tl/Cl 19.0/4.2-20.0/4.8), 15 females (15.0/3.4-23.0/5.0), Hastings, Port Western, Victoria, dredging in the bay, down to about 18 m, Th. Mortensen's Pacific Expedition, leg. Th. Mortensen, 06.ix.1914.

Diagnosis. — Mxp3 ischium-merus suboperculiform, merus distinctly oblique, declined on distal margin, showing a protruded distomesial angle, male Plp1 uniramous, two-segmented, male Plp2 absent, telson truncate on posterior margin, bearing a median spine.

Type locality. — Illawarra District, New South Wales, Australia.

Distribution. — Townsville to Port Phillip Bay, eastern Australia, intertidal (Poore & Griffin, 1979).

Callianassa australis Kensley, 1974

Callianassa subterranea australis Kensley, 1974: 271, figs. 3-5.
Callianassa australis; De Saint Laurent & Le Loeuff, 1979: 51, fig. 9a, b, d; Sakai, 1999c: 40.
Biffarius australis; Tudge et al., 2000: 142.

Diagnosis. — Mxp3 ischium-merus subpediform, merus obliquely truncate on distomesial margin, male Plp1 uniramous, two-segmented, male Plp2 uniramous, unsegmented, telson truncate on posterior margin, bearing a median spine (Kensley, 1974: 275, figs. 3B, G, 4H, G).

Type locality. — South West Africa, Lüderitz Bay, 180 m.

Distribution. — Lüderitz Bay; Orange River mouth, 10-180 m; South West Africa (16°46'E 22°41'S).

Callianassa bangensis sp. nov.
(figs. 17-18)

Material examined. — ZMUC CRU-3785, holotype, 1 male (Tl/Cl 16.0/3.8, damaged), 6 miles (10.8 km) E. of Port Banga, Mindanao, Philippines, 27 fms (= 49 m) depth, Th. Mortensen's Pacific Expedition, leg. Th. Mortensen, 07.iii.1914.

Diagnosis. — Rostrum triangular, distally pointed in dorsal view; frontal margin of carapace devoid of anterolateral projections. Eyestalks apically rounded, shorter than rostrum, overreaching basal segment of antennular peduncle. Mxp3 ischium-merus broad, oval in shape, merus rounded on distomesial margin; carpus, propodus, and dactylus pediform. Male larger cheliped massive, ischium ventral margin denticulate, dorsal margin unarmed; merus oval, dorsal margin arched and smooth, ventral margin arched and denticulate; carpus short and high, entirely arched on proximoventral margin; chela stout. P3 propodus ovate. Plp1-2 uncertain. Telson shield-like in outline, slightly longer than broad, convergent posteriorly, posterior margin straight, lacking median spine. Uropodal endopod subquadrate, longer than telson, distally rounded; uropodal exopod broad, longer than endopod, distally obliquely truncate.

Description of male holotype (fig. 17A). — Rostrum (fig. 17B) triangular and pointed distally in dorsal view, overreaching eyes; frontal margin of carapace smooth, devoid of anterolateral projections; dorsal oval present, transversely incomplete posterior to rostrum, conspicuous cervical groove in posterior fourth of carapace. Linea thalassinica present at full length. Eyes (fig. 17A-B) apically rounded; dorsally widely descending distally from rostrum, slightly overreaching antennular basal segment; cornea distinct, located in distal half; colour brown. Antennular peduncle slightly shorter than antennal peduncle, terminal segment 2.5 times as long as penultimate segment. A2 scaphocerite vestigial; terminal segment about three-fourths length of penultimate segment; antennal flagellum about 5 times as long as antennular flagella. Mxp3 (fig. 17C) without exopod; endopodal ischium-merus broadened, of oval shape; ischium subrectangular, about as long as broad; merus subrectangular, 0.7 times as long as ischium, 1.6 times as broad as long, largely convex on mesiodistal angle; carpus, propodus, and dactylus pediform.

Larger cheliped (fig. 17D) massive; ischium rod-like, 2.0 times as long as broad; dorsal margin slightly undulate and unarmed, ventral margin straight and armed with row of distinct, irregular denticles; merus spindle-shaped, slightly longer than ischium, about 1.3 times as long as high, dorsal margin arcuate and smooth, ventral margin also arcuate and armed with row of triangular denticles, enlarged proximal lobe absent; exterior surface swollen with median carina. Carpus broadened, 2.0 times as high as long and about two-thirds length of merus, entirely arched on proximoventral margin. Chela heavy, 4 times as long as carpus; palm 2.8 times as long as carpus, about 1.2 times as long as high, dorsal and ventral margins smooth, distal gap convex and armed with denticulate protrusion at its lower corner, running to triangular proximal concavity of fixed finger; fixed finger one-third length of palm, prehensile margin entirely concave, incurved downward with row of triangular denticles in proximal half, and smooth in distal half; dactylus massive, incurved downward distally on dorsal margin, prehensile margin distinctly concave proximally, distally with single denticle, remainder of margin with row of minute denticulations. Smaller cheliped absent. P2 (fig. 18A) chelate; ischium thick, 1.4-1.6 times as long as broad, merus broadened, 2.2 times as long as broad, 2.5 times as long as ischium, flexor margin with closely set setae; carpus triangular, 0.6 times as long as merus; chela slightly longer than carpus, setose on margins; both fingers 1.3 times as long as palm. P3 (fig. 18B) simple; ischium rectangular, 1.6-1.8 times as long as broad, merus 2.0 times as long as ischium, carpus triangular, 2.0 times as long as broad, two-thirds length of merus, propodus oval, 0.8 times as long as carpus and 1.3 times as long as high, bearing strong spine near ventrodistal angle, dactylus slender, two-thirds length of propodus; missing on right side. P4 (fig. 18C) simple; ischium rectangular, 2.0 times as long as broad, merus 3.0 times as long as broad, two-thirds length of ischium, unarmed; carpus about 0.8 times as long as merus, ventral margin en-

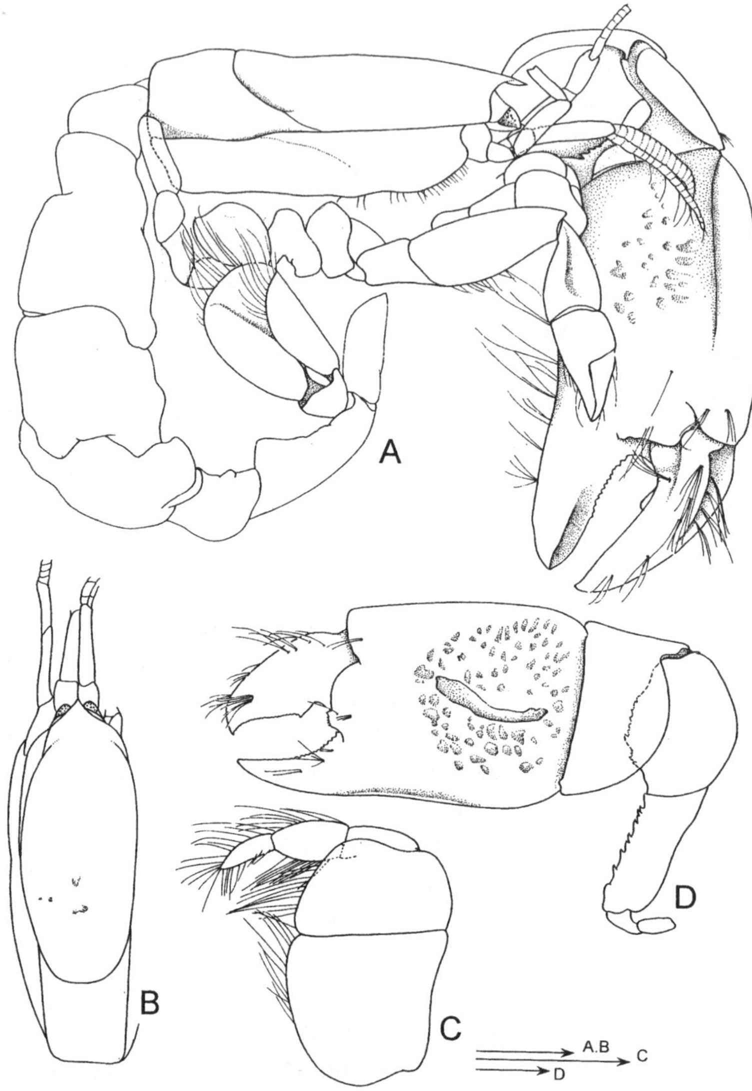

Fig. 17. *Callianassa bangensis* sp. nov. A, whole body, lateral view; B, carapace, dorsal view; C, Mxp3, lateral view; D, larger cheliped, lateral view. A-D, ZMUC 38, holotype male, 6 miles (11 km) off Port Banga, Mindanao, 27 fms (49 m) depth. Scales 1 mm.

tirely arched downward; propodus rectangular, slightly shorter than carpus, lateral surface scattered with soft setae, ventrodistal corner not protruded; dactylus slender, slightly less than half length of propodus and setose on external surface. P5 damaged on right side and missing on left side. Abdominal somites

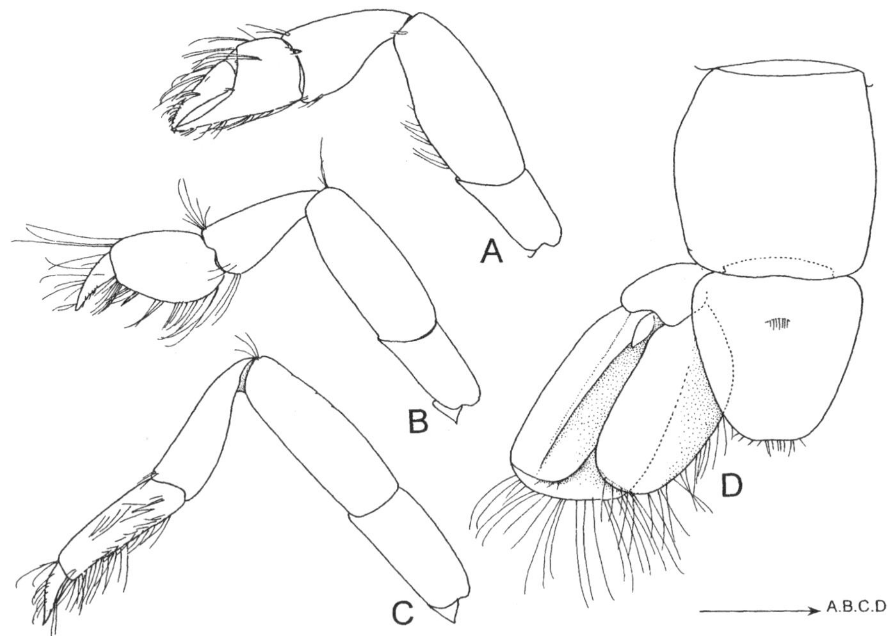

Fig. 18. *Callianassa bangensis* sp. nov. A, pereiopod 2, lateral view; B, pereiopod 3, lateral view; C, pereiopod 4, lateral view; D, abdominal somite 6 and telson, with uropod on left side, dorsal view. A-D, ZMUC 38, holotype male, 6 miles (11 km) off Port Banga, Mindanao, 27 fms (49 m) depth. Scale 1 mm.

in poor condition due to evaporation of alcohol. Telson (fig. 18D) shield-shaped, slightly longer than broad; lateral margins converging posteriorly to rounded posterolateral corners; posterior margin slightly setose, without median spine; dorsal surface with transverse row of setae medially in proximal third. Uropodal endopod rectangular; distally truncate, distinctly longer than telson, dorsal surface with median carina; uropodal exopod broadened and obliquely truncate on distal margin, dorsal surface with a longitudinal medial carina continuing distally to second setose lobe.

Plp1-5 unknown.

Etymology. — Named after the type locality, Port Banga, Mindanao. The specific name is an adjective agreeing in gender with the (feminine) generic name.

Remarks. — *Callianassa bangensis* sp. nov. is small in size (16 mm), and similar to *C. parva* Edmondson, 1944, from Oahu, Hawaii in the form of Mxp3, in having an ovate P3 propodus bearing a stout spine at the distoventral angle, the shape of the tail-fan, and the antennal peduncle being slightly longer than the antennular peduncle. *C. bangensis* sp. nov., however, differs in having a triangular, distally pointed rostrum that extends anterior to the eyestalks, in contrast to *C. parva*, in which the rostrum is scarcely developed.

Type locality. — Six miles off Port Banga, Mindanao, Philippines, 50 m.

Distribution. — Known only from the type locality.

Callianassa bouvieri Nobili, 1904

Callianassa (Trypaea) Bouvieri Nobili, 1904: 236; Nobili, 1906b: 105, pl. 6 fig. 3; De Man, 1928b: 27, 107, 146.

Callianassa maldivensis Borradaile; 1904: 753, pl. 58 fig. 3a, 3b; Tudge et al., 2000: 143. [Type locality: Hulule, Male Atoll, Maldives.]

Callianassa (Trypaea) maldivensis; Borradaile, 1903: 546 (nomen nudum); Borradaile, 1904: 753, pl. 58 fig. 3a, 3b; Pearson, 1905: 90; De Man, 1928a: 22; De Man, 1928b: 28, 107, 134, 146. [Type locality: Hulule, Male Atoll, Maldives.]

Callianassa bouvieri; Holthuis, 1958: 37, 38, fig. 15; Sakai, 1970a: 46; Sakai, 1987a: 303; Dworschak & Pervesler, 1988: 3, fig. 3; Dworschak, 1992: 192. Sakai, 1999c: 40, fig. 6a-c; Tudge et al., 2000: 143; Sakai & Apel, 2002: 276.

Callianassa rectangularis Ngoc-Ho, 1991: 292, fig. 5. [Type locality: Atoll de Surprise, New Caledonia, 36 m depth.]

Cheramus rectangularis; Tudge et al., 2000: 145.

Material examined. — ZMUC CRU-3786, 1 female (Tl 21.0, damaged), 08°46'S 115°14'E, S. of Bali, sand, 30 m depth, "Galathea" Exped. 1950-1952, Station 482, leg. R/V "Galathea", 12.ix.1951; ZMUC CRU-3787, 1 male (Tl/Cl 22.8/5.0), both chelipeds present; 1 male (26.5/6.0, no chelipeds); 1 male (16.0/3.5, no chelipeds), 1 ovig. female (23.0/5.0, no chelipeds), Puerto Galera sand coast, Mindanao, Philippines, Th. Mortensen's Pacific Expedition, leg. Th. Mortensen, 03.ii.1914; ZMUC CRU-3788, 1 ovig. female, Banda, Java Sea, Danish Exped. to Kei Islands, 1922, leg. Th. Mortensen, 06.viii.1922; ZMUC CRU-3789, 2 males (Tl 20.0-21.0), 04°33'S 129°55'E, Lontor, Banda Is., low tide, Danish Exped. to Kei Islands, 1922, leg. Th. Mortensen, 06.vi.1922; ZMUC CRU-3790, 1 male (Tl 23.0), Puerto Galera, Mindoro, Philippines, Th. Mortensen's Pacific Expedition, leg. Th. Mortensen, 01.ii.1914.

Diagnosis. — Mxp3 ischium-merus suboperculiform, merus rounded on distomesial margin, male Plp1-2 absent, telson convex on posterior margin, slightly concave medially with a distinct median spine.

Type locality. — Djibouti.

Distribution. — Egypt, Red Sea; Djibouti, Gulf of Aden; Gulf of Mannar, India; Maldives; Sri Lanka; Mindanao, Philippines; Bali and the Kei Islands, Indonesia; Amami-Ohshima, Amakusa Island and Tsushima, Japan. Here recorded from Mindanao, Philippines and Bali and the Kei Islands, Indonesia.

Callianassa brachytelson Sakai, 2002

Callianassa brachytelson Sakai, 2002: 499, fig. 21A-E.

Diagnosis. — Mxp3 ischium-merus subsquare, merus straight distally, convex on mesial margin, male Plp1-2 absent, telson slightly convex on posterior margin, bearing a distinct median spine.

Type locality. — Andaman Sea, 07°45.452'N 97°57.951'E, 70.0 m, coarse and fine sand.

Distribution. — Andaman Sea, 7°45.452'N 97°57.951'E; 08°00.016'N 97°53.927'E, 70.0-76.1 m, muddy sand, coarse and fine sand.

Callianassa brevirostris Sakai, 2002

Callianassa brevirostris Sakai, 2002: 514, figs. 30A-E, 31A-I.

Diagnosis. — Mxp3 ischium-merus suboperculiform, merus rounded on distomesial margin, male Plp1 uniramous, two-segmented, male Plp2 absent, telson convex on posterior margin, medially concave with a median spine.

Type locality. — Andaman Sea, 7°51.862'N 98°47.584'E, 20.9 m, sand with shell fragments.

Distribution. — Andaman Sea, 6°59.913'N 99°23.822'E; 9°30.351'N 97°57.168'E, 17.0-60.7 m, mud, fine sand, sand with shell fragments.

Callianassa ceramica Fulton & Grant, 1906

Callianassa ceramica Fulton & Grant, 1906: 12, pl. 5; Hale, 1927: 86; Poore, 1975: 205; Poore
    & Griffin, 1979: 257, figs. 22, 23; Sakai, 1988: 57; Sakai, 1999c: 41.
Callianassa (Trypaea) ceramica; De Man, 1928b: 27, 93, 104.
Biffarius ceramica; Tudge et al., 2000: 143.
Biffarius ceramicus; Davie, 2002: 457.

Material examined. — ZMUC CRU-3791, 2 males (Tl/Cl 28.0/5.2-38.0/8.0), 1 female (34.0/7.0), Plimmerton, New Zealand, fine sand, at low tide, leg. H. Tarqubar, 1912; ZMUC CRU-3792, 4 males (35.0/7.0-33.0/7.2), 3 females (31.0/6.2-43.0/9.0), New Zealand, x.1937.

Diagnosis. — Mxp3 ischium-merus suboperculiform, merus rounded on distomesial margin, male Plp1 uniramous, two-segmented, male Plp2 uniramous, unsegmented, telson slightly convex on posterior margin, bearing a distinct median spine.

Remarks. — In the present specimens from New Zealand, the antennular peduncle is slightly longer than the antennal one as figured by Fulton & Grant (1906, pl. 5). However, Poore & Griffin (1979: 259, fig. 22a) mentioned: "Peduncle of antenna 1 reaching midway along last segment of antenna 2", showing the antennular peduncle as distinctly shorter than the antennal peduncle. It is most probable that Poore & Griffin (1979) misidentified their specimens.

Type locality. — Port Phillip and Western Port, Victoria, Australia.

Distribution. — Victoria to the south of Western Australia, intertidal to shallow subtidal (Poore & Griffin, 1979); Plimmerton, North Island, New Zealand.

Callianassa chakratongae Sakai, 2002

Callianassa chakratongae Sakai, 2002: 513, figs. 28A-D, 29A-G; Sakai, 2004: 574-581, figs. 9-12.

Diagnosis. — Mxp3 ischium-merus pediform, male Plp1-2 unknown, form of telson uncertain.

Type locality. — Andaman Sea, 9°00.062'N 97°53.366'E, 65.4, muddy sand.

Distribution. — Known only from the type locality.

Callianassa contipes Sakai, 2002

Callianassa contipes Sakai, 2002: 523, fig. 35A-E.

Diagnosis. — Mxp3 ischium-merus pediform, merus straight distally, continuous with convex mesial margin, male Plp1-2 unknown, telson convex on posterior margin, medially concave with a median spine.

Type locality. — Andaman Sea, 7°30.006'N 98°56.633'E, 37.5 m, mud.

Distribution. — Andaman Sea, 7°29.921'N 99°00.977'E; 7°44.638'N 98°16.496'E, 20.5-30.5 m, mud.

Callianassa exilimaxilla sp. nov.
(figs. 19-20)

Material examined. — ZMUC CRU-3793, holotype, 1 ovig. female (Tl/Cl 11.0/2.3), South China Sea, 5°09'N 106°47'E, clay, 63 m depth, "Galathea" Exped., 1950-1952, Station 404, leg. R/V "Galathea", 30.vi.1951; ZMUC CRU-3794, paratypes, 3 males (Tl 9-10), 3 ovig. females (Tl 10-11), same data as holotype.

TABLE IV

Branchial formula of *Callianassa exilimaxilla* sp. nov.

	Maxillipeds			Pereiopods				
	1	2	3	1	2	3	4	5
Exopods	1	1	–	–	–	–	–	
Epipods	1	–	–	–	–	–	–	–
Podobranchs	–	r	–	–	–	–	–	–
Arthrobranchs	–	–	2	2	2	2	2	–
Pleurobranchs	–	–	–	–	–	–	–	–

(r = rudimentary)

Diagnosis. — Rostrum spike-like; frontal margin of carapace without ante-rolateral protuberances. Eyestalks evidently descending from rostrum distally. Mxp3 ischium-merus pediform, entirely slender and elongate; carpus triangular; propodus subquadrate; dactylus digitiform. P1 unequal and dissimilar; female larger cheliped with ischium bearing a subterminal spine on ventral margin; merus with a small proximal spine on ventral margin; carpus 1.5 times as long as high, declined on ventroproximal margin; chela 1.8 times as long as carpus; dactylus slender. Smaller cheliped with ischium and merus unarmed; carpus 1.8 times as long as merus; chela 0.8 times as long as carpus, dactylus slender and incurved. P3 propodus bean-shaped, distally increasing in height. Male Plp1 uniramous, unsegmented; Plp2 absent. Telson trapezoid, broader than long; lateral margins convergent in distal two-thirds, continuous with posterior margin, posterior margin convex without posteromedian spine. Uropodal endopod oval; uropodal exopod expanded, truncate distally, larger than endopod.

Description of female holotype (fig. 19A). — Rostrum (fig. 19B-C) spike-like; frontal margin of carapace smooth, lacking anterolateral protuberances; dorsal oval conspicuous; cervical groove located in posterior fourth of carapace. Linea thalassinica extending at full length. Eyestalks triangular, longer than broad, dorsal surface descending distally from the level of the rostrum, convex on dorsal surface; almost reaching distal end of antennular basal segment, cornea poorly delimited. Antennular peduncle slightly longer than antennal peduncle, terminal segment about 2.5 times as long as penultimate. Antennal scale scarcely developed; terminal segment 0.8 times as long as penultimate; A2 flagellum about 2.5 times as long as antennular flagella. Mxp3 (fig. 19D) without exopod; ischium-merus entirely slender and elongate; ischium subrectangular, 2.3 times as long as broad; crista dentata with a row of sparse, distinct denticles; merus slender, 1.8 times as broad as long, connecting with carpus over full width; carpus triangular, 1.8 times as long as broad; propodus subquadrate, 2.0 times as long as broad; dactylus digitiform, 0.7 times as long as propodus. Branchial formula as shown in table IV.

P1 unequal in size and dissimilar in shape. Larger cheliped (fig. 19E) with ischium slender, dorsal margin straight and unarmed, ventral margin bearing a single subdistal spine; merus about as long as ischium, about 2.0 times as long as high, dorsal margin slightly arcuate and smooth, ventral margin bearing a small proximal spine, exterior surface slightly convex. Carpus broadened, 1.5 times as high as long, about 1.2 times as long as merus, entirely declined on ventroproximal margin. Chela heavy, 1.7 times as long as carpus; palm as long as carpus, about 1.2 times as long as high, dorsal margin smooth and convex, ventral margin smooth, distal gap swollen, unarmed, extending to fixed finger; fixed finger armed with two subterminal tubercles on prehensile margin; dactylus slender, unarmed, incurved downward. Smaller cheliped (fig. 19F) slender and less massive than larger cheliped; ischium narrow, 3.2 times as long as broad, unarmed on dorsal and ventral margins; merus rectangular, 2.1 times as long as ischium, slightly shorter than ischium, unarmed on dorsal and ventral margins; carpus narrow and elongate, 1.7 times as long as merus, 3.8 times as long as high, entirely declined on proximoventral margin. Chela 0.8 times as long as carpus; palm subsquare, about 0.4 times as long as carpus and 0.8 times as long as high; fixed finger about as long as palm, unarmed on prehensile margin; distal gape incurved and continuous with fixed finger; fixed finger slightly longer than palm; dactylus slender and incurved, prehensile margin unarmed, crossing with fixed finger distally. P2 (fig. 19A) chelate; ischium 1.3 times as long as broad; merus broadened, 2.6 times as long as ischium and 2.5 times as long as broad, armed with sparse long setae on flexor margin; carpus 0.6 times as long as merus; chela slightly longer than carpus, setose on margins; chela about as long as carpus, both fingers 2.0 times as long as palm. P3 (fig. 19G) simple, ischium 1.5 times as long as broad; merus 2.3 times as long as ischium, and 3.0 times as long as high; carpus triangular, 2.0 times as long as high and 0.7 times as long as merus; propodus showing bean shape, 2.3 times as long as high, increased in height distally, roundly protruded at posteroventral angle; dactylus digitiform, as long as propodus. P4 (fig. 19A) simple, ischium 2.5 times as long as broad; merus 1.6 times as long as ischium; carpus 0.7 times as long as merus; propodus rectangular, 1.2 times as long as carpus, ventrodistal part densely setose; dactylus two-fifths length of propodus; missing on right side. P5 (fig. 19A) chelate; propodus protruded ventro-

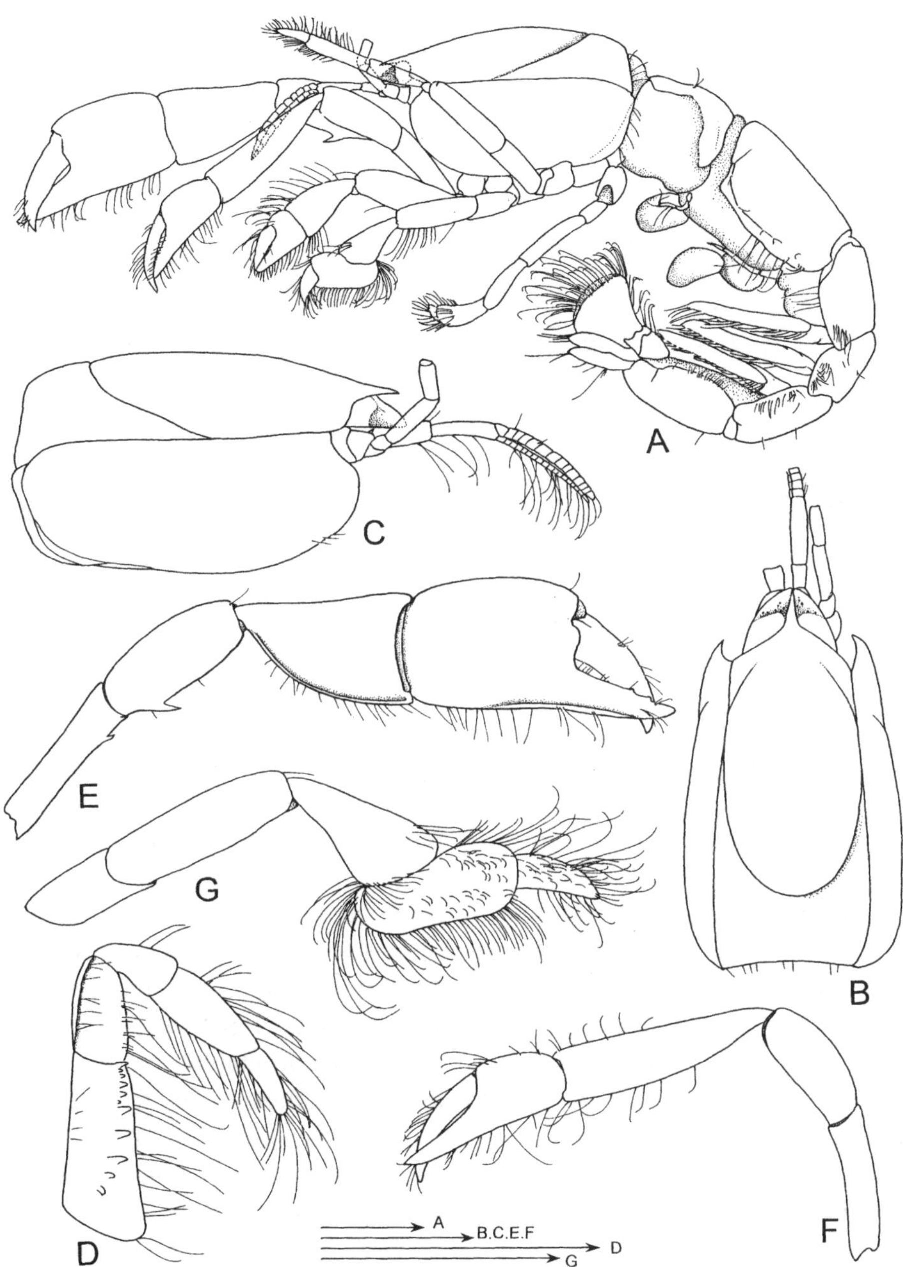

Fig. 19. *Callianassa exilimaxilla* sp. nov. A, whole body, lateral view; B, carapace, dorsal view; C, same, lateral view; D, Mxp3, mesial aspect; E, larger cheliped; F, smaller cheliped; G, pereiopod 3. A-C, E-G, ZMUC 21a, holotype, ovig. female, 5°09'N 106°47'E, South China Sea, 63 m depth. D, ZMUC 21, paratype male, 5°09'N 106°47'E, South China Sea, 63 m depth. Scales 1 mm.

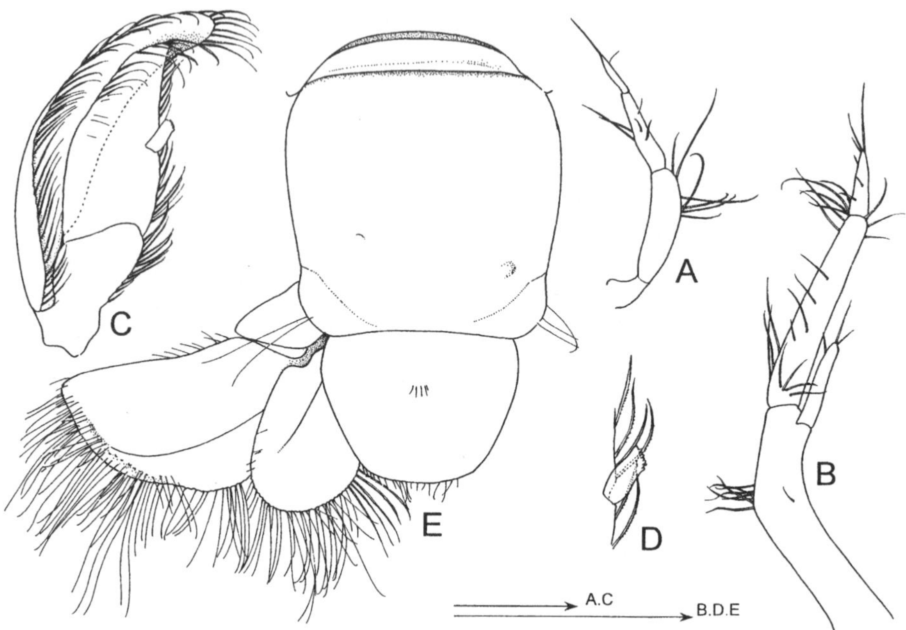

Fig. 20. *Callianassa exilimaxilla* sp. nov. A, male Plp1, B, male Plp2; C, Plp3; D, appendix interna on Plp3; E, abdominal somite 6 and tail-fan. A-D, ZMUC 21, paratype male, 5°09'N 106°47'E, South China Sea, 63 m depth; E, ZMUC 21a, holotype, ovig. female. Scales for A, 0.25 mm; B, D, 0.5 mm; for C, E, 1.0 mm.

distally to form chela with dactylus, dactylus hooked towards external side of fixed finger and deflected distally.

Abdominal somites smooth, dorsally glabrous; pleura 2-5 each with lateral tuft of setae; abdominal somite 6 (fig. 20E) smooth on lateral margins, about as long as broad in dorsal view. Plp1 (fig. 20A) simple, 3-segmented. Plp2 (fig. 20B) biramous, slender, endopod of two segments. Plp3 (fig. 20C) to Plp5 biramous, narrowly foliaceous, each bearing a small, stubby, projecting appendix interna (fig. 20D) on mesial margin of endopod. Telson (fig. 20E) trapezoid, slightly broader than long; lateral margins parallel over proximal third, then clearly convergent posteriorly, posterior margin slightly setose and without median spine; dorsal surface medially with short, transverse row of setae in proximal third. Uropodal endopod oval; distal margin rounded; dorsal surface shortly carinate medially. Uropodal exopod broadly expanded and larger than endopod, about 1.8 times as long as broad, truncate distally; anterior margin slightly concave, dorsal surface with a medial carina.

Male Plp1 simple, minute; Plp2 absent.

Etymology. — The species name is derived from the Latin, *exilis*, meaning slim, and also the Latin *maxilla*, meaning jaw. The specific name thus is a noun in apposition with the generic name.

Remarks. — The female type specimen of *C. exilimaxilla* is small, measuring no more than 10 mm in total length. The present new species is closely similar to *C. chakratongae* Sakai, 2002 from the Andaman Sea off Phuket, in having a spiny rostrum and similar chelipeds, but differs clearly in the shape of the P3 propodus and Mxp3. In *C. exilimaxilla* the P3 propodus is broad, with a distally diverging ventral margin; the crista dentata of Mxp3 bears rather obtuse teeth, whereas in *C. chakratongae* the P3 propodus is broad and entirely parallel, and the denticles of the crista dentata on Mxp3 are sharp in form. The tail-fan of *C. exilimaxilla* seems similar to that of *C. chakratongae*, though it is difficult to compare them with each other, because this part is damaged in the *C. chakratongae* specimens available.

Type locality. — South China Sea, 5°09'N 106°47'E.

Distribution. — Known only from the type locality.

Callianassa filholi A. Milne-Edwards, 1878

Callianassa Filholi A. Milne-Edwards, 1878: 112; Filhol, 1886: 491, pl. 53 figs. 10-12; Borradaile, 1903: 548.

Callianassa filholi; Chilton, 1907: 461-464, pl. 16 figs. 1-5; Miller & Batt, 1973: 110; Berkenbusch et al., 1998: 55; Sakai, 1999c: 43, fig. 7a-c; Berkenbusch et al., 2000: 397; Tudge et al., 2000: 143.

Callianassa (Trypaea) Filholi; De Man, 1928b: 27, 101, 104.

Diagnosis. — Mxp3 ischium-merus operculiform, merus rounded on distomesial margin, male Plp1 uniramous, two-segmented, Plp2 absent, telson convex on posterior margin, bearing a median spine (Sakai, 1999c: 43, fig. 7b, c).

Type locality. — Stewart Island.

Distribution. — Timaru; Oamaru; Stewart Is., New Zealand.

Callianassa gravieri Nobili, 1905

Callianassa (Trypaea) Gravieri Nobili, 1905: 396 (not 395); Nobili, 1906b: 107, pl. 6 fig. 4; Balss, 1915: 2; De Man, 1928a: 23, pl. 6 fig. 11-11e; De Man, 1928b: 27, 107.

Callianassa (Trypaea) cristata Borradaile, 1910: 263, pl. 16 fig. 7; De Man, 1928b: 27, 107. [Type locality: Salomon Atoll; Chagos Archipelago.]

Callianassa (Trypaea) gravieri; Holthuis, 1953b: 51.

Callianassa gravieri; Sakai, 1999c: 43, fig. 6d-f; Tudge et al., 2000: 143.
Callianassa cristata; Tudge et al., 2000: 143.

Diagnosis. — Mxp3 ischium-merus subsquare, merus straight distally with convex mesial margin, male Plp1-2 probably absent (De Man, 1928a: 24), telson straight on posterior margin, bearing a median spine.

Type locality. — Obock, Gulf of Aden.

Distribution. — Harmil Is., Red Sea; Obock and Djibouti, Gulf of Aden; Salomon Atoll; Chagos Archipelago.

Callianassa gruneri Sakai, 1999

Callianassa mucronata; Tirmizi, 1977 (partim): 21, fig. 1b. [Not *Callianassa mucronata* Strahl, 1862.]
Callianassa gruneri; Sakai, 1999c: 44, fig. 8a-g.

Diagnosis. — Mxp3 ischium-merus subsquare, merus obliquely straight distally in damaged type, male Plp1-2 uncertain, telson straight on posterior margin, lacking a median spine.

Type locality. — Luzon, Philippines.

Distribution. — Luzon, Philippines.

Callianassa intermedia De Man, 1905

Callianassa intermedia De Man, 1905: 609; Sakai, 1999c: 46; Tudge et al., 2000: 143.
Callianassa (*Cheramus*) *intermedia*; De Man, 1928b: 26, 98, 143, pl. 14 fig. 21-21d.

Not *Callianassa subterranea intermedia* forma; Czerniavsky, 1884: 80 [nomen dubium].

Diagnosis. — Mxp3 ischium-merus unknown, male Plp1 uniramous, two-segmented, Plp2 absent, telson concave on posterior margin, bearing a median spine.

Remarks. — Lewinsohn & Holthuis (1986: 21) cited *Callianassa subterranea intermedia* forma Czerniavsky, 1884 as an unavailable name.

Type locality. — Indonesia, Bali Sea (7°46'S 114°30.5'E), 330 m.

Distribution. — Bali Sea (7°46'S 114°30.5'E), Indonesia, 330 m.

Callianassa japonica Ortmann, 1891

Callianassa subterranea var. *japonica* Ortmann, 1891: 56, pl. 1 fig. 10a; Bouvier, 1901: 332-334; Doflein, 1902: 644; Balss, 1914: 91; Nakazawa, 1927: 1039, fig. 1999; Yokoya, 1930: 543; Kikuchi, 1932: 7; Yokoya, 1933: 52; Miyazaki, 1936: 317-320, figs. 1-3; Kamita, 1957: 107-109, fig. 49.

Callianassa californiensis var. *japonica* Bouvier, 1901: 332. [Type locality: Japan.]

Callianassa Harmandi Bouvier, 1901: 332-334. [Type locality: Japan.]

Callianassa (Trypaea) Harmandi; Borradaile, 1903: 546; Parisi, 1917: 24, fig. 7; De Man, 1928a: 13-15, fig. 6-6j; De Man, 1928b: 27, 102-103; Yü, 1931: 92-93, fig. 3.

Callianassa (Trypaea) japonica; Borradaile, 1903: 546; De Man, 1928a: 19-22, pl. 5 fig. 10-10a; De Man, 1928b: 27, 93, 106; Yü, 1931: 95-96, fig. 5; Makarov, 1938: 69-71, fig. 25.

Callianassa (Trypaea) californiensis; Parisi, 1917: 23. [Not *Callianassa californiensis* Dana, 1854.]

Callianassa hermandi; Nakazawa, 1927: 1039, fig. 2000.

Callianassa (Trypaea) californiensis var. *japonica*; De Man, 1928a: 18-19, pl. 4 fig. 9-9e; De Man, 1928b: 27, 105; Yü, 1931: 94, fig. 4.

Callianassa (Trypaea) harmandi; Makarov, 1938: 66-67, figs. 22-23.

Callianassa (Trypaea) californiensis var. *bouvieri* Makarov, 1938: 71-72, fig. 26. [Replacement name for *C. californiensis* var. *japonica*.]

Callianassa japonica; Nakazawa & Kubo, 1947: 754, fig. 2174; Miyake, 1965: 633, fig. 1037 (illustrated by K. Sakai); Sakai, 1968: 2-3, fig. 8; Sakai, 1969: 232, pls. 9-12; Miyake, 1982: 92, pl. 31 fig. 4; Sakai, 1987a: 303; Holthuis, 1991: 246, figs. 449, 450; Dworschak, 1992: 198; Liu & Zhong, 1994: 562 (list); Tamaki et al., 1996: 675, tabs.; Tamaki et al., 1997: 223; Miyabi et al., 1998: 101; Sakai, 1999c: 46; Sakai, 2001: 937-948, figs. 1-4, tab. 1; Sakai, 2002, fig. 15D-F.

Callianassa harmandi; Nakazawa & Kubo, 1947: 754, fig. 2173; Liu, 1955: 63, pl. 23, figs. 1-5; Utinomi, 1956: 63, pl. 32 fig. 2; Miyake et al., 1962: 124.

Callianassa petalura; Liu, 1955, pl. 23 figs. 6-9; Holthuis, 1991, fig. 453 (larger cheliped of male and female). [Not *Callianassa petalura* Stimpson, 1860.]

Nihonotrypaea japonica Manning & Tamaki, 1998: 889, fig. 1; Tamaki et al., 1999: 37; Tamaki & Miyabe, 2000: 182; Tudge et al., 2000: 133, 143; Sakai, 2001: 946.

Callianassa ?japonica; Yamaguchi & Holthuis, 2001: 112-113, figs. 2, 3.

Nihonotrypaea harmandi; Tudge et al., 2000: 143.

Material examined. — ZMUC CRU-3795, 1 male (Tl/Cl 27.0/5.5), 1 female (48.0/8.1), at river mouth, Peitaiho, China, vertical holes in mud, dry at ebb tide, leg. Dr. (?) Klumingsen, 07.vi.1942.

Diagnosis. — Mxp3 ischium-merus suboperculiform, merus rounded on distomesial angle, male Plp1 uniramous, two-segmented, Plp2 absent, telson straight on posterior margin, bearing a median spine.

Remarks. — *Callianassa japonica* was once placed in *Nihonotrypaea* Manning & Tamaki, 1998. It is here confirmed to belong in the genus *Callianassa* (cf. Sakai, 2001: 937). The material reported here was collected from a muddy tidal habitat.

Type locality. — Bay of Tokyo, Japan.

Distribution. — Japan from Nemuro, Funka Bay, Hokkaido to Kagoshima, Kyushu Is., both on Pacific side and in Japan Sea; Hou Hai, Yellow Sea (Shantung Peninsula to southwestern coast of Korean peninsula); Peter the Great Bay. Intertidal mud flats in bays and estuaries, to 192 m.

Callianassa joculatrix De Man, 1905

Callianassa joculatrix De Man, 1905: 610; Poore & Griffin, 1979: 266, figs. 28-29; Sakai, 1988: 53 (key); Ngoc-Ho, 1991: 287, fig. 3; Liu & Zhong, 1994: 562; Ngoc-Ho, 1994: 51; Sakai, 1999c: 47; Tudge et al., 2000: 143; Davie, 2002: 458.
Callianassa (Cheramus) joculatrix; De Man, 1928b: 18, 26, 93, 95, 98, 130-137, 141, 146, 148, 151, 153, pl. 12 fig. 19, 19c, pl. 13 fig. 19a, d-m (not pl. 12 fig. 19b); McNeil, 1968: 26.

Material examined. — Philippines: ZMUC CRU-3796, 2 ovig. females (Tl 14-17), 09°36'N 125°46'E, Candos Bay, Mindanao, anchorage, greenish mud, 22 m depth, "Galathea" Exped., 1950-1952, Station 428, leg. R/V "Galathea", 30.vii.1951.

Indonesia: ZMUC CRU-3815, 1 male (10.5/2.4); 2 ovig. females (13.0/3.2-11.0/ 2.7), 4 females (12.5/3.2-11.5/2.8), Java, 05°36'S 106°13'E, mud, 52 m depth, Danish Exped. to Kei Islands, 1922, Station 112, 06.viii.1922; ZMUC CRU-3816, 1 ovig. female (13.5/3.0), Java Sea, 05°23'S 116°02'E, coral-clay, 60 m depth, "Galathea" Exped., 1950-1952, Station 454, leg. R/V "Galathea", 25.viii.1951; ZMUC CRU-3817, 1 left cheliped, detached, 10°43'S 139°17'E, Arafura Sea, coral-sand and gravel, 54 m depth, "Galathea" Exped., 1950-1952, Station 502, leg. R/V "Galathea", 27.ix.1951; ZMUC CRU-3818, 1 male (10.0/2.5), 1 female (13.0/3.0), 10°43'S 139°17'E, Arafura Sea, coral-sand and gravel, 54 m depth, "Galathea" Exped., 1950-1952, Station 501, leg. R/V "Galathea", 27.ix.1951; ZMUC CRU-3819, 1 female (10.0/2.4), 5°45'S 108°19'E, Java Sea, coral-clay, 44 m depth, "Galathea" Exped., 1950-1952, Station 457, leg. R/V "Galathea", 27.viii.1951.

South China Sea: ZMUC CRU-3797, 1 ovig. female (Tl/Cl 10.2/2.5), 05°09'N 106°47'E, South China Sea, clay, 63 m depth, "Galathea" Exped., 1950-1952, Station 404, leg. R/V "Galathea", 30.vi.1951.

Sunda Strait: ZMUC CRU-3820, 1 female (10.0/2.7), Sunda Strait, 47 m depth, Danish Exped. to Kei Islands, 1922, Station 79, leg. Th. Mortensen, 29.iii.1922; ZMUC CRU-3821, 1 ovig. female (13.0/3.1), 06°25'S 105°41'E, Sunda Strait, 30 m depth, Th. Mortensen's Pacific Expedition, leg. Th. Mortensen, 27.vii.1922.

Gulf of Thailand: ZMUC CRU-3808, 1 damaged specimen, 07°00'N 103°18'E, Gulf of Thailand, 54 m depth, "Galathea" Exped., 1950-1952, Station 381, leg. R/V "Galathea", 08.vi.1951; ZMUC CRU-3806, 1 male (Tl 8.0), 07°00'N 103°18'E, Gulf of Thailand, muddy sand and shells, 55 m depth, "Galathea" Exped., 1950-1952, Station 381, leg. R/V "Galathea", 08.vi.1951; ZMUC CRU-3803, 1 male (10.5/2.5, with only smaller cheliped), 1 male (11.0/2.8, with only larger cheliped), 1 female (12.0/3.0, no chelipeds), carapace (Cl 2.8), carapace (Cl 2.2), 07°00'N 103°18'E, Gulf of Thailand, muddy sand and shells, 54 m depth, "Galathea" Exped., 1950-1952, Station 381, leg. R/V "Galathea", 08.vi.1951; ZMUC CRU-3814, 2 males (Tl 10.0-11.0), 2 females (Tl 9.0-11.0), 10°25'N 101°29'E, Gulf of Thailand, muddy clay, a little sand and shells, 72 m depth, "Galathea" Exped., 1950-1952, Station 386, leg. R/V "Galathea", 10.vi.1951. ZMUC CRU-3807, 1 female (17.0/3.8), 10°33'N 101°24'E, Gulf of Thailand, muddy clay, a little sand and shells, 70 m depth, "Galathea" Exped., 1950-1952, Station 388, leg. R/V "Galathea", 10.vi.1951; ZMUC CRU-3798, 1 male (12.8/3.1), 13°02'N 100°33'E, Gulf of Thailand, muddy sand with shells, 22 m depth, "Galathea" Exped., 1950-1952, Station 390, leg. R/V "Galathea", 11.vi.1951; ZMUC CRU-3799, 1 ovig. female (14.0/3.5), with only larger cheliped, 4-6 miles S. of Koh Samit, Th. Mortensen's Pacific Expedition, leg. Th. Mortensen, 01.ii.1900; ZMUC CRU-3804, 1 male (7.0/1.8), 3 places between Koh Chuen and Koh Chang, 27 m depth, mud and shells, Th. Mortensen's Pacific Expedition, leg. Th. Mortensen, 03.iii.1900; ZMUC CRU-3801, 1 male (13.5/3.3), 8 miles N. W. of Koh Chang, 18 m depth, Th. Mortensen's Pacific Expedition, leg. Th. Mortensen, 24.ii.1900; ZMUC CRU-3802, 2 ovig. females (14.2/3.5-13.0/3.0), with only smaller cheliped, W. of Koh Chang, 27 m depth, Th. Mortensen's Pacific Expedition, leg. Th. Mortensen, 29.i.1900; ZMUC CRU-3800, 1 male (13.0/3.2), without larger cheliped, 12 miles E. of Koh Mak, 37 m depth, Th. Mortensen's Pacific Expedition, leg. Th. Mortensen, 28.i.1900; ZMUC CRU-3809, Koh Mak, S. of Koh Chang, Th. Mortensen's Pacific Expedition, leg. Th. Mortensen, 09.i.1900; ZMUC CRU-3810, 1 male (13.0/3.2), 3 ovig. females (15.5/3.6-15.0/3.6), N. of Koh Kut, Th. Mortensen's Pacific Expedition, leg. Th. Mortensen, 23.i.1900; ZMUC CRU-3813, 1 male (10.0/2.0), 15 miles (27 km) S. of Koh Kut, 17-20 fms (31-37 m) depth, Th. Mortensen's Pacific Expedition, leg. Th. Mortensen, 28.i.1900; ZMUC CRU-3805, 1 ovig. female (14.0/3.0), 15 miles (27 km) S. of Koh Kut, 17-20 fms (31-37 m) depth, Th. Mortensen's Pacific Expedition, leg. Th. Mortensen, 28.i.1900; ZMUC CRU-3811, 3 males (15.0/3.5-7.2/1.7), 1 female (11.3/3.2), 4 ovig. females (15.0/3.1-12.1/2.9), fishing grounds, W. of Koh Kong, 11 m depth, Th. Mortensen's Pacific Expedition, leg. Th. Mortensen, 25.i.1900.

West Malay Peninsula: ZMUC CRU-3822, 26 males (15.0/3.4-7.0/2.0), 11 ovig. females (13.5/3.5-9.5/2.5), 33 females (12.5/3.0-8./2.2), 07°00'N 99°22'E, Southern Thailand, 27 m depth, Thai/Danish Exped. 1966, Station 1052, 10.ii.1966; ZMUC CRU-3825, 1 female (13.1/2.8), 07°00'N 99°22'E, Southern Thailand, 27 m depth, Thai/Danish Exped. 1966, Station 1052, 10.ii.1966. ZMUC CRU-3823, 2 males (10.9/2.7-10.3/2.5), 2 females (12.0/3.1-11.0/2.8), 09°43'N 98°20'E, Southern Thailand, 22 m depth, Thai/Danish Exped. 1966, Station 1168, 06.iii.1966; ZMUC CRU-3824, 1 male (10.0/2.5), 1 female (13.0/3.2), 09°42'N 98°21'E, sandy clay, 22 m depth, Thai/Danish Exped. 966, Station 1164, 06.iii.1966.

East Africa: ZMUC CRU-3812, 1 male (Tl 16.0), 04°03'S 039°37'E, Mombasa, Port Reitz, grey mud, 12 m depth, "Galathea" Exped., 1950-1952, Station 254, leg. R/V "Galathea", 20.iii.1951.

Diagnosis. — Mxp3 ischium-merus pediform, merus not protruded on distomesial angle, male Plp1 uniramous, two-segmented, Plp2 absent, telson straight on posterior margin, lacking a median spine.

Type locality. — Bay of Labuan Tring (8°44.5'S 116°02.5'E), west coast of Lombok, Indonesia, 18-27 m.

Distribution. — N. South China Sea; Taiwan; Philippines; Indonesia (Lombok, Java Sea), Arafura Sea, northern Queensland; north-west and north-east Australia; New Caledonia; South Vietnam; Mombasa; 15-300 m depth.

Callianassa lewtonae Ngoc-Ho, 1994

Callianassa lewtonae Ngoc-Ho, 1994: 52, fig. 1; Sakai, 1999c: 47.
Biffarius lewtonae; Tudge et al., 2000: 143; Davie, 2002: 457.

Diagnosis. — Mxp3 ischium-merus subsquare, merus rounded on distomesial margin, male Plp1-2 unknown, telson convex on posterior margin, lacking a median spine.

Type locality. — Queensland, Britomart Reef, reef front (18°17'S 146°38'E), 15 m.

Distribution. — Queensland, Britomart Reef, 18°17'S 146°38'E.

Callianassa lignicola Alcock & Anderson, 1899

Callianassa lignicola Alcock & Anderson, 1899: 288; Alcock, 1900, pl. 42 fig. 2, 2a, 2b; Alcock, 1901: 200; Borradaile, 1903: 545; Balss, 1925: 212; Sakai, 1999c: 48; Tudge et al., 2000: 143.
Callianassa (*Calliactites*) *lignicola*; De Man, 1928b: 25, 97.

Diagnosis. — Mxp3 ischium-merus suboperculiform, merus rounded on distomesial angle, male Plp1-2 unknown, telson straight on posterior margin, lacking a median spine (Alcock & Anderson, 1899: 289; Alcock, 1900, pl. 42 fig. 2, 2b).

Type locality. — Andaman Sea, 185-244 m.

Distribution. — Andaman Sea, 180-445 m.

Callianassa limosa Poore, 1975

Callianassa limosa Poore, 1975: 201-205, figs. 4, 5; Poore & Griffin, 1979: 270, figs. 32-33.
Neocallichirus limnosa; Sakai, 1988: 61. (Erroneous spelling of *N. limosus.*)
Neocallichirus limosus; Sakai, 1999c: 103.
Biffarius limosa; Tudge et al., 2000: 143.
Biffarius limosus; Davie, 2002: 457.

Diagnosis. — Mxp3 ischium-merus subsquare, merus rounded on distomesial angle, male Plp1 uniramous, two-segmented, male Plp2 minute, medially lobed, tapering papilla, telson convex on posterior margin, lacking a median spine (Poore, 1975: 201-204, figs. 4b, 5g, i, j).

Remarks. — Sakai (1999c) placed this species in *Neocallichirus*, because the Mxp3 propodus is swollen, but returned it to *Callianassa* by the form of Plp1-2, which are different from those of *Neocallichirus*.

Type locality. — Port Phillip Bay, Victoria, Australia.

Distribution. — Central New South Wales to Tasmania, shallow water to 100 m, Australia.

Callianassa lobetobensis De Man, 1905

Callianassa lobetobensis De Man, 1905: 607; Sakai, 1999c: 48; Tudge et al., 2000: 143.
Callianassa (Cheramus) lobetobensis; De Man, 1928b: 26, 93, 98, 137, pl. 13 fig. 20, pl. 14 fig. 20a-d.

Diagnosis. — Mxp3 ischium-merus subpediform, merus declined straightly on distal margin, male Plp1-2 unknown, telson distinctly notched medially on posterior margin, bearing a median spine (De Man, 1928b: 138, 140, pl. 14 fig. 20, 20a).

Type locality. — Indonesia, Banda Sea, Lobetobi Strait between Flores and Solor (8°27'S 122°54.5'E), 247 m.

Distribution. — Banda Sea, Lobetobi Strait between Flores and Solor, Indonesia (8°27'S 122°54.5'E).

Callianassa longicauda Sakai, 1967

Callianassa (Calliactites) longicauda Sakai, 1967: 324, figs. 3, 4, pl. 11C.
Callianassa longicauda; Sakai, 1987a: 303; Sakai, 1999c: 48.
Cheramus longicaudatus; Tudge et al., 2000: 145.

Diagnosis. — Mxp3 ischium-merus subsquare, merus straight with a distinct median tooth on distal margin, male Plp1 uniramous, two-segmented, Plp2 biramous, telson convex distally, bearing a median spine (Sakai, 1967: 324-326, figs. 3C, F, 4A, B).

Type locality. — East China Sea (32°N 122°30'E).

Distribution. — East China Sea (32°N 122°30'E).

Callianassa malaccaensis Sakai, 2002

Callianassa malaccaensis Sakai, 2002: 492, figs. 18A-D, 19A-K.

Material examined. — ZMUC CRU-3826, 2 males (Tl/Cl 16.0/3.9, chelipeds of both sides and P2-3 on right side absent; 18.0/4.2 mm, chelipeds absent and P2 of right side detached), 10°43'S 139°17'E, Arafura Sea, coral-sand and gravel, 54 m depth, "Galathea" Exped., 1950-1952, Station 502, leg. R/V "Galathea", 27.ix.1951; ZMUC CRU-3827, 1 male (30.0/7.0 mm, chelipeds, P2 on right side, and P4 on right side absent; and detached cheliped not belonging to the present species), 10°43'S 139°17'E, Arafura Sea, coral-sand and gravel, 54 m depth, "Galathea" Exped., 1950-1952, Station 502, leg. R/V "Galathea", 27.ix.1951; ZMUC CRU-3828, 1 male (Tl 11.0 mm, damaged), 07°00'N 103°18'E, Gulf of Thailand, muddy sand and shells, 54 m depth, "Galathea" Exped., 1950-1952, Station 381, leg. R/V "Galathea", 08.vi.1951.

Diagnosis. — [Revised after Sakai, 2002.] Rostrum sharply pointed. No anterolateral projections on carapace. Eyestalks descending downward distally, separated from level of rostrum. A1 peduncle reaching proximal quarter of A2 peduncle, distal segment. Mxp3 ischium-merus subpediform, merus obliquely truncate on distal margin. P1 unequal; meri with a single median marginal tooth; fingers unarmed. P3 propodus oval, ventral margin entirely rounded at ventroproximal angle and transferred distally to dactylus at the same level with dactylus. Abdominal somite 6 rectangular, 1.2 times as long as wide. Male Plp1 uniramous, two-segmented, male Plp2 biramous. Telson convergent posteriorly, lateral margin with two pairs of spines and posterior margin slightly concave, with middle tooth. Uropodal exopod broadened distally, with truncate distal margin.

Remarks. — This species was originally considered to be small, but the male specimen from the Arafura Sea is 30 mm in total length and, therefore, the species is not particularly small. The male pleopod 2 is biramous, with the endopod consisting of two segments, not one as shown by Sakai (2002, fig. 19G). The known range of *C. malaccaensis* is here extended to the Arafura Sea.

Type locality. — Andaman Sea, 6°45.045'N 99°20.766'E, 38.2 m.

Distribution. — Andaman Sea, 6°45.045'N 99°20.766'E, 38.2 m; 6°45.019'N 98°44.832'E, 82.7 m; 9°30.351'N 97°57.168'E, 60.7 m; muddy sand, sandy mud, fine sand and shell fragments.

Callianassa matzi Sakai, 2002

Callianassa matzi Sakai, 2002: 506, figs. 26A-C, 27A-I.

Material examined. — ZMUC CRU-3829, 1 male (Cl 2.1, without pereiopods 1-5, tail-fan, and larger cheliped), 10°11'N 101°37'E, Gulf of Thailand, muddy clay with a little sand, 72 m depth, "Galathea" Exped., 1950-1952, Station 384, leg. R/V "Galathea", 10.vi.1951. ZMUC CRU-3830, 1 male (Tl 13.0), 2 ovig. females (Tl 12.0-13.0); 1 male (Tl 11.0), male larger cheliped, Java Sea, mud, 32 m depth, Danish Exped. to Kei Islands, 1922, Station 121, leg. Th. Mortensen, 08.viii.1922; ZMUC CRU-4270, 3 males (9.0/1.9-15.0/3.1), 3 ovig. females (12.0/2.7-15.0/3.1), 1 female (14.0/3.1), S. of Koh Kut, Th. Mortensen's Pacific Expedition, leg. Th. Mortensen, 25.i.1900; PMBC 15501, paratypes, 11 males, 4 females, Sta. BIOSHELF B2, 09°15'N 097°54'E-09°15'N 097°25'E, 58 m depth.

Diagnosis. — Mxp3 ischium-merus pediform, merus not protruded on distomesial angle, male Plp1 uniramous, two-segmented, Plp2 absent, telson convex distally, bearing a minute median spine.

Type locality. — Andaman Sea, 7°59.839'N 98°13.625'E, 41.8 m, sandy mud.

Distribution. — Andaman Sea, 6°43.303'N 99°03.304'E; 6°45.961'N 99°20.968'E; 9°14.939'N 97°54.212'E; 9°00.091'N 97°43.147'E; 20.5-83.3 m, mud, fine sand, muddy sand, with shell fragments.

Callianassa mocambiquensis Sakai, 2004

Callianassa mocambiquensis Sakai, 2004: 585-592, figs. 15-17.

Type locality. — Mozambique Channel, 26.00 m.

Callianassa modesta De Man, 1905

Callianassa (Calliactites) modesta De Man, 1905: 604 (partim); Balss, 1925: 212; De Man, 1928b: 26, 97, 118, pl. 10 fig. 16-16b, pl. 11 fig. 16c-e.
Callianassa modesta; Liu & Zhong, 1994: 562; Sakai, 1999c: 48; Tudge et al., 2000: 143.

Material examined. — ZMUC CRU-3831, 1 female (Tl 11.0, chelipeds and P3 absent), 10°11'N 101°37'E, Gulf of Thailand, muddy clay with a little sand, 72 m depth, "Galathea" Exped., 1950-1952, Station 384, leg. R/V "Galathea", 10.vi.1951; ZMUC CRU-3832, 1 male (Tl/Cl 11.0/2.5), 07°00'N 103°18'E, Gulf of Thailand, muddy sand and shells, 54 m depth, "Galathea" Exped., 1950-1952, Station 381, leg. R/V "Galathea", 08.vi.1951; ZMUC CRU-3833, 1 male (9.0/2.2), 35 miles (63 km) W. of Koh Chang, 55 m depth, Th. Mortensen's Pacific Expedition, leg. Th. Mortensen, 31.i.1900; ZMUC CRU-3834, 2 males (11.0/2.6-9.0/2.0), 07°00'N 103°18'E, Gulf of Thailand, muddy sand and shells, 54 m depth, "Galathea" Exped., 1950-1952, Station 381, leg. R/V "Galathea", 08.vi.1951; ZMUC CRU-3835, 1 ovig. female (11.0/2.9), 1 female (11.0/2.9), 07°00'N 103°18'E, Gulf of Thailand, muddy sand and shells, 55 m depth, "Galathea" Exped., 1950-1952, Station 381, leg. R/V "Galathea", 09.vi.1951; ZMUC CRU-3836, 1 male (9.0/2.2), 07°00'N 103°18'E, Gulf of Thailand, muddy sand and shells, 55 m depth, "Galathea" Exped., 1950-1952, Station 381, leg. R/V "Galathea", 08.vi.1951; ZMUC CRU-3837, 1 abdomen lacking carapace, 07°00'N 103°18'E, Gulf of Thailand, muddy sand and shells, 55 m depth, "Galathea" Exped., 1950-1952, Station 381, leg. R/V "Galathea", 08.vi.1951; ZMUC CRU-3838, 1 male (10.0/2.5), 07°00'N 103°18'E, Gulf of Thailand, muddy sand and shells, 55 m depth, "Galathea" Exped., 1950-1952, Station 381, leg. R/V "Galathea", 08.vi.1951; ZMUC CRU-3839, 1 male (Tl 10.0), 1 female (Tl 10.0), 04°43'N 103°43'E, off Kerteh, Trengganu, mud, 48 m depth, "Galathea" Exped., 1950-1952, Station 380, leg. R/V "Galathea", 07.vi.1951; ZMUC CRU-3840, 2 females (9.0/2.3-11.0/2.5), 10°25'N 101°29'E, Gulf of Thailand, muddy clay, a little sand and shells, 72 m depth, "Galathea" Exped., 1950-1952, Station 386, leg. R/V "Galathea", 10.vi.1951.

Diagnosis. — Mxp3 ischium-merus subpediform, merus straight with a median tooth on distal margin, male Plp1 uniramous, two-segmented, Plp2 biramous, slender, telson convex distally, bearing a median spine.

Type locality. — Elat, Kepulauan Kai, Indonesia, 27 m.

Distribution. — N. South China Sea (Liu & Zhong, 1994); W. of Kwandang Bay; Bay of Bima (0°58.5'N 122°42.5'E), 27-310 m depth; Great Kei Island, Indonesia (5°40'S 132°26'E), 27-310 m depth.

Callianassa nieli Sakai, 2002

Callianassa nieli Sakai, 2002: 518, fig. 32A-E.

Diagnosis. — Mxp3 ischium-merus subsquare, merus rounded on distomesial margin, male Plp1-2 undescribed, telson slightly concave medially, bearing a median spine (Sakai, 2002: 518, fig. 32C, D).

Type-locality. — Andaman Sea, 7°36.100'N 98°19.257'E, 55.4 m.

Distribrution. — Andaman Sea, 7°34.598'N 98°16.250'E; 7°36.100'N 98°19.257'E; 7°35.995'N 98°25.119'E; 48.9-68.8 m, muddy sand.

Callianassa nigroculata Sakai, 2002

Callianassa nigroculata Sakai, 2002: 525, figs. 36A-D, 37A-H.

Diagnosis. — Mxp3 ischium-merus subpediform, merus convex distally with rounded distomesial angle, male Plp1-2 undescribed, telson slightly concave medially, bearing a median spine (Sakai, 2002: 525-528, figs. 36D, 37A).

Type locality. — Andaman Sea, 7°30.221'N 98°22.005'E-7°35.932'N 98°25.348'E, 43.2-74.8 m.

Distribution. — Andaman Sea, 7°29.786'N 98°17.348'E; 7°15.061'N 98°34.181'E; 7°35.932'N 98°25.348'E; 60.4-78.7 m, muddy sand.

Callianassa orientalis (Bate, 1888)

Cheramus orientalis Bate, 1888: 30, pl. 1 fig. 2.
Callianassa (Cheramus) orientalis; Borradaile, 1903: 546; De Man, 1928a: 9, fig. 2, 2a; De Man, 1928b: 26, 93, 98, 119, 132, 137.
Callianassa orientalis; Sakai, 1999c: 49, fig. 5a-c.
Cheramus orientalis; Holthuis, 1991: 239 [invalid selection as type species of the genus *Cheramus*]; Tudge et al., 2000: 145.

Diagnosis. — Mxp3 ischium-merus subpediform, merus obliquely convex on distal margin, male Plp1-2 unknown, telson slightly concave medially, lacking a median spine (Bate, 1888: 31).

Remarks. — *Callianassa stenomastaxa* Sakai, 2002, from the Andaman Sea, is a species closely similar to *C. orientalis* in the form of the Mxp3 and the relative length of A1-2 peduncle. However, in *C. stenomastaxa*, the appendices internae of Plp3-5 are large and triangular, whereas in *C. orientalis* they are stubby (see Remarks on *Call. stenomastaxa*).

Type locality. — Arafura Sea (9°59'S 139°42'E), 28 m depth.

Distribution. — Arafura Sea (9°59'S 139°42'E), 28 m depth.

Callianassa parva Edmondson, 1944

Callianassa (Calliactites) parva Edmondson, 1944: 45, fig. 5a-j; Edmondson, 1946: 261.
Callianassa parva; Sakai, 1999c: 50; Tudge et al., 2000: 143.

Diagnosis. — Mxp3 ischium-merus subsquare, merus rounded on distomesial margin, male Plp1 uniramous, two-segmented, Plp2 undescribed, telson straight on posterior margin, lacking a median spine (Edmondson, 1944: 45, fig. 5b, g, i).

Type locality. — Hanauma Bay, Oahu.

Distribution. — Hanauma Bay and Palos Bay, Hawaii; Celebes (= Sulawesi), Indonesia.

Callianassa parvula Sakai, 1988

Callianassa parvula Sakai, 1988: 59, fig. 3; Sakai, 1999c: 50; Tudge et al., 2000: 143; Davie, 2002: 458.

Diagnosis. — Mxp3 ischium-merus unknown, male Plp1 uniramous, two-segmented, Plp2 biramous, telson slightly concave medially on posterior margin, lacking a median spine (Sakai, 1988: 59, fig. 3D, E, F).

Type locality. — Western Australia, North West Shelf (19°04.4'S 118°47.35'E), 83 m.

Distribution. — North West Shelf, Western Australia, 83 m.

Callianassa petalura Stimpson, 1860

Callianassa petalura Stimpson, 1860: 23; A. Milne-Edwards, 1870: 88, 101; Bouvier, 1901: 332-334; De Man, 1928b: 115; Yokoya, 1939: 277-278; Liu, 1955: 65, pl. 23 figs. 6-9 (= *Callianassa japonica*); Miyake et al., 1962: 124; Miyake, 1965: 633, fig. 1036 (illustrated by K. Sakai); Sakai, 1968: 2-3, fig. 9; Sakai, 1969: 233, pls. 13-15; Miyake, 1982: 91, pl. 31 fig. 3; Sakai, 1987a: 303; Holthuis, 1991: 249, 264 (not figs. 453, 454 = *C. japonica*); Liu & Zhong, 1994: 562 (list); Sakai, 1999c: 50.
Callianassa (*Trypaea*) *petalura*; Borradaile, 1903: 546; De Man, 1928b: 28, 115.
Callianassa subterranea var. *japonica*; Balss, 1914: 91.
Callianassa subterranea japonica; Kikuchi, 1932: 7.
Callianassa (*Trypaea*) *gigas* var. *japonica* Makarov, 1935: 323-324, fig. 4. [Type locality: Patrocle Bay, Peter the Great Bay.]
Callianassa (*Trypaea*) *gigas* var. *eoa* Makarov, 1938, 67-69, fig. 24. [New name for *Callianassa* (*Trypaea*) *gigas* var. *japonica* Makarov, 1938.]
Nihonotrypaea petalura; Tudge et al., 2000: 143.

Material examined. — ZMUC CRU-3841, 1 ovig. female (Tl/Cl 40.0/8.5), 2 females (44.0/8.9-46.0/9.5), Misaki, Sagami Bay, Japan (coast under stone), sandy coast, Th. Mortensen's Pacific Expedition, leg. Th. Mortensen, 30.iv.1914; ZMUC CRU-3842, 1 male (25.0/5.5), 1 ovig. female (50.0/8.9), 2 females (41.0/8.5.-50.0/9.0), Misaki, Sagami Bay, Japan (coast under stone), sandy coast, Th. Mortensen's Pacific Expedition, leg. Th. Mortensen, 28.vi.1914.

Diagnosis. — Mxp3 ischium-merus suboperculiform, merus rounded on distomesial angle, male Plp1 uniramous, two-segmented, Plp2 absent, telson slightly concave medially on posterior margin, bearing a median spine.

Remarks. — The labels attached to the two vials show that the specimens are collected under coastal sands, but the list of stations (Dr. Th. Mortensen's Expeditions 1899-1930, List of Stations, unpubl.) shows different localities: Misaki, Sagami Bay, Japan, 200 fms (366 m) depth, under stone, 30.vi.1914; and Okinose, Sagami Bay, Japan, 300 fms (549 m) depth, 28.vi.1914.

Type locality. — Shimoda, Japan.

Distribution. — Japan from Hokkaido to Kyushu Is. both on Pacific side and in Japan Sea, on sandy mud flats, facing open sea; Ho Hai, Yellow Sea, shallow water (Liu & Zhong, 1994).

Callianassa persica sp. nov.
(fig. 21)

Material examined. — ZMUC CRU-3843, holotype, 1 male (Tl/Cl 20.0/4.0) 28°54'N 50°11'E, Persian Gulf, 56 m depth, Thorson's Expedition, Station 22A, leg. G. Thorson, 16.iii.1937; ZMUC CRU-3844, paratype, 1 female (20.0/4.7), 29°04'N 49°56'E, Persian Gulf, 50 m depth, Thorson's Expedition, Station 23D, leg. G. Thorson, 13.iii.1937; ZMUC CRU-3845, 1 male, (Tl 14 mm, damaged), 29°04'N 49°56'E, Persian Gulf, Thorson's Expedition, Station 23D?, leg. G. Thorson, 20.iv.1937.

Diagnosis. — Rostrum of a broad, triangular shape; anterolateral projections of carapace obtuse. A1 peduncle slightly shorter than that of A2. Mxp3 ischium-merus narrowly oblong, merus obliquely straight on distal margin and slanting on distal margin; propodus oval; dactylus digitiform. P1 unknown. Male Plp1-2 absent. Telson shield-shaped, slightly longer than broad; lateral margins convergent distally, posterior margin slightly convex and with a small median spine. Uropodal endopod oval distally; uropodal exopod broadly truncate distally.

Description of male holotype. — Rostrum (fig. 21A-B) of a broad, triangular shape in dorsal view; frontal margin of carapace almost smooth and with obtuse anterolateral projections; dorsal oval conspicuous; cervical groove located in posterior fourth of carapace. Linea thalassinica extending at full length. Eyestalks triangular, longer than broad, descending distally downward, separated from rostrum, scarcely shorter than distal end of antennular basal segment; cornea poorly delimited, located distomedially, faintly pigmented in alcohol. A1 peduncle slightly shorter than that of A2, terminal segment about 2.5 times as long as penultimate. A2 terminal segment distinctly shorter than penultimate; scaphocerite minute. Mxp3 (fig. 21C) ischium-merus narrowly oblong, obliquely straight on distal margin; ischium subrectangular, 1.8 times as long as broad; crista dentata with row of sparse denticles; merus subtriangular, about as long as broad, half length of ischium; distal margin slanting entirely, and continuous with short mesial margin by rounded corner; carpus triangular, 1.2 times as long as broad; propodus oval, 1.3 times as long as broad, entirely rounded on ventral margin; dactylus digitiform, 0.7 times as long as propodus; exopod not present. Mxp3 on right side missing.

P1 absent on both sides. P2 chelate; merus broadened, 2.6 times as long as ischium; carpus 0.7 times as long as merus; chela slightly longer than carpus, setose on margins; both fingers 1.8 times as long as palm; missing on right side. P3 (fig. 21D) ischium-merus oblong; merus about three times as long as

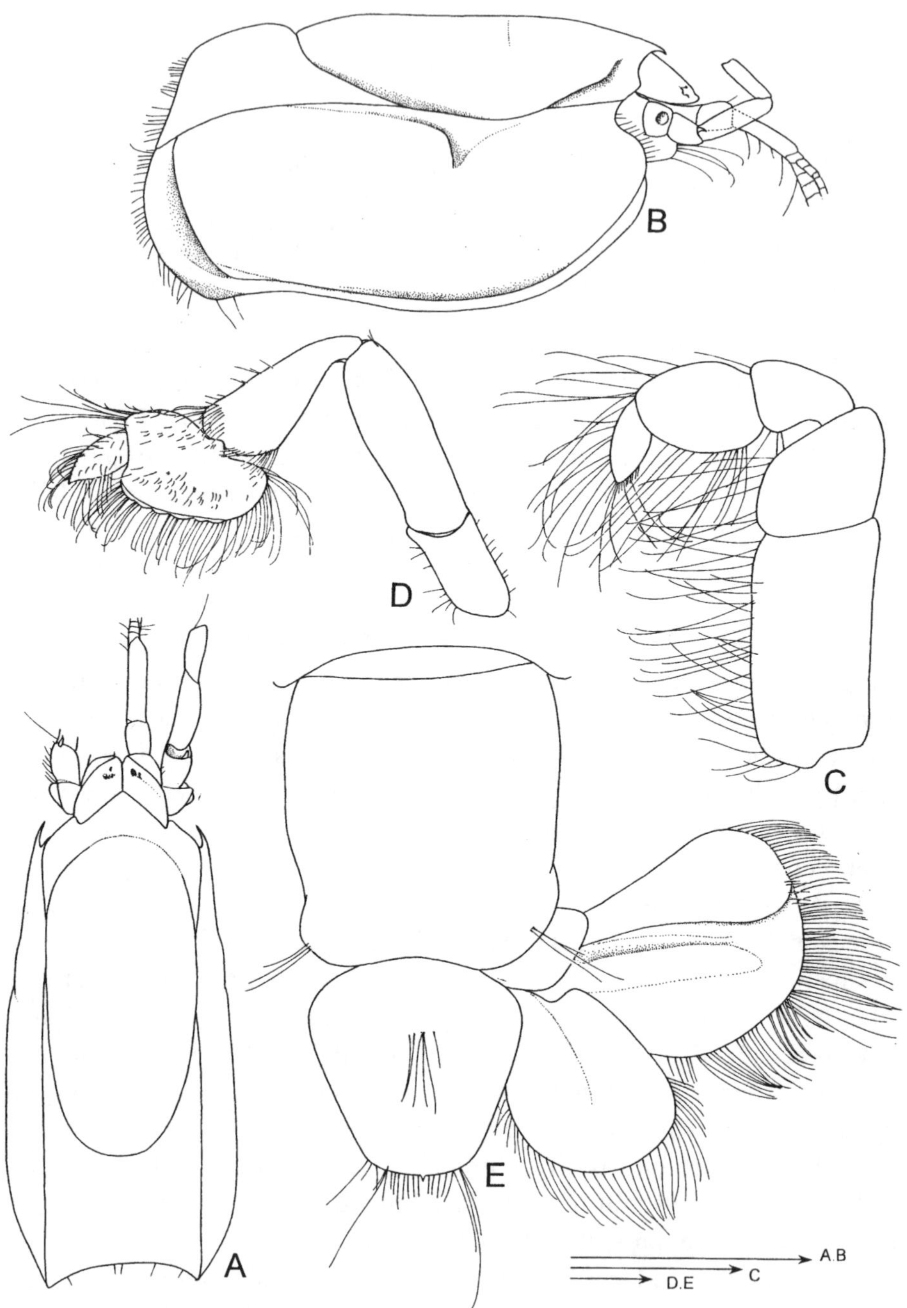

Fig. 21. *Callianassa persica* sp. nov. A, carapace, dorsal view; B, same, lateral view; C, Mxp3, lateral view; D, pereiopod 3, lateral view; E, abdominal somite 6 and telson, with uropod on right side. A-E, ZMUC 199, holotype male, Persian Gulf, Sta. 22A, 28°54'N 50°11'E, 56 m depth. Scales 1 mm.

broad and 2.0 times as long as ischium; carpus elongate and triangular, 0.8 times as long as merus; propodus shoe-shaped, narrow at ventroproximal part and broadened distally, ventral margin broadly convex and crenulate, with setae; dactylus triangular; missing on right side. P4 slender, simple; merus 1.2 times as long as ischium: carpus 0.7 times as long as merus; propodus about as long as carpus, ventral margin with thick setae on distal half, ventrodistal corner not protruded; dactylus less than half length of propodus and setose on external surface; missing on right side. P5 chelate; merus slender, 2.0 times as long as ischium; carpus shorter than merus; propodus forming a broad fixed finger ventrodistally, ventral surface with dense setation; dactylus hooked towards external side of fixed finger, tip deflected.

Abdominal somites smooth, glabrous dorsally; pleura 2-5 each with a tuft of setae laterally; abdominal somite 6 about as long as broad, almost parallel on lateral margins. Telson (fig. 21E) simple shield-shape, slightly longer than broad; lateral margins convergent distally, posterior margin narrow and straight, setose and with a small median spine; dorsal surface with transverse row of setae medially. Uropodal endopod broad and rounded on posteromesial margin; dorsal surface with longitudinal median carina. Uropodal exopod broadly truncate distally, larger than endopod, slightly longer than broad, broadly truncate on distal margin; dorsal surface with a longitudinal medial carina. Plp1-2 absent.

Remarks. — The present species, *Callianassa persica* is similar to *C. australis* Kensley, 1974 from Lüderitz Bay, South West Africa, in body shape, in the relative size of the P3 propodus, in the Mxp3, the telson, and the rostrum. However, they differ as follows: in the present species, the anterolateral lobes on the frontal margin of the carapace are obtuse; the A2 peduncle is slightly longer than the A1 peduncle; and abdominal somite 6 is longer than broad; whereas in *C. australis*, the anterolateral lobes are scarcely developed; the A2 peduncle is distinctly longer than the A1 peduncle; and abdominal somite 6 is broader than long.

Type locality. — Persian Gulf, 28°54'N 50°11'E.

Distribution. — Persian Gulf (= Arabian Gulf).

Callianassa plantei Sakai, 2004

Callianassa plantei Sakai, 2004: 592-599, figs. 18-23.

Type locality. — Mozambique Channel, 30.00 m.

Distribution. — Mozambique Channel, 16-64 m.

Callianassa poorei Sakai, 1999

Callianassa poorei Sakai, 1999b: 373, figs. 1-2.
Biffarius poorei; Davie, 2002: 457.

Diagnosis. — Mxp3 ischium-merus suboperculiform, merus rounded on distomesial margin, male Plp1 uniramous, two-segmented, Plp2 absent, telson truncate on posterior margin, lacking a median spine (Sakai, 1999b: 373, figs. 1C, 2A).

Type locality. — East coast of Tasmania, Great Oyster Bay (42°09'S 148°10'E).

Distribution. — Tasmania and Victoria, Australia.

Callianassa praedatrix De Man, 1905

Callianassa praedatrix De Man, 1905: 607; Sakai, 1988: 59, fig. 4; Ngoc-Ho, 1994: 54, fig. 2; Sakai, 1999c: 51.
Callianassa (Cheramus) praedatrix; De Man, 1928b: 26, 97, 99, 146, pl. 15 fig. 22-22d.
Cheramus praedatrix; Tudge et al., 2000: 145; Davie, 2002: 459.

Diagnosis. — Mxp3 ischium-merus subsquare, merus armed with a sharp spine on distal margin, rounded on distomesial margin, male Plp1 uniramous, two-segmented, Plp2 biramous and slender, telson straight with a shallow triangular notch on posterior margin, bearing a median spine (De Man, 1928b: 147, 148, 149, pl. 15 fig. 22-22a, b).

Type locality. — Between Bowoni and Boetoeng (= Buton), Indonesia (4°20'S 122°58'E), 75-94 m.

Distribution. — Between Bowoni and Boetoeng (= Buton), Indonesia, 75-94 m depth; North West Shelf, Western Australia; sublittoral, benthic, continental shelf, 41-51 m depth.

Callianassa propinqua De Man, 1905

Callianassa propinqua De Man, 1905: 609; Ngoc-Ho, 1991: 290, fig. 4; Ngoc-Ho, 1994: 54, fig. 2; Sakai, 1999c: 51.
Callianassa (Cheramus) propinqua; De Man, 1928b: 27, 98, 127, pl. 12 fig. 18-18d.
Cheramus propinquus; Tudge et al., 2000: 145; Davie, 2002: 459.

Diagnosis. — Mxp3 ischium-merus subpediform, merus armed with a sharp spine on distal margin, rounded on distomesial margin, male Plp1 uniramous, two-segmented, Plp2 biramous, telson convex with a median spine (Ngoc-Ho, 1991: 290, 291, fig. 4d, f, k).

Type locality. — Teluk Kwandang, Sulawesi, Indonesia (0°58.5'N 122°55'E), 75 m.

Distribution. — Kwandang Bay, Indonesia; New Caledonia; North West Shelf, Australia; benthic, continental shelf, continental slope; shelly sand to mud substrates, 75-300 m depth.

Callianassa propriopedis Sakai, 2002

Callianassa propriopedis Sakai, 2002: 522, fig. 34A-F.

Diagnosis. — Mxp3 narrow, ischium-merus subpediform, merus rounded on distomesial margin, male Plp1-2 undescribed, telson convex, with a median spine (Sakai, 2002: 522, fig. 34C, D).

Type locality. — Andaman Sea, 7°59.956'N 98°38.587'E, 17.0 m, sand with shell fragments.

Distribution. — Known only from the type locality.

Callianassa pugnatrix De Man, 1905

Callianassa pugnatrix De Man, 1905: 611; Sakai, 1999c: 51; Tudge et al., 2000: 143.
Callianassa (Cheramus) pugnatrix; De Man, 1928b: 27, 93, 99, 138, 146, 151, pl. 15 fig. 23-23a, pl. 16 fig. 23b-e.

Material examined. — ZMUC 19, larger and smaller chelipeds, and P3 present, 11°10'N 079°59'E, off Tranquebar, coarse sand and mud, 75 m depth, "Galathea" Exped., 1950-1952, Station 295, leg. R/V "Galathea", 22.iv.1959.

Diagnosis. — Mxp3 ischium-merus subpediform, merus concave on distal margin, rounded on mesial margin, male Plp1 uniramous, two-segmented, Plp2 absent, telson notched medially, with a median spine.

Remarks. — The single specimen examined was damaged with the body missing; the larger and smaller chelipeds and P3 on both sides are separately preserved. This species is recorded for the first time from Tranquebar, India.

Type locality. — Anchorage of Djangkar, Java, Indonesia, 7°46'S 114°30.5'E, "Siboga" Exped., Sta. 5, 330 m.

Distribution. — Anchorage of Djangkar, Java, Indonesia (De Man, 1928); 330 m depth; off Tranquebar, India, 75 m depth.

Callianassa pygmaea De Man, 1928

Callianassa (Cheramus) pygmaea De Man, 1928b: 1, 18, 27, 93, 99, 155, pl. 16 fig. 24-24g.
　[Type locality: Ambon (= Amboina) Anchorage, 54 m.]
Callianassa amboinae; Sakai, 1999c: 37 [not *C. amboinae* Bate, 1888].
Callianassa pygmaea; Tudge et al., 2000: 143.

Diagnosis. — Antennular and antennal peduncle equal in length. Mxp3 ischium-merus subpediform, merus concave on distal margin, convex on lateral margin; P3 propodus oval. Male Plp1 uniramous, two-segmented, male Plp2 absent, telson subsquare, convex on posterior margin, medially concave with a minute median spine, uropodal exopod and endopod, respectively, each with a spine on anterior margin (De Man, 1928b, pl. 16 fig. 24-24b, c, g).

Remarks. — This species was synonymized with *C. amboinae* (Bate, 1888) by Sakai (1999: 37), but it is different in the form of the P3 propodus and in the relative length of antennular and antennal peduncles; in *C. amboinae*, the P3 propodus is elongate, whereas in *C. pygmaea* it is shortly oval, so in the present paper *C. amboinae* and *C. pygmaea* are separated as good species.

Type locality. — Ambon (= Amboina) Anchorage, 54 m.

Distribution. — Known only from the type locality.

Callianassa rotundicaudata Stebbing, 1902

Callianassa rotundicaudata Stebbing, 1902: 41, pl. 8; Pearson, 1905: 90; Stebbing, 1910: 369;
　Kensley, 1974: 277 (key); Dworschak, 1992: 202, fig. 11a-f; Sakai, 1999c: 52; Tudge et al.,
　2000: 143.
Callianassa (Calliactites) rotundicaudata; Borradaile, 1903: 545; De Man, 1928b: 26, 92, 94,
　97, 123, 136, 149, 150; Barnard, 1950: 512, fig. 95i-l.
Callianassa (Calliactites) rotundicaudata; Stebbing, 1910: 369.

Diagnosis. — Mxp3 ischium-merus subsquare, merus straight on distal margin, convex on mesial margin, male Plp1-2 absent, telson rounded, without a median spine (Barnard, 1950: 513, fig. 95j; Dworschak, 1992, fig. 11e, f).

Type locality. — 34°02.45'S 25°10'E, St. Francis Bay, South Africa.

Distribution. — Orange River mouth, Saldanha Bay, St. Francis Bay, Algoa Bay, 10-35 m, Kowie, Port Alfred, South Africa; Cheval Paar, Sri Lanka.

Callianassa sibogae De Man, 1905

Callianassa Sibogae De Man, 1905: 613.

Callianassa (?*Cheramus*) *sibogae*; De Man, 1928b: 124.
Callianassa (*Cheramus*) *sibogae*; De Man, 1928b: 27, 98, pl. 11 fig. 17-17e.
Callianassa sibogae; Ngoc-Ho, 1994: 54, fig. 3; Sakai, 1999c: 52.
Cheramus sibogae; Tudge et al., 2000: 145; Davie, 2002: 460.

Diagnosis. — Mxp3 ischium-merus subpediform, merus truncate on distal margin, convex on mesial margin, male Plp1 uniramous, two-segmented, male Plp2 absent, telson convex with a median spine (Ngoc-Ho, 1994: 56, fig. 3c, h).

Type locality. — Indonesia, 7°46'S 114°30'E, 330 m.

Distribution. — Indonesia (7°46'S 114°30.5'E), 330 m; North West Shelf, Australia; benthic, continental shelf, continental slope; 134-330 m depth.

Callianassa spinophthalma Sakai, 1970

Callianassa (*Cheramus*) *spinophthalma* Sakai, 1970a: 40, figs. 2, 3, 4a-b; Sakai, 1987a: 306.
Callianassa spinophthalma; Sakai, 1999c: 52.
Cheramus spinophthalmus; Tudge et al., 2000: 145.

Diagnosis. — Mxp3 ischium-merus subsquare, merus truncate on distal margin, convex on mesial margin, male Plp1-2 unknown, telson concave with a median spine (Sakai, 1970a: 40, figs. 3b, 4a).

Type locality. — Tsushima Island, Japan (34°58'N 129°26.9'E), 210 m.

Distribution. — Known only from the type locality.

Callianassa spinoculata sp. nov.
(figs. 22-23)

Material examined. — ZMUC CRU-3847, holotype, 1 female (Tl/Cl 16/3.4 mm), 07°00'N 99°22'E, W. Malay Peninsula, 27 m depth, Thai/Danish Exped. 1966, Station 1052, 10.ii.1966.

Diagnosis. — Small in size. Rostrum triangular and pointed distally in dorsal view; frontal margin of carapace with anterolateral projections slightly developed. Antennular peduncle slightly shorter than antennal peduncle. Mxp3 ischium-merus subsquare, broadened, merus rounded on distomesial margin; ischium subsquare; merus entirely rounded on mesiodistal angle and with a row of 12 teeth on mesial surface. Larger cheliped merus with a simple proximal tooth on ventral margin. P3 propodus oval, paddle-shaped. Plp3-5 narrow, with thumb-like appendices internae. Abdominal somite 6 subsquare. Male

Plp1-2 unknown. Telson trapezoid, about as long as wide, not convex proximally, posterior margin slightly concave medially, and with a median tooth.

Description of female holotype (fig. 22A). — Rostrum triangular in dorsal view, slightly failing to reach middle of eyestalks; frontal margin of carapace smooth, bearing anterolateral projections that are slightly developed; dorsal oval distinct; cervical groove located at posterior fifth of carapace including rostrum. Linea thalassinica present entirely. Eyestalks (fig. 22A-B) oval, slightly longer than broad, almost reaching distal end of A1 basal segment, bearing an obtusely angular protrusion distally; dorsal surface convex, touching rostrum, descending forward in distal quarter; cornea rounded, located in middle part. A1 peduncle slightly shorter than peduncle of A2, terminal segment more than 2.0 times as long as penultimate. A2 without scaphocerite; terminal segment slightly less than half length of penultimate segment. Mxp3 (fig. 22A-B) ischium-merus broadened, about 1.6 times as long as wide, setose on mesial margin; ischium slightly longer than broad, with crista dentata consisting of a row of 12 teeth on mesial surface; merus half as long as broad and half length of ischium, entirely rounded at mesiodistal angle; three terminal segments slender and with long setae on flexor margins. Exopod absent.

P1 (fig. 23A) massive on right side; ischium rod-like, dorsal margin slightly undulate and unarmed, ventral margin armed with a row of three anteriorly directed denticles in middle part; merus about as long as ischium, about 1.8 times as long as high, dorsal margin entirely arched, ventral margin with a small, simple proximal tooth, exterior surface medially swollen; carpus as high as long and about as long as merus, dorsal margin almost straight, ventral margin smooth and carinate, and largely divergent downward on proximoventral margin; chela broadened, about 1.8 times as long as high; palm about as long as carpus, dorsal margin smooth and slightly incurved distally, ventral margin with smooth keel extending to base of fixed finger; distal gap largely convex and unarmed; fixed finger armed with row of denticles on prehensile margin and distally incurved; dactylus slightly shorter than palm, slightly overreaching tip of fixed finger, prehensile margin sinuous, unarmed, and incurved distally. Smaller cheliped absent. P2 chelate, ischium unarmed; merus more than 2.0 times as long as ischium; carpus 0.7 times as long as merus; chela slightly longer than carpus, with long setae on dorsal and ventral margins; dactylus longer than palm. P3 (fig. 23B) simple, ischium 1.8 times as long asbroad, unarmed; merus rectangular, 2.0 times as long as ischium, 2.5 times as long as high; carpus broadened distally, slightly shorter than merus and 2.0 times as

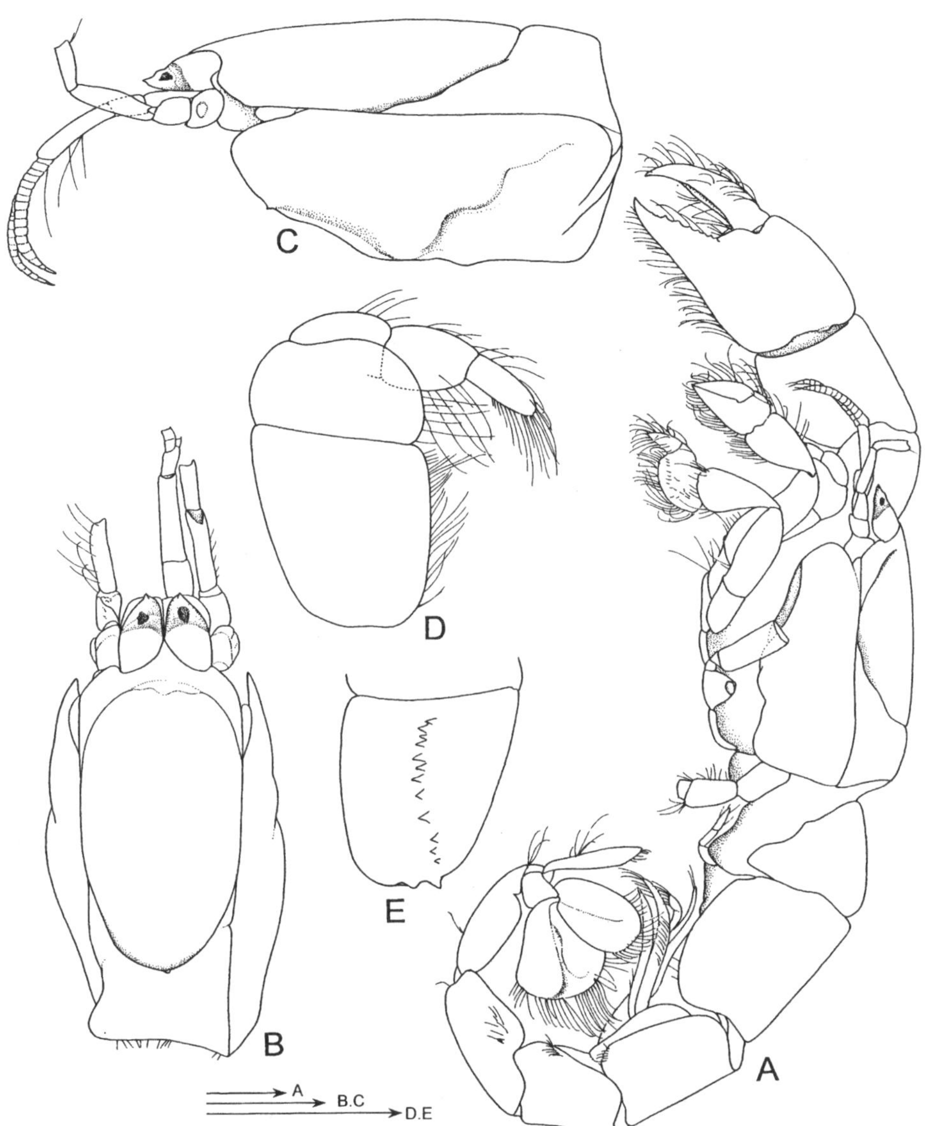

Fig. 22. *Callianassa spinoculata* sp. nov. A, whole body, lateral view; B, carapace, dorsal view; C, same, lateral view; D, Mxp3, lateral view; E, ischium of same Mxp3, mesial view. A-D, ZMUC 90, holotype female, 7°00'N 99°22'E, West Malay Peninsula, 27 m depth. Scales 1 mm.

long as high; propodus oval, paddle-shaped, ventral margin rounded entirely, with long marginal setae; dactylus triangular in shape, external surface densely setose. P4 absent on both sides. P5 chelate; merus 4.5 times as long as wide;

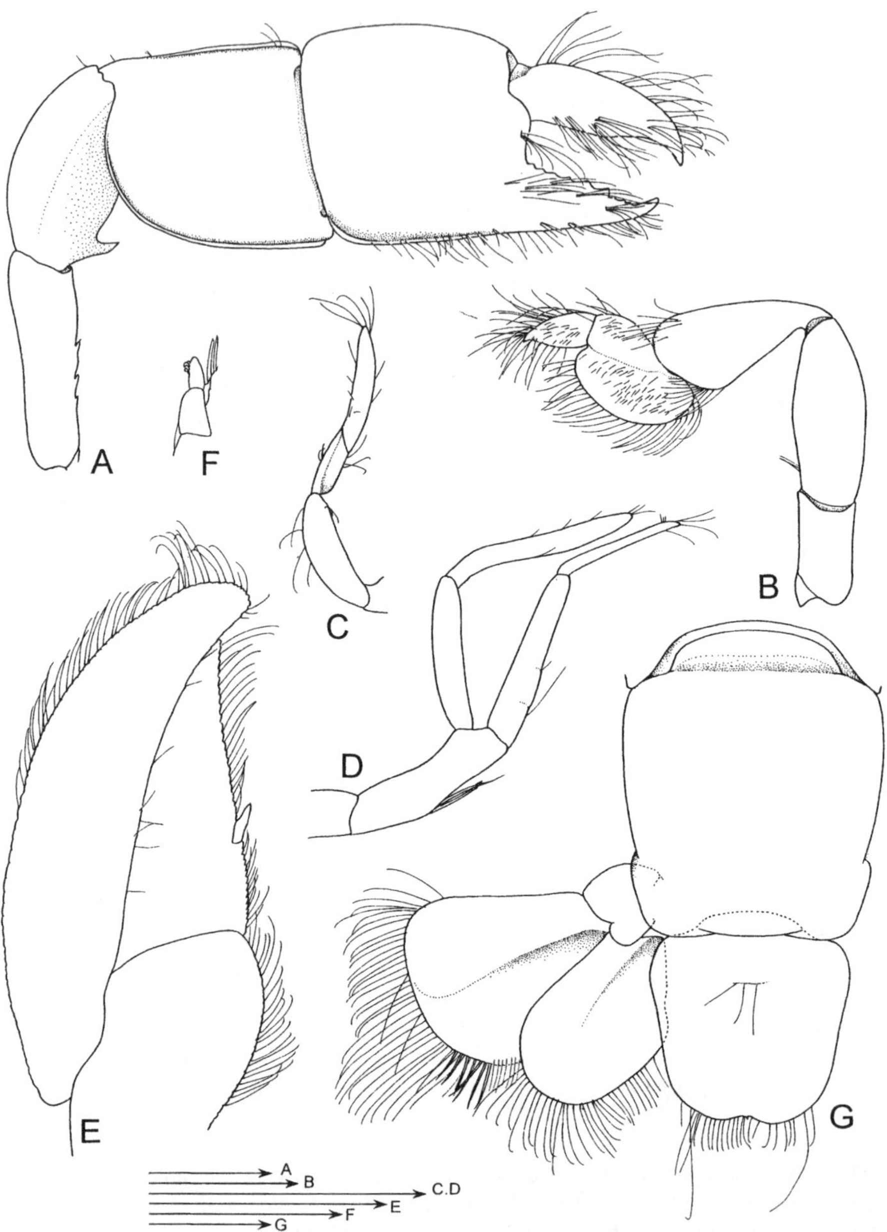

Fig. 23. *Callianassa spinoculata* sp. nov. A, larger cheliped, lateral view; B, pereiopod 3; C, female Plp1; D, female Plp2; E, Plp3; F, appendix interna on same Plp3; G, abdominal somite 6 and telson, with uropod on left side. A-G, ZMUC 90, holotype female, 7°00'N 99°22'E, West Malay Peninsula, 27 m depth. Scales 1 mm.

TABLE V

Morphological differences between *Callianassa spinoculata* sp. nov. and *C. brevirostris* Sakai, 2002

C. spinoculata sp. nov.	*C. brevirostris* Sakai, 2002
A1 peduncle slightly shorter than A2 peduncle	A1 peduncle about as long as A2 peduncle
Eyestalks with obtusely angular distal protrusion	Eyestalks without a distal protrusion
Mxp3 ischium with crista dentata, consisting of 12 teeth on internal surface	Mxp3 ischium with crista dentata, consisting of 9 teeth on internal surface
P3 propodus oval, ventral margin entirely rounded	P3 propodus oval, ventral margin distally truncate

carpus 0.8 times as long as merus; propodus about as long as carpus; dactylus chelate with distal part of propodus. P5 on left side missing.

Abdominal somites smooth, glabrous dorsally; pleura 3-5 each with tuft of setae laterally; abdominal somite 6 (fig. 23G) rectangular, parallel on lateral margins. Plp1 (fig. 23C) uniramous, 3-segmented. Plp2 (fig. 23D) biramous, slender. Plp 3 (fig. 23E) to Plp5 biramous, each bearing 2-segmented, thumb-like appendix interna (fig. 23F) on mesial margins of endopods. Telson (fig. 23G) trapezoid, slightly longer than wide and slightly convex proximally; lateral margin convergent posteriorly, posterior margin concave medially, setose, and with a median tooth; dorsal surface weakly elevated, medially with transverse row of setae. Uropodal endopod subquadrate, slightly longer than telson, 1.6 times as long as broad and rounded distally; dorsal surface slightly elevated in lateral half, bordered by faint longitudinal median, carina. Uropodal exopod broadened distally, about 1.2 times as long as wide, setose on distal margin; dorsal surface elevated in lateral half.

Etymology. — The species name, *spinoculata* is the combination of the Latin words "spina" and "oculatus", meaning "thorn" and "-eyed", respectively, based on the characteristic form of the eye peduncles. It is an adjective agreeing in gender with the (feminine) generic name.

Remarks. — The present new species is closely similar to *Callianassa brevirostris* Sakai, 2002 from the Andaman Sea, off Phuket, Thailand, in the shape of Mxp3, the female larger cheliped, and the telson bearing a median tooth on the concave posterior margin. The differences between *C. spinoculata* and *C. brevirostris* are shown in table V.

Type locality. — Thai/Danish Exped., St. 1052, 7°00'N 99°22'E, W. Malay Peninsula, 27 m.

Distribution. — Known only from the type locality.

Callianassa stenomastaxa Sakai, 2002

Callianassa stenomastaxa Sakai, 2002: 496, fig. 20A-G.

Diagnosis. — Mxp3 ischium-merus narrow, subpediform, merus rounded on distomesial margin, male Plp1-2 unknown, telson notched medially without a median spine (Sakai, 2002: 496, fig. 20C, D).

Remarks. — The present species is closely similar to *Callianassa orientalis* (Bate, 1888) from the Arafura Sea, because the Mxp3 is elongate, and the A1 peduncle is shorter than the A2 peduncle, but it differs in the form of the appendix interna. In *Ch. orientalis* Bate, 1888 it is described as stubby (Bate, 1888, fig. 2mr), while in *C. stenomastaxa* it is triangular (Sakai, 2002, fig. 20G).

Type locality. — Andaman Sea, 8°00.046'N 98°29.383'E, 19.0 m, sand with shell fragments.

Distribution. — Known only from the type locality.

Callianassa tenuipes Sakai, 2002

Callianassa tenuipes Sakai, 2002: 528, fig. 38A-D.

Diagnosis. — Mxp3 ischium-merus narrow, pediform, male Plp1 uniramous, thick protrusion, male Plp2 absent, telson convergent distally, straight on posterior margin, bearing a median spine (Sakai, 2002: 528, fig. 38C, D).

Type locality. — Andaman Sea, 41-61 m depth.

Distribution. — Known only from the type locality.

Callianassa thailandica sp. nov.
(figs. 24-25)

Material examined. — ZMUC CRU-3848, holotype, male (Tl/Cl 18.0/4.2, right branchial cavity with swelling by infection of bopyrids, larger and smaller cheliped detached), 13°13'N 100°34'E, Gulf of Thailand, some fine sandy mud with mostly clay, 20 m depth, "Galathea" Exped., 1950-1952, Station 394, leg. R/V "Galathea", 11.vi.1951; ZMUC CRU-3849, paratypes, 3 males, (Tl 9.0-19.0, missing larger cheliped), 1 ovig. female (16.0/3.5, missing larger cheliped),

13°13'N 100°34'E, Gulf of Thailand, mud with some fine sand, mud with clay, 20 m depth, "Galathea" Exped., 1950-1952, Station 394, leg. R/V "Galathea", 11.vi.1951; ZMUC CRU-3850, two carapaces, two abdominal parts, male larger cheliped of left side, 13°13'N 100°34'E, Gulf of Thailand, mud with some fine sandy mud with clay, 20 m depth, "Galathea" Exped., 1950-1952, Station 394, leg. R/V "Galathea", 11.vi.1951.

Diagnosis. — Rostrum barely developed; frontal margin of carapace with obtuse anterolateral projections. Eyestalks subsquare, truncate apically, shorter than A1 basal segment. A1 peduncle distinctly shorter than A2 peduncle. Mxp3 ischium-merus subpediform, merus rounded on distomesial margin. P1 merus distally convergent in width; ventral margin with distinct proximal tooth and denticulate distal to proximal teeth; dorsal margin slightly arched and with a subproximal denticle. P3 propodus square. Plp1 uniramous, weakly 2-segmented. Plp2 absent. Telson square, lateral margins convergent to rounded posterolateral angles, posterior margin concave in middle part and without median spine. Uropodal endopod rounded distally; uropodal exopod square-ended distally, larger than endopod.

Description of male holotype. — Rostrum (fig. 24A, B) barely developed in dorsal view; frontal margin of carapace with obtuse anterolateral projections; dorsal oval conspicuous; cervical groove located at posterior fourth of carapace. Linea thalassinica present at full length. Eyestalks subsquare, broader than long proximally; tip truncate distally, and shorter than A1 basal segment; cornea small, located at middle part, pigmented black in alcohol specimens. A1 slightly overreaching A2 penultimate segment, terminal segment about two-thirds length of penultimate. A2 scaphocerite vestigial. Mxp3 (fig. 24C) merus-ischium narrow; ischium subrectangular, 1.8 times as long as broad; crista dentata (fig. 24D) with a row of fine and slender denticles; merus subtriangular, two-thirds length of ischium, mesial margin broadly convex, distal margin entirely facing to carpus distally; carpus triangular, 1.5 times as long as broad; propodus subquadrate, longer than broad, with entirely convex ventral margin; dactylus digitiform, almost as long as propodus. Exopod not present. Branchial formula as shown in table VI.

P1 unequal and dissimilar. Larger cheliped (fig. 24E, F) massive; ischium slender, dorsal margin sinuous and unarmed, ventral margin almost straight and armed with a distinct denticle on middle part; merus slightly longer than ischium, about 2.0 times as long as high, dorsal margin slightly arched and with a subproximal tubercle, ventral margin with an arcuate proximal tooth, of which the ventral margin bears a subdistal tooth, and distal to the proximal tooth straight and armed with a row of triangular denticles; lateral surface dis-

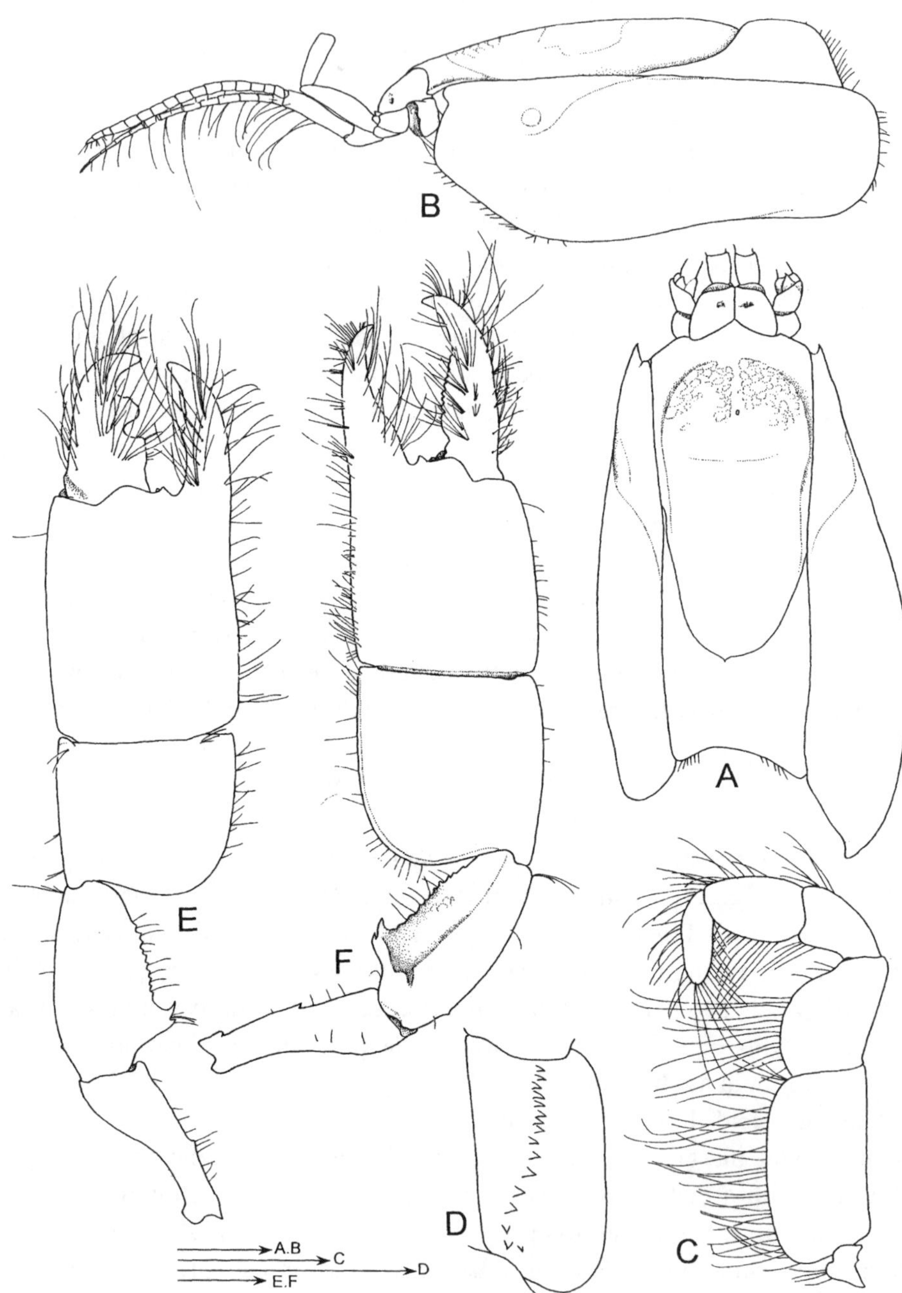

TABLE VI

Branchial formula of *Callianassa thailandica* sp. nov.

	Maxillipeds			Pereiopods				
	1	2	3	1	2	3	4	5
Exopods	1	1	–	–	–	–	–	
Epipods	1	–	–	–	–	–	–	–
Podobranchs	–	r	–	–	–	–	–	–
Arthrobranchs	–	–	2	2	2	2	2	–
Pleurobranchs	–	–	–	–	–	–	–	–

(r = rudimentary)

tinctly engraved in ventral half. Carpus broadened, as high as long, slightly shorter than merus, largely convex on posteroventral angle. Chela heavy, 2.0 times as long as carpus; palm as long as carpus and slightly longer than high, dorsal and ventral margins smooth, distal gap protruded and armed with a denticle above ventral concavity to fixed finger; fixed finger two-thirds length of palm and smooth on prehensile margin; dactylus slightly longer than fixed finger, dorsal margin incurved downward distally, prehensile margin entirely concave and denticulate over its proximal half. Smaller cheliped (fig. 25A) slender and less massive than larger cheliped; ischium narrow, dorsal and ventral margins unarmed; merus rectangular, unarmed, slightly shorter than ischium; carpus elongate-triangular, 3.0 times as long as high, broadly descending distally on proximoventral margin. Chela shorter than carpus; palm subsquare, about 1.3 times as long as high; distal gap slanting distally to cutting margin of fixed finger; fixed finger 1.5 times as long as palm and unarmed on prehensile margin. Dactylus slender, as long as fixed finger, crossing distally with fixed finger, prehensile margin unarmed. P2 (fig. 25B) chelate; merus broadened, slightly less than three times as long as wide; carpus triangular, more than half length of carpus; chela 1.3 times as long as carpus, both fingers 2.3 times as long as palm along midline. Missing on the left side. P3 (fig. 25C) simple; merus 2.8 times as long as broad; carpus divergent distally, 0.8 times as long as merus; propodus subsquare, half length of carpus and 0.8 times as long as broad, distoventral angle slightly protruded forward and setose on lat-

Fig. 24. *Callianassa thailandica* sp. nov. A, carapace, dorsal view. B, same, lateral view; C, Mxp3, lateral view; D, ischium of same Mxp3, mesial view; E, male larger cheliped, detached, lateral view; F, male larger cheliped, detached, lateral view. A-D, F, holotype male, ZMUC 10A, 13°13'N 100°34'E, Gulf of Thailand, mud with a little fine sand, mud with clay, 20 m depth; E, ZMUC 10D, non-type, same locality. Scales 1 mm.

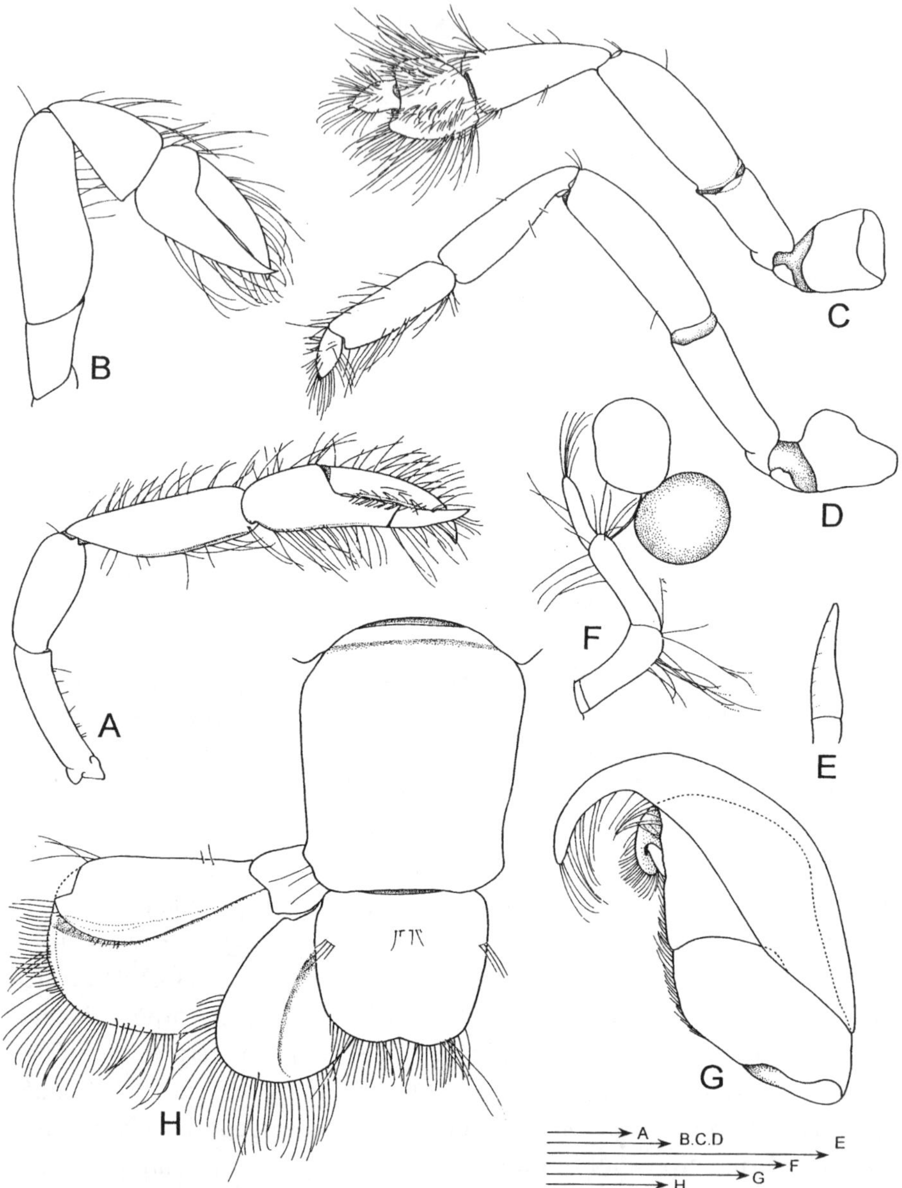

Fig. 25. *Callianassa thailandica* sp. nov. A, smaller cheliped, detached from body, lateral view; B, pereiopod 2, right lateral view; C, pereiopod 3; D, pereiopod 4; E, male Plp1; F, female Plp1 with eggs; G, Plp3; H, abdominal somite 6 and tail fan, dorsal view. A-E, holotype, male, ZMUC 10A, 13°13'N 100°34'E, Gulf of Thailand, mud with a little fine sand, mud with clay, 20 m depth; F-G, female paratype, ZMUC 10C, same locality; H, non-type, ZMUC 10D, same locality. Scales 1 mm.

eral surface; dactylus triangular, 0.8 times as long as propodus. P4 (fig. 25D) slender, simple; merus 1.3 times as long as ischium; carpus 0.8 as long as merus; propodus rectangular in lateral view, 0.8 times as long as carpus, lateral surface scarcely setose, ventrodistal corner obtusely protruded; dactylus less than half length of propodus. Missing on the right side. P5 chelate; propodus with a broad fixed finger ventrodistally, ventral surface with dense setation; dactylus hooked outside towards fixed finger, tip deflected.

Abdominal somites smooth, glabrous dorsally; pleura 2-5 each with a tuft of setae laterally; abdominal somite 6 smooth, 1.2 times as long as broad, unarmed on lateral margin. Telson (fig. 25H) square in dorsal view, slightly broader than long; lateral margins smooth, slightly convergent to rounded posterolateral angles; posterior margin shallowly concave in middle part, lacking in a median spine; dorsal surface with transverse row of setae medially at anterior fourth. Uropodal endopod oval on distal margin and longer than telson; anterior margin slightly convex, extending to rounded anterodistal angle; dorsal surface with a weak longitudinal medial carina. Uropodal exopod broadened, about 1.5 times as long as broad, larger and longer than endopod, squared distally; dorsal surface–with three longitudinal medial carinae. Plp1 (fig. 25E) uniramous, weakly 2-segmented. Plp2 absent. Plp3 (fig. 25G) to Plp5 biramous, broadly foliaceous, each bearing a stubby, projecting appendix interna.

Female. — Plp1 (fig. 25F) uniramous and 3-segmented. Plp2 absent in the female material (ZMUC CRU-3849).

Etymology. — The species is named after the type locality. The name is an adjective agreeing in gender with the (feminine) generic name.

Remarks. — *Callianassa thailandica* is closely similar to *Callianassa stenomastaxa* Sakai, 2002 in the form of Mxp3, P3 propodus, and abdominal somite 6, but *C. thailandica* differs in that the rostrum is weakly triangular in shape, the P3 propodus is subsquare with a distally protruding distoventral angle, the uropodal exopod is distally truncate, while the telson is slightly concave and lacks a median tooth. In *C. stenomastaxa*, the rostrum is narrowly triangular, the P3 propodus subsquare with the distoventral angle not protruding, the uropodal exopod is distally rounded, and the telson is unarmed and distinctly concave on the posterior margin.

In the present species, *C. thailandica*, the larger chelipeds (figs. 24E, F) are variable in the cutting edge of the male larger cheliped, as observed in *Callianassa japonica* (cf. Sakai, 2002).

Type locality. — Gulf of Thailand, 13°13'N 100°34'E.

Distribution. — Gulf of Thailand.

Callianassa thorsoni sp. nov.
(figs. 26-27)

Material examined. — ZMUC CRU-3851, holotype, 1 ovig. female (Tl/Cl 14.0/3.0, P2 on right side and P3-4 on left side present, but missing both chelipeds and P2 on left side as well as P3-5 on right side), 27°03'N 51°02'E, N. of Bahrein lightship, Persian Gulf, clay mixed with sand, 71 m depth, Thorson's Expedition, Station 38D, leg. G. Thorson, 27.iii.1937; ZMUC CRU-3852, paratype, 1 female (11.0/2.0, wanting P1-2, Plp2 on right side, Plp 3 on left side, and Plp4-5), 27°31'N 52°07'E, N. E. of Bahrein lightship, Persian Gulf, 58 m depth, Thorson's Expedition, Station 43B, leg. G. Thorson, 31.iii.1937; ZMUC CRU-3853, paratype, 1 male (Tl 10.0), 27°03'N 51°02'E, N. of Bahrein lightship, Persian Gulf, clay mixed with sand, 71 m depth, Thorson's Expedition, Station 38C, leg. G. Thorson, 27 iii, 1937; ZMUC CRU-3854, paratype, 1 female (13.0/2.5, chelipeds lacking, P2-5 present), Persian Gulf, 27 m depth, Thorson's Expedition, Station 33C, leg. G. Thorson, 08.iv.1937; ZMUC CRU-3855, paratype, 1 male (12.0/2.8, with smaller cheliped present, but missing larger cheliped and P5), 27°00'N 52°53'E, S. W. of Bustani, Persian Gulf, clay strongly mixed with sand, 72 m depth, Thorson's Expedition, Station 48A, leg. G. Thorson, 06.iv.1937; ZMUC CRU-3856, paratype, 1 female (Tl/Cl ca. 11.0/2.0, chelipeds on both sides absent, P2-5 on left side, abdomen detached from carapace), 27°03'N 51°02'E, N. of Bahrein lightship, Persian Gulf, clay mixed with sand, 71 m depth, Thorson's Expedition, Station 38A, leg. G. Thorson, 27.iii.1937.

Diagnosis. — Rostrum triangular and pointed distally; frontal margin of carapace with rounded anterolateral projections. Eyestalks triangular, characteristically with slender distal spine extending beyond antennular basal segment. Mxp3 ischium-merus broadened, subsquare, merus straight distally and rounded on distomesial margin; ischium rectangular, 1.5 times as long as broad; merus 1.5 times as broad as long, distal margin broadened, slightly concave, and largely rounded on mesiodistal angle. P3 propodus rhombic, as long as broad, ventral margin distinctly deflexed medially. Plp1 uniramous, two-segmented; Plp2 absent. Telson trapezoid, slightly broadened proximally on lateral margins, posterior margin concave medially and with a distinct median spine.

Description of female holotype (fig. 26A). — Rostrum (fig. 26B) triangular and pointed apically; frontal margin of carapace almost smooth, bearing paired obtuse anterolateral projections; dorsal oval conspicuous; cervical groove located in posterior fourth to fifth of carapace. Linea thalassinica extending at full length. Eyestalks (fig. 26B-C) triangular, broader than long, convex dorsally, armed characteristically with a slender distal spine extending beyond antennular basal segment; cornea located medially, pigmented in alcohol specimen. A1 peduncle slightly longer than A2 peduncle, terminal segment three times as long as penultimate. A2 scaphocerite small and triangular; terminal

TABLE VII

Branchial formula of *Callianassa thorsoni* sp. nov.

	Maxillipeds			Pereiopods				
	1	2	3	1	2	3	4	5
Exopods	1	1	–	–	–	–	–	
Epipods	1	–	–	–	–	–	–	–
Podobranchs	–	–	–	–	–	–	–	–
Arthrobranchs	–	–	2	2	2	2	2	–
Pleurobranchs	–	–	–	–	–	–	–	–

segment two-thirds length of penultimate. Mxp3 (fig. 26D-E) ischium-merus broadened, ischium rectangular, 1.5 times as long as broad; crista dentata with a row of sharp denticles; merus oblong, 1.5 times as broad as long, lateral margin divergent distally, distal margin slightly concave and distomesial angles largely rounded; carpus triangular, 1.5 times as long as broad; propodus rectangular, 1.8 times as long as broad; dactylus digitiform, 0.8 times as long as propodus. Exopod absent. Branchial formula as shown in table VII.

P1 absent on both sides. P2 chelate; merus broadened, 2.0 times as long as ischium, ventral margin with closely set setae; carpus slightly longer than half length of merus, setose on dorsal and flexor margins; chela longer than carpus, setose on both margins; both fingers 1.2 times as long as palm, corneous on prehensile margins and tips; missing on left side. P2 chelate. P3 (fig. 27B) simple; merus short, 2.5 times as long as ischium; carpus three-fourths length of merus, divergent distally on ventral margin; propodus rhombic, as long as broad, ventral margin distinctly deflexed medially; dactylus digitiform; missing on right side. P4 (fig. 26A) simple; merus as long as ischium; carpus slightly shorter than merus; propodus 1.2 times as long as carpus; dactylus slightly less than half length of propodus; missing on right side. P5 absent.

Abdominal somites smooth, glabrous dorsally; pleurites 2-5 each with a tuft of setae laterally; abdominal somite 6 (fig. 27F) slightly longer than broad and parallel laterally. Plp1 (fig. 27C) uniramous, 3-segmented. Plp2 (fig. 27D) biramous; exopod lanceolate; endopod 2-segmented, distal segment simple and without an appendix interna. Plp3 (fig. 27E) to Plp5 biramous, narrowly foliaceous, each bearing a small appendix interna on mesial margin of endopod. Telson (fig. 27F) trapezoid, slightly broader than long proximally; lateral margins slightly broadened over proximal third, then clearly convergent posteriorly and with two pairs of spinules at posterolateral angles; posterior margin concave medially and with a distinct median spine; dorsal surface with trans-

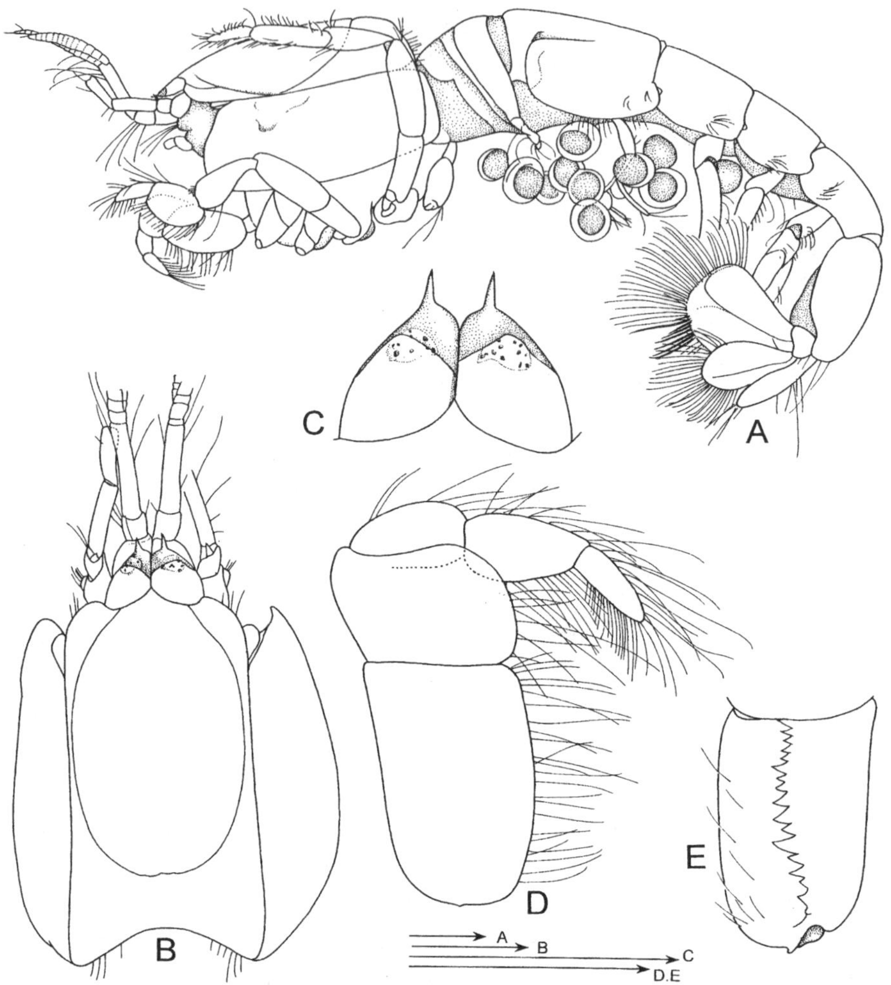

Fig. 26. *Callianassa thorsoni* sp. nov. A, ovigerous female, lateral view; B, carapace, dorsal view; C, eyes, dorsal view; D, Mxp3, lateral view; E, ischium of Mxp3, mesial view. A, D-E, ZMUC 156, holotype ovigerous female, Persian Gulf, north of Bahrein lightship, 27°03'N 51°02'E, clay mixed with sand, 71 m depth; B-C, ZMUC 159, paratype male, Persian Gulf, S. W. of Bustani, 27°00'N 52°53'E, clay strongly mixed with sand, 72 m depth. Scales 1 mm.

verse row of setae proximomedially. Uropodal endopod leaf-shaped; entire anterior margin slightly convex, oval and narrow distally, dorsal surface with a median longitudinal carina. Uropodal exopod square distally, larger than endopod, about 1.5 times as long as broad; anterior margin slightly divergent distally; dorsal surface with a longitudinal medial carina.

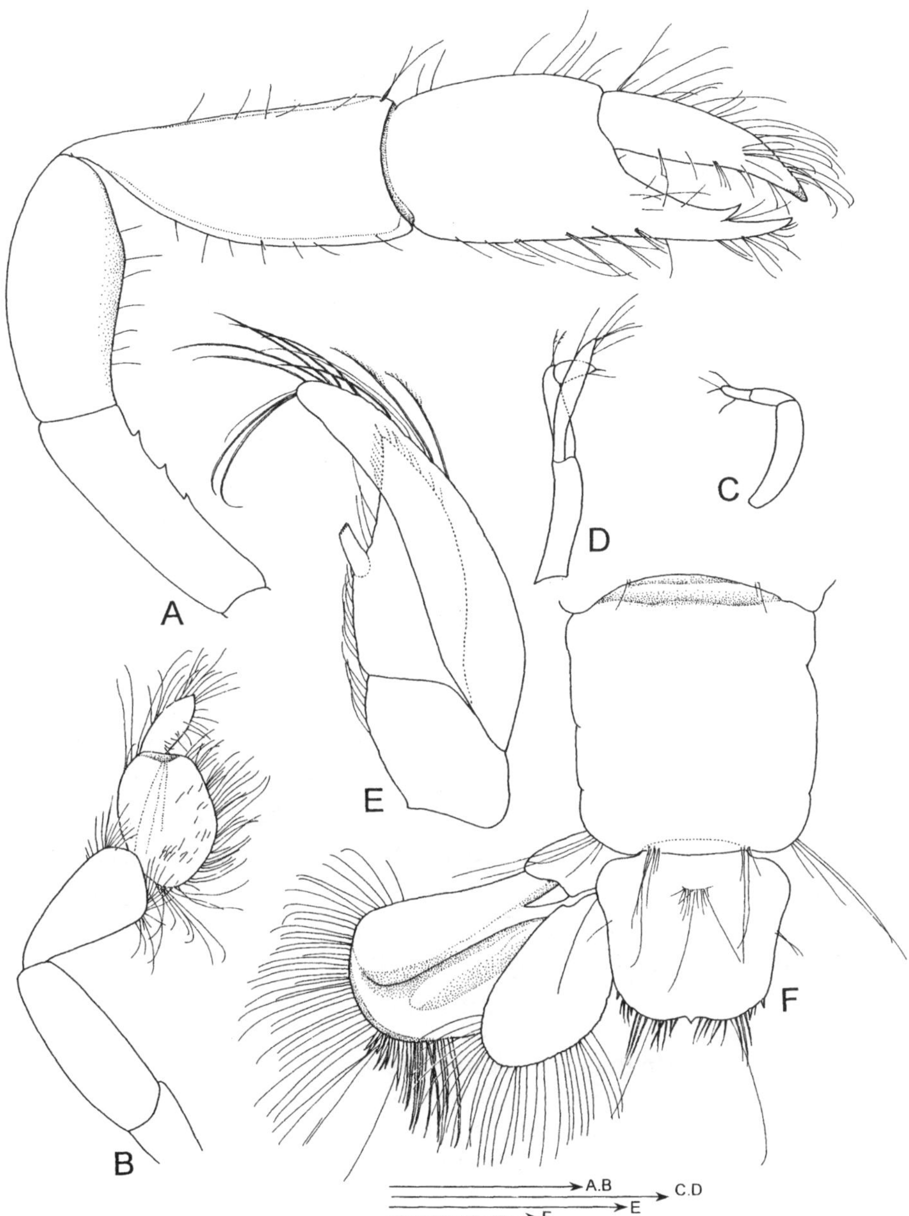

Fig. 27. *Callianassa thorsoni* sp. nov. A, smaller cheliped, lateral view; B, pereiopod 3, lateral view; C, male pleopod 1; D, male pleopod 2; E, female Plp3, with appendix interna; F, abdominal somite 6 and telson, with uropod on left side, dorsal view. A, C-D, ZMUC 159, paratype male, Persian Gulf, S. W. of Bustani, 27°00'N 52°53'E, clay strongly mixed with sand, 72 m; B, E, ZMUC 155, paratype female, G. Thorson, Persian Gulf, Sta. 33C, 08.iv.1937, 27 m; F, ZMUC 156, holotype ovigerous female, Persian Gulf, north of Bahrein lightship, 27°03'N 51°02'E, clay mixed with sand, 71 m. Scales 1 mm.

Description of male. — Larger cheliped unknown. Smaller cheliped (fig. 27A) with ischium three times as long as broad, ventral margin with three denticles in its distal half and dorsal margin straight and unarmed; merus spindle-shaped, slightly longer than ischium, unarmed on both margins; carpus 1.2 times as long as merus, divergent distally on ventral margin; chela 1.3 times as long as carpus; palm 0.8 times as long as carpus and 1.2 times as long as broad, distal gap divergent to cutting margin of fixed finger; fixed finger with a subdistal tooth on cutting edge; dactylus about as long as palm, unarmed on cutting edge. Plp1 uniramous, 2-segmented; Plp2 absent.

Etymology. — The species name, *thorsoni* is chosen in memory of the Danish Scientist, Gunar Thorson, who made a great contribution to the Persian (= Iranian) Expedition. The name is derived from a personal name, hence treated as a noun in the genitive singular.

Remarks. — *Callianassa thorsoni* is similar to the Australian species, *C. acutirostella* Sakai, 1988 from the North West Shelf, Western Australia, in the shape of the rostrum, the Mxp3, abdominal somite 6, Plp1-2, tail-fan, and the antennular peduncle being slightly longer than the antennal peduncle. The P3 propodus in *C. acutirostella* has not been described; in *C. acutirostella* the eye lacks a distal spine and the telson is not armed with two pairs of spinules at the posterodistal corners, as in *C. thorsoni*. The present species is also similar to *C. propinqua* De Man, 1905 in the shape of the P3 propodus, but it is otherwise very different: in *C. propinqua*, the Mxp3 merus bears a distinct spine on the distal margin, and there is no distal spine on the eyestalks; also, the antennular peduncle is shorter than the antennal one.

Type locality. — Persian Gulf, 27°03'N 51°02'E.

Dstribution. — Persian Gulf (= Arabian Gulf).

Callianassa tonkinae Grebenjuk, 1975

Callianassa (Scallasis) tonkinae Grebenjuk, 1975: 302, fig. 3.
Callianassa tonkinae; Sakai, 1999c: 52; Tudge et al., 2000: 143; Sakai, 2002: 503, figs. 24A-C, 25A-G.

Diagnosis. — Mxp3 ischium-merus subpediform, merus straight on distal margin, convex on lateral margin, male Plp1 uniramous, two-segmented, male Plp2 absent, telson truncate, medially concave on posterior margin, bearing a median spine.

Type locality. — Tonkin Bay.

Distribution. — Tonkin Bay, South China Sea; Andaman Sea, 7°45.452'N 97°57.951'E; 9°00.062'N 97°53.366'E, 65.4-70.0 m, muddy sand, coarse and fine sand; East Lagoon, New Caledonia, 21 m.

Callianassa sp. Haswell, 1882

Callianassa sp. Haswell, 1882: 167; Sakai, 1999c: 53.

Locality. — Molle Island, Whitsunday Passage, Queensland.

Callianassa sp. De Man, 1928

Callianassa (*Calliactites*) sp. De Man, 1928b: 26, 93, 97, 116, pl. 10 fig. 15-15c.
Callianassa sp.; Sakai, 1999c: 53.

Locality. — Anchorage off Donggala, Sulawesi (Celebes), Palos Bay, Indonesia, 36 m.

Callianassa sp. Sakai, 1970

Callianassa (*Cheramus*) sp. Sakai, 1970a: 45, fig. 4c-f.
Callianassa sp.; Sakai, 1999c: 53.

Locality. — Around Tsushima Is., Japan, 34°37.5'N 129°50.7'E, 110 m, coarse sand and mud.

Callianassa sp. 1 Sakai, 2002

Callianassa sp. 1, Sakai, 2002: 488, fig. 16A-D.

Distribution. — Andaman Sea, 8°30.039'N 98°00.051'E, 62.7 m, muddy sand.

Callianassa sp. 2 Sakai, 2002

Callianassa sp. 2, Sakai, 2002: 498.

Distribution. — Andaman Sea, 9°30.351'N 97°57.168'E, 60.7 m, sandy mud, fine sand and shell fragments.

Callianassa sp. 3 Sakai, 2002

Callianassa sp. 3, Sakai, 2002: 520, fig. 33A-C.

Distribution. — Andaman Sea, 9°30.351'N 97°57.168'E, 60.7 m, sandy mud, fine sand and shell fragments.

Callianassa sp. 4 Sakai, 2002

Callianassa sp. 4, Sakai, 2002: 528, fig. 38A-D.

Distribution. — Andaman Sea, St. B18 OS; 7°15.066'N 98°51.212'E, 31-61.1 m.

Subfamily CALLICHIRINAE Manning & Felder, 1991

Callichirinae Manning & Felder, 1991: 775; Tudge et al., 2000: 136.

Not: Callichirinae, Ngoc-Ho, 2003: 486.

Definition. — Rostrum developed or reduced, lacking rostral carina. Carapace with dorsal oval; cardiac prominence and hepatic sulcus(i) absent. Eyestalks dorsoventrally flattened and contiguous. Abdominal somite 6 lacking lateral projections. A2 scaphocerite developed as a small process. Mxp3 ischium-merus variform, from pediform to operculiform; propodus broadened; dactylus digitiform. P1 chelate, unequal or subequal, and similar or dissimilar; larger cheliped with or without meral hook. P2 chelate. P3 propodus broadened. Male Plp1 slender and uniramous; female Plp1 slender or blade-like, and uniramous. Male Plp2 present or absent and, when present, slender or blade-like, and uniramous or biramous, bearing or lacking appendix interna and appendix masculina; female Plp2 slender or blade-like and biramous, bearing or lacking appendix interna. Plp3-5 larger than Plp1-2 in size and more robust in shape, biramous, foliaceous, and with appendices internae in both sexes. Uropodal exopod with a secondary setal lobe.

Type genus. — *Callichirus* Stimpson, 1866.

Genera included. — *Callichirus* Stimpson, 1866; *Glypturus* Stimpson, 1866; *Lepidophthalmus* Holmes, 1904; *Michaelcallianassa* Sakai, 2002; *Neocallichirus* Sakai, 1988; *Podocallichirus* Sakai, 1999.

Remarks. — The subfamily Callichirinae Manning & Felder, 1991 is based on the type genus, *Callichirus* Stimpson, 1866, and includes 6 genera: *Callichirus* Stimpson, 1866, *Glypturus* Stimpson, 1866 (syns.: *Corallianassa* Manning, 1987 and *Corallichirus* Manning, 1992), *Lepidophthalmus* Holmes, 1904, *Michaelcallianassa* Sakai, 2002, *Neocallichirus* Sakai, 1988 (syn.: *Sergio* Manning & Lemaitre, 1994), and *Podocallichirus* Sakai, 1999. The genera *Michaelcallianassa* Sakai, 2002 and *Podocallichirus* Sakai, 1999 are added to the Callichirinae in the present revision.

In my previous revision (Sakai, 1999c), the Callichirinae were synonymized with the Callianassinae sensu Manning & Felder, 1991, because in *Callichirus major*, the type species of *Callichirus*, Plp2 is not typical in shape for the Callichirinae in both sexes: in males it is biramous, foliaceous, and lacking its appendix interna and appendix masculina, and in females it is biramous, pediform, and lacking the appendix interna. However, in other genera of the Callichirinae, *Glypturus*, *Lepidophthalmus*, *Neocallichirus*, and *Podocallichirus*, the male Plp2 is biramous, foliaceous, and bears the appendix interna and appendix masculina, though the appendix interna is often reduced, fused, or absent, and the female Plp2 is also biramous, foliaceous, and bearing the appendix interna. In *Michaelcallianassa*, Plp2 is uniramous in both sexes, and lacks the appendix interna and appendix masculina in males, while it bears the appendix interna in females. Taking into account that all the examined males in *Michaelcallianassa indica* are small in size, it is considered that the appendix interna and appendix masculina in males are undeveloped in juvenile forms. Thus, in the present review, the genus *Michaelcallianassa* is included in the Callichirinae. The Callichirinae are to be separated from the Callianassinae, for the reason that in the Callianassinae the Mxp3 propodus is slender, and the male Plp2 is biramous and pediform, whereas in the Callichirinae the Mxp3 propodus is broadened and the male Plp2 is usually biramous and foliaceous. In *Callianassa limosa* Poore, 1975, the Mxp3 propodus is slightly broadened, so it was located in *Neocallichirus* in the previous revision (Sakai, 1999c: 103), but the Plp2 is pediform in both sexes, as in the Callianassinae, so it has been removed to *Callianassa*.

Tudge et al. (2000) admitted *Corallianassa* Manning, 1987 as a valid genus in the Callichirinae. However, it is to be synonymized with the genus *Glypturus*. Manning & Felder (1991) mentioned that: *"Corallianassa* (the type species is *Callianassa longiventris* A. Milne-Edwards, 1870) resembles *Glypturus* (type species is *Glypturus acanthochirus* Stimpson, 1866) in having three strong anterior spines on the carapace. It differs from *Glypturus* in having a much larger, subglobular cornea and in lacking [a] dorsal spine on the propodus of the cheliped." (Manning & Felder, 1991: 777). However, in *Corallianassa longiventris*, *Glypturus articulatus*, *G. borradailei*, *G. collaroy*, *G. coutierei*, *G. hartmeyeri*, *G. martensi*, and *Corallichirus xuthus*, the type species of the genus *Corallichirus*, the cornea is large, subglobular, and located distally; in *G. acanthochirus*, *G. haswelli*, *G. intesi*, *G. lanceolata*, and *G. winslowi*, the cornea is distinct, but not so large as in the species just mentioned above, and located subdistally. As a result, it is difficult to separate *Corallianassa* from *Glypturus* by the size and the location of the cornea. In *G. acanthochirus*, the dorsal spine of the propodus of the chelipeds is observed as in the definition, but in the other species of *Glypturus*, no such spines are present. The dorsal spine of the propodus of the chelipeds is not a generic character differentiating one from the other of those two genera, so the genus *Corallianassa* is to be synonymized with *Glypturus*.

Corallichirus Manning, 1992 is based on the type species, *Corallichirus xuthus*, which was formerly placed in *Corallianassa* for the reason that in the type species of that genus, *Corallianassa longiventris*, abdominal somite 2 is distinctively longer than somite 6, while in *Corallichirus xuthus*, abdominal somite 2 is subequal to somite 6 (Manning, 1992: 571). Tudge et al. (2000) included *Glypturus hartmeyeri* (Schmitt, 1935), *Glypturus coutierei* (De Man, 1905) (= junior synonym of *G. coutierei*), *Lepidophthalmus tridentatus* (Von Martens, 1869), and *Glypturus xuthus* (Manning, 1988) in *Corallichirus*. Though *G. xuthus* shows that abdominal somite 2 is subequal to somite 6, this character is not decisive enough to separate the genus *Corallichirus* from the genus *Glypturus*, and *G. acanthochirus*, the type species of *Glypturus*, also shows the same character of abdominal length as does *G. xuthus*. *Corallianassa longiventris* is characterized by an elongated abdominal somite 2, but it is difficult to separate *Corallianassa* from *Glypturus* only by the elongated abdominal somite 2.

As shown in the remarks on the Callianassinae, Tudge et al. (2000) applied their heuristic search analysis to the thalassinid taxa, and Ngoc-Ho (2002: 548) followed them and mentioned only Callichirinae sensu Tudge et al., 2000, without presenting any particular reasons to separate the Callichirinae from the Callianassinae.

Tudge et al. (2000: 144) listed the species of the genera in the Callichirinae, but the species defined as Callichirinae by Tudge et al. (2000) are to be revised as follows:

Subfamily Callichirinae Manning & Felder, 1991:

Genus *Callichirus* Stimpson, 1866: *C. adamas* (Kensley, 1974); *C. balssi* (Monod, 1933) (removed to *Podocallichirus*); *C. foresti* Le Loeuff & Intès, 1974 (removed to *Podocallichirus*); *C. garthi* (Retamal, 1975) (syn. of *Callianassa seilacheri*); *C. guineensis* (De Man, 1928) (removed to *Podocallichirus*; not clearly distinguished in *C. guineensis* of Le Loeuff & Intès, 1974, which belongs to either *C. guineensis* (De Man, 1928) or *C. foresti* Le Loeuff & Intès, 1974); *C. intesi* De Saint Laurent & Le Loeuff, 1979 (removed to *Glypturus*); *C. islagrande* (Schmitt, 1935); *C. kraussi* (Stebbing, 1910); *C. major* (Say, 1818); *C. monodi* De Saint Laurent & Le Loeuff, 1979 (removed to *Neocallichirus*); *C. pentagonocephala* (Rossignol, 1962) (removed to *Neocallichirus*); *C. seilacheri* (Bott, 1955); and *C. tenuimanus* De Saint Laurent & Le Loeuff, 1979 (removed to *Podocallichirus*). However, the present author reconfirmed that *C. adamas* (Kensley, 1974), *C. islagrande* (Schmitt, 1935), *C. kraussi* (Stebbing, 1910), *C. major* (Say, 1818), and *C. seilacheri* (Bott, 1955) (syn.: *C. garthi* (Retamal, 1975)) are included in *Callichirus*, while *C. balssi* (Monod, 1933), *C. foresti* Le Loeuff & Intès, 1974, *C. guineensis* (De Man, 1928), and *C. tenuimanus* De Saint Laurent & Le Loeuff, 1979 are not included in *Callichirus* but instead in *Podocallichirus* Sakai, 1999, because there is no ornament on abdominal somites 3-5 as in *Callichirus*, and the Mxp3 ischium-merus is elongate as in *Podocallichirus*. *C. intesi* is instead included in *Glypturus*; and *C. monodi* and *C. pentagonocephala* are instead included in *Neocallichirus*.

Genus *Corallianassa* Manning, 1987: *C. articulata* (Rathbun, 1906) (removed to *Glypturus*); *C. borradailei* (De Man, 1928) (syn. of *Glypturus coutierei*); *C. collaroy* (Poore & Griffin, 1979) (removed to *Glypturus*); *C. longiventris* (A. Milne-Edwards, 1870) (removed to *Glypturus*). All these species are removed to *Glypturus*.

Genus *Corallichirus* Manning, 1992: *C. hartmeyeri* (Schmitt, 1935); *C. placidus* (De Man, 1905) (syn. of *C. coutierei* (Nobili, 1904)); *C. tridentatus* (Von Martens, 1869); *C. xuthus* (Manning, 1988). All these species are also removed to *Glypturus*, except *C. tridentatus*, which belongs in *Lepidophthalmus*.

Genus *Glypturus* Stimpson, 1966: *G. acanthochirus* Stimpson, 1988; *G. armatus* (A. Milne-Edwards, 1870); *G. karumbus* (Poore & Griffin, 1979) (removed to *Neocallichirus*); *G. laurae* (De Saint Laurent, 1984) (syn. of *Glypturus armatus* (A. Milne-Edwards, 1870)); *G. martensi* (Miers, 1884); *G. motupore* Poore & Suchanek, 1988 (removed to *Neocallichirus*); *G. mucronatus* (Strahl, 1862) (removed to *Neocallichirus*). Most of these species are also removed to *Glypturus*, except *G. karumbus*, *G. motupore*, and *G. mucronatus*, which belong in *Neocallichirus*.

Genus *Lepidophthalmus* Holmes, 1904: *L. bocourti* (A. Milne-Edwards, 1870); *L. eiseni* (Holmes, 1904) (syn. of *L. bocourti*); *L. jamaicense* (Schmitt, 1935); *L. louisianensis* (Schmitt, 1935); *L. rafai* Felder & Manning, 1998; *L. ranongensis* (Sakai 1983) (removed to *Neocallichirus*); *L. richardi* Felder & Manning, 1997; *L. sinuensis* Lemaitre & Rodrigues, 1991; *L. siriboius* Felder & Rodrigues, 1993; *L. turneranus* (White, 1861). All these species are confirmed to be included in *Lepidophthalmus*, except *L. ranongensis*, which belongs to *Neocallichirus* because the A1 peduncle is shorter than the A2 peduncle.

Genus *Neocallichirus* Sakai, 1988: *N. cacahuate* Felder & Manning, 1995; *N. caechabitator* Sakai, 1988; *N. darwinensis* Sakai, 1988; *N. denticulatus* Ngoc-Ho, 1994; *N. grandimanus* (Gibbes, 1850); *N. horneri* Sakai, 1988; *N. indicus* (De Man, 1905); *N. jousseaumei* (Nobili, 1904); *N. lemaitrei* Manning, 1993; *N. manningi* Kazmi & Kazmi, 1992 (syn. of *N. indicus*); *N. moluccensis* (De Man, 1905); *N. natalensis* (Barnard, 1946) (syn. of *N. indicus*); *N. nickellae* Manning, 1993; *N. pachydactylus* (A. Milne-Edwards, 1870), *N. rathbunae* (Schmitt, 1935); *N. sassandrensis* (Le Loeuff & Intès, 1974); and *N. taiaro* Ngoc-Ho, 1995 (syn. of *N. indicus*). All these species should belong in *Neocallichirus* and three of them, *N. manningi*, *N. natalensis*, and *N. taiaro*, are synonymous with *N. indicus*.

Genus *Paraglypturus* Türkay & Sakai, 1995. Tudge et al. (2002) included *Paraglypturus* in the Callichirinae, but it should be removed to the subfamily Eucalliacinae because in *Paraglypturus calderus* Türkay & Sakai, 1995, the type species of the genus *Paraglypturus*, the Mxp3 dactylus is ovoid as in the Eucalliacinae.

The genus *Sergio* Manning & Lemaitre, 1994 was re-examined and synonymized again with *Neocallichirus* Sakai, 1988. Manning & Lemaitre (1994: 40) mentioned that "The species of *Sergio* differ from members of *Neocallichirus* in that in *Sergio* the posterior margin of the telson is divided by an armed or an unarmed median cleft and have a uropodal endopod that is distinctly longer than broad, tapering distally. In *Neocallichirus* the posterior margin of the telson is entire, broadly rounded, and the uropodal endopod is distinctly broader than long, widening distally". However, it is not reasonable that they put stress on differences in the form of the tail-fan and regard these as generic characters, by which *Sergio* could be considered to be different from members of *Neocallichirus*, without taking into account any such differences in members of other genera, because the tail-fan is not considered at the same level with the other taxonomic characters, such as Mxp3, rostrum, A1-2, and Plp, but rather treated as subordinate to those. So, in the species assigned to *Sergio*, the character of the tail fan should be defined as a specific character within the genus *Neocallichirus*, pertaining to a limited number of its members only. This means that *Sergio* can be handled as a synonym of *Neocallichirus*.

The genus *Sergio* Manning & Lemaitre, 1994, then, is based on *Callianassa guassutinga* Rodrigues, 1971, and includes eight species: *S. guaiqueri* Blanco Rambla & Liñero Arana, 1994; *S. guara* (Rodrigues, 1971); *S. guassutinga* (Rodrigues, 1971); *S. mericeae* Manning & Felder, 1995 (syn. of *Neocallichirus guassutinga*); *S. mirim* (Rodrigues, 1971); *S. monodi* (De Saint Laurent & Le Loeuff, 1979); *S. sulfureus* Lemaitre & Felder, 1996; and *S. trilobatus* (Biffar, 1970) (Tudge et al., 2000: 144). However, all these species are to be removed to *Neocallichirus* by the character of the broadened Mxp3 propodus. In *Neocallichirus audax*, *N. mucronatus*, and *N. ranongensis*, the posterior margin of the telson is entire; in *N. guara* (*Sergio* by Manning & Lemaitre, 1994), *N. guaiqueri* (*Sergio* by Manning & Lemaitre, 1994), *N. guassutinga* (removed to *Sergio* by Manning & Lemaitre, 1994), *N. mericeae* (removed to *Sergio* by Manning & Lemaitre, 1994), *N. pentagonocephalus*, *N. sulfureus* (removed to *Sergio* by Manning & Lemaitre, 1994), and *N. trilobatus* (removed to *Sergio* by Manning & Lemaitre, 1994), it is largely concave, lacking a median spine; in *N. sassandrensis* it is slightly concave entirely, bearing a small median spine; in *N. mirim* (removed to *Sergio* by Manning & Lemaitre, 1994) and *N. monodi* (removed to *Sergio* by Manning & Lemaitre, 1994) it bears an armed median cleft; in *N. denticulatus*, *N. maxima*, *N. motupore*, *N. karumba*, and *N. kempi*, it is convex. Those above-mentioned species of *Sergio* and *Neocallichirus* show a uropodal exopod that is longer than broad, tapering distally, while in *N. sulfureus* (removed to *Sergio* by Manning & Lemaitre, 1994), and *N. mericeae* (removed to *Sergio* by Manning & Lemaitre, 1994) it is broad as in *N. cacahuate*, *N. grandimanus*, *N. jousseaumei*, *N. lemaitrei*, *N. manningi*, *N. moluccensis*, *N. nickellae*, *N. pachydactylus*, and *N. rathbunae*; in *N. mirim*, the uropodal exopod shows an intermediate form as in *N. caechabitator*, *N. calmani*, *N. darwinensis*, *N. horneri*, *N. indicus*, *N. mauritianus*, and *N. vigi-*

lax. As a result, it is difficult to separate *Sergio* from *Neocallichirus* by the form of the telson and of the uropods, respectively.

The genera cited above are also polymorphic in the form of Plp1-2 in both sexes. In the type species of those genera, *Callichirus major* (Say, 1818), *Glypturus acanthochirus* Stimpson, 1866, *Corallianassa longiventris* (A. Milne-Edwards, 1870), *Lepidophthalmus bocourti* (A. Milne-Edwards, 1870), *Neocallichirus horneri* (Sakai, 1988), and *Podocallichirus madagassus* (Lenz & Richters, 1881), Plp1-2 are fundamentally not different in both sexes: in *Callichirus major* (Say, 1818) the male Plp1 is uniramous, 2-segmented, distal segment terminally simple; male Plp2 endopod without appendix interna and appendix masculina; female Plp1 uniramous, 3-segmented, distal segment foliaceous; female Plp2 biramous (Sakai, 1999c: 62). In *Glypturus acanthochirus* Stimpson, 1866, the male Plp1 is uniramous and 2-segmented, distal segment chelate distally; male Plp2 biramous, endopod with appendix masculina and appendix interna; female Plp1 biramous, endopod with appendix interna; and the female Plp2 is biramous, bearing an appendix interna. In *Corallianassa longiventris* (A. Milne-Edwards, 1870) the male Plp1 is uniramous and 2-segmented, distal segment chelate distally; male Plp2 biramous, endopod with appendix masculina and appendix interna; female Plp1 uniramous; female Plp2 biramous, with appendix interna. In *Neocallichirus horneri* (Sakai, 1988), the male Plp1 is uniramous and 2-segmented, distal segment chelate distally; male Plp2 biramous, endopod with appendix masculina fused with appendix interna; female Plp1 uniramous; female Plp2 biramous, with or without appendix interna. In *Podocallichirus madagassus* (Lenz & Richters, 1881), the male Plp1 is uniramous and 2-segmented, distal segment not chelate; male Plp2 biramous, endopod with distally demarcated appendix masculina, laterally with small appendix interna; female Plp1 3-segmented; female Plp2 biramous, exopod pediform, endopod with appendix interna.

The recent genus, *Michaelcallianassa* Sakai, 2002, is to be placed in the Callichirinae by the form of Mxp3, while the genus, *Grynaminna* Poore, 2000 is a synonym of *Podocallichirus*.

Genus **Callichirus** Stimpson, 1866

Callichirus Stimpson, 1866: 47; De Man, 1928b: 96; Edmondson, 1944: 51; Gurney, 1944: 83; De Saint Laurent, 1973: 514; Le Loeuff & Intès, 1974: 40; De Saint Laurent & Le Loeuff, 1979: 55; Manning & Felder, 1986: 439; Manning, 1987: 397; Manning & Felder, 1991: 775; Poore, 1994: 102; Sakai, 1999c: 59.

Definition. — [Revised from Sakai, 1999c: 59.] Carapace with dorsal oval; rostral spine present or absent; no rostral carina, hepatic sulcus, or prominence developed. Abdominal somites 3-5 dorsally ornamented. A1 peduncle longer and more robust than A2 peduncle. Mxp3 ischium-merus broadened, operculiform, propodus broadened, and dactylus digitiform; exopod usually absent. P1 chelate, equal, or unequal; in male ischium of larger cheliped much elongated, merus much elongated or normal in length, bearing a meral hook or not, and carpus much longer than chela. P2 chelate. P3 propodus broadened. P4 subchelate. P5 subchelate or chelate. Male Plp1 uniramous and blade-like, uni-, bi-, or triarticulate, distal segment simple distally; male Plp2 biramous and blade-like, usually without an appendix interna and masculina, but when appendix interna is present, it is embedded on the appendix masculina (Sakai, 1999c, fig. 13e of *Callichirus kraussi* (Stebbing, 1900)). Female Plp1 uniramous, bi- or triarticulate, female Plp2 biramous, without appendix interna. Plp3-5 foliaceous with appendices internae in both sexes. Uropodal endopod longer than broad, and strap-shaped.

Type species. — *Callianassa major* Say, 1818, by original designation and monotypy. Gender of generic name, *Callichirus*, masculine.

Species included. — East Atlantic species: *C. adamas* (Kensley, 1974). West Atlantic species: *Callichirus islagrande* (Schmitt, 1935); *C. major* (Say, 1818). East Pacific species: *C. seilacheri* (Bott, 1955). Indo-West Pacific species: *C. kraussi* (Stebbing, 1900).

East Atlantic species

Callichirus adamas (Kensley, 1974)

Callianassa adamas Kensley, 1974: 266, 277, figs. 1-2; De Saint Laurent & Le Loeuff, 1979: 67, figs. 14f, 16a, 17a, 19f, 20e-g, 23f-i.
Callichirus adamas; Manning & Felder, 1986: 439; Sakai, 1999c: 63; Tudge et al., 2000: 144.

Diagnosis. — Mxp3 ischium-merus showing an inverse triangle, merus convex on distal margin, male Plp1 uniramous, two-segmented, male Plp2 biramous, endopod without appendix interna, telson shallowly concave medially on posterior margin, bearing a tiny median notch (Kensley, 1974: 269).

Type locality. — Orange River mouth, South Africa.

Distribution. — Orange River mouth and Lambert's Bay, South Africa; Senegal; Cape Verde Islands.

West Atlantic species

Callichirus islagrande (Schmitt, 1935)

Callianassa (Callichirus) islagrande Schmitt, 1935b: 5, pl. 1 fig. 3, pl. 2 fig. 1, pl. 3 fig. 2, pl. 4 fig. 5.
Callianassa islagrande; Biffar, 1971a: 654; Phillips, 1971: 165-196, figs. 3B, D, F, 4; Felder, 1973: 24, pl. 2 figs. 12-14; Rabalais et al., 1981: 105; Manning & Felder, 1986: 438, fig. 2.
Callichirus islagrande; De Saint Laurent & Le Loeuff, 1979: 79; Abele & Kim, 1986: 27; Manning & Felder, 1986: 439; Manning, 1987: 397; Williams et al., 1989: 28; Manning & Felder, 1991: 775; Dworschak, 1992: 208; Sakai, 1999c: 60, fig. 11b-f; Strasser et al., 2000: 100; Tudge et al., 2000: 144.

Diagnosis. — Mxp3 ischium-merus subsquare, merus obliquely straight on distal margin, male Plp1 uniramous, two-segmented, male Plp2 biramous, endopod without appendix interna (Sakai, 1999c, fig. 12f), telson divergent posteriorly, and truncate, shallowly concave medially on posterior margin, lacking a median spine.

Type locality. — Grand Isle, Louisiana, U.S.A.

Distribution. — Gulf of Mexico, common in shallow subtidal of sandy beaches.

Callichirus major (Say, 1818)

Callianassa major Say, 1818: 238; White, 1847: 70; Gibbes, 1850: 194; A. Milne-Edwards, 1870: 86, 101; Stimpson, 1871: 122; Schmitt, 1935b: 3; Lunz, 1937: 1-15, figs. 1-3; Pearse et al., 1942: 153, 155, 156, 185, figs. 10, 14; Willis, 1942: 2; Gurney, 1944: 83; Pohl, 1946: 71-80, figs. 7-28; Hoyt & Weimer, 1963: 10; Williams, 1965: 100-102; Frankenberg et al., 1967: 113-120; Holthuis, 1969: 12; Biffar, 1971a: 651-653; Coelho & Ramos, 1973: 161; Rabalais et al., 1981: 105; Williams, 1984: 183, fig. 127.
Callichirus major; Stimpson, 1866: 47; Stimpson, 1871: 122; Kingsley, 1878: 327; Hay & Shore, 1917: 407, pl. 29 fig. 10; De Saint Laurent, 1973: 514; Rodrigues, 1983: 25, figs. 23-52; Manning & Felder, 1986: 439, fig. 1; Manning, 1987: 397; Rodrigues & Hödl, 1990: 50, fig. 1; Manning & Felder, 1991: 775, figs. 1, 3-6; Dworschak, 1992: 208; Sakai, 1999c: 61, fig. 11a; Strasser et al., 1999a: 211; Strasser et al., 1999b: 844; Tudge et al., 2000: 144.
Callianassa subterranea major var.; Czerniavsky, 1884: 76, 80.
Callianassa (Callichirus) major; Borradaile, 1903: 547; De Man, 1928a: 30, pl. 7 fig. 14-14b, pl. 8 fig. 14c, 14d; De Man, 1928b: 29, 91, 94, 111 (key); Williams, 1965: 100, fig. 78; Rodrigues, 1971: 191, figs. 1-20.

Material examined. — SMF 23579, 1 ovig. female (Tl/Cl 18.2/1.9), Blackbeard Island, Sapelo Island, Georgia, U.S.A., sandband, vii.1969, leg. J. Doerjes.

Diagnosis. — Mxp3 ischium-merus rhombic, merus concave and declined on distal margin, male Plp1 uniramous, two-segmented, male Plp2 biramous, endopod without appendix interna and appendix masculina (Sakai, 1999c, fig. 11a; Rodrigues, 1971, fig. 17); female Plp1 uniramous, female Plp2 biramous, lacking appendix interna. Telson rounded, shallowly concave medially on posterior margin, lacking a median spine.

Remarks. — The type species of the genus *Callichirus*, *C. major*, is observed again as follows: Female P1 equal, chelate. P4 propodus setose on mesial surface, forming a triangular distoventral angle; dactylus located laterally on ventrodistal angle of propodus; P5 chelate. Female Plp1 2-segmented, distal segment tapering distally; Plp2 biramous, bearing no appendices internae. Plp3 appendices internae fused with the respective endopods.

Type locality. — Coast of southern [United] States and eastern Florida, St. Johns River.

Distribution. — Cape Lookout, Barden Inlet, Ferry Landing, North Carolina, intertidal; Gulf of Mexico; Praia de Jose Menino, Santos, São Paulo, Brazil (Dworschak, 1992).

East Pacific species

Callichirus seilacheri (Bott, 1955)

Callianassa seilacheri Bott, 1955: 47, fig. 7a-g.
Callichirus seilacheri; Manning & Felder, 1986: 439, fig. 3; Manning & Felder, 1991: 775; Hendrickx, 1995: 390; Sakai, 1999c: 62, fig. 12a-f; Tudge et al., 2000: 144.
Callianassa garthi Retamal, 1975: 178, figs. 1-8. [Type locality: Playa Negra, Chile, 36°45'S 73°10'W.]
Callichirus garthi; Tudge et al., 2000: 144.

Material examined. — SMF 4941, 1 male, S. Salverry, La Libertad, Peru.

Diagnosis. —Mxp3 ischium-merus rhombic, merus concave and oblique on distal margin, male Plp1 uniramous, two-segmented (Sakai, 1999c, fig. 12e), male Plp2 biramous, endopod without appendix interna (Sakai, 1999c, fig. 11a), telson rounded, shallowly concave medially on posterior margin, lacking a median spine.

Remarks. — The male specimen was observed again as follows: P4 subchelate; propodus forming a thick tooth ventrodistally; P5 subchelate. Male Plp1 uniramous, 3-segmented, distal segment showing thick flagellum; Plp2 biramous, no appendix interna or appendix masculina.

Type locality. — El Salvador.

Distribution. — El Salvador; W. of Tubul a Playa Negra, Chile (36°45'S).

Indo-West Pacific species

Callichirus kraussi (Stebbing, 1900)

Callianassa kraussi Stebbing, 1900: 39, pls. 2, 3; Kensley, 1974: 277 (key); Kensley, 1975: 57; Holthuis, 1991: 248, 264, figs. 451, 452; Dworschak, 1992: 198; Wynberg et al., 1997: 139.
Callianassa (Callichirus) kraussi; Borradaile, 1903: 547; De Man, 1928b: 28, 94, 95, 113, 179, 182, 183; Barnard, 1950: 506, fig. 94.
Callichirus kraussi; Stebbing, 1910: 369; Sakai, 1999c: 64, fig. 13a-e; Tudge et al., 2000: 144.

Diagnosis. — Mxp3 ischium-merus operculiform, merus straight on distal margin, male Plp1 uniramous, three-segmented, male Plp2 biramous, endopod with appendix interna embedded on appendix masculina (Sakai, 1999c, fig. 13e), telson concave on posterior margin, medially convex (Holthuis, 1991, fig. 451).

Type locality. — Cape of Good Hope, Gordon's Bay, South Africa.

Distribution. — Saldanha Bay, False Bay to Zululand, a little below high water mark, sandy littoral zone in bays and estuaries.

Genus **Glypturus** Stimpson, 1866

Glypturus Stimpson, 1866: 46; Borradaile, 1903: 548; Boone, 1927: 85; Manning, 1987: 390, 398; Sakai, 1988: 61; Manning & Felder, 1991: 778; Sakai, 1999c: 72; Tudge et al., 2000: 144; Davie, 2002: 460.
Corallianassa Manning, 1987: 392, 397; Manning & Felder, 1991 (partim): 776, figs. 1, 2, 5; Poore, 1994: 102; Tudge et al., 2000: 144; Davie, 2002: 460. [Type species: *Callianassa longiventris* A. Milne-Edwards, 1870, by original designation.]
Corallichirus Manning, 1992: 571, figs. 1b, 2; Poore, 1994: 102; Tudge et al., 2000: 144. [Type species: *Corallianassa xutha* Manning, 1968, by original designation.]

Definition. — [Revised from Sakai, 1999c: 72.] Carapace with dorsal oval, lacking rostral carina, cardiac prominence, and transverse cardiac sulci; front with spinous rostrum and with anterolateral projections bearing a non-calcified proximal area. Antennular peduncle not longer or stouter than antennal peduncle. Mxp3 ischium-merus broadened, subpediform, propodus broadened and ovate; dactylus narrow, digitiform; exopod absent. P1 chelate, unequal, and dissimilar; male larger cheliped with or without meral hook. P2 chelate. P3 propodus broadened. P4 subchelate. P5 chelate.

Male Plp1 uniramous, 2-segmented, distal segment simple or chelate distally; male Plp2 biramous, only with appendix interna or with both appendix masculina and appendix interna. Female Plp1 uniramous, 2- or 3-segmented; Plp2 biramous and with an appendix interna. Plp3-5 foliaceous and with or without appendices internae in both sexes. Telson broader than long, the posterior margin usually with a medial, broadly convex lobe. Uropodal endopod distinctly longer than broad and oval or triangular in shape. Uropodal exopod entirely bent posteriorly in distal half.

Remarks. — The genus *Glypturus* Stimpson, 1866 is based on the type species, *G. acanthochirus* Stimpson, 1866, and includes *G. armatus* (A. Milne-Edwards, 1870), *G. laurae* (De Saint Laurent, 1984) (syn. of *G. armatus*), *G. motupore* Poore & Suchaneck, 1988 (removed to *Neocallichirus*) (Manning & Felder, 1991: 778). Tudge et al. (2000: 144) listed other species in the genus *Glypturus* than the species recognized by Manning & Felder (1991): *G. karumbus* (Poore & Griffin, 1979) (removed to *Neocallichirus*); *G. martensi* (Miers, 1884); and *G. mucronatus* (Strahl, 1862) (removed to *Neocallichirus*). Most of these species are thus also removed to *Neocallichirus*.

Manning (1988: 883) created the genus *Corallichirus* for the type species, *Corallianassa xutha* Manning, 1988, thus formerly assigned to the genus *Corallianassa*. Kensley (2001: 331) considered that *Corallianassa* characteristically has abdominal somite 2 equal in length to abdominal somite 6 (Kensley, 2001: 331), and by this character, the following species were included in the genus *Corallichirus*: *Corallichirus hartmeyeri* (Schmitt, 1935b); *Corallichirus placidus* (De Man, 1905); *Corallichirus xuthus* (Manning, 1988); and *Corallichirus tridentatus* (Von Martens, 1869). However, *Corallichirus hartmeyeri*, *Corallichirus placidus* (junior synonym of *Corallichirus coutierei* (Nobili, 1904)), and *Corallichirus xuthus* are characterized by having a pair of anterolateral spines separated from the front by a non-calcified membrane; and the A1 peduncle not longer or stouter than the A2 peduncle. Because of this, they are assigned to the genus *Glypturus* Stimpson, 1866. *Corallichirus tridentatus* is characterized by the carapace lacking a pair of anterolateral spines, and the A1 peduncle being extremely long, very much longer than the A2 peduncle (contrary to Poore's (1994) definition for *Corallichirus*, in which that author defined the A1 peduncle as **not** longer than the A2 peduncle), so that this species obviously is assigned to *Lepidophthalmus*.

Glypturus hartmeyeri (Schmitt, 1935b), *G. coutierei* (Nobili, 1904) (= *G. placidus*), *G. xuthus* (Manning, 1988), and *Lepidophthalmus tridentatus* (Von Martens, 1869) were thus examined in regard to the relative lengths of abdominal somite 2 and abdominal somite 6. In *G. hartmeyeri*, somite 2 is in fact 0.9 times as long as somite 6, according to Manning (1988, fig. 2g, h); yet Manning (1988) mentioned that abdominal somite 2 is exactly as long as somite 6; in *G. coutierei*, somite 2 is 0.8 times as long as somite 6, though in *G. placidus* somite 2 is almost the same length as somite 6, about 4.25 mm (De Man, 1928b: 171); in *G. xuthus*, somite 2 is as long as somite 6; and in *G. longiventris* somite 2 is 1.5 times as long as somite 6 (Borradaile, 1904, pl. 58 fig. 2a). However, in *G. armatus* somite 2 is 0.7 times as long as somite 6 (A. Milne-Edwards, 1870, pl. 1-1); in *G. martensi*, somite 2 is 0.8-1.0 times as long as somite 6 (Tirmizi, 1974, fig. 1A; Sakai, 1999c, fig. 19a); in *G. intesi*, somite 2 is 1.3 times as long as somite 6 (De Saint Laurent & Le Loeuff, 1979: 17b); and in *G. acanthochirus*, somite 2 is 1.2 times as long as somite 6 (Manning, 1987, fig. 3g). These results clearly show that the relative lengths of abdominal somites 2 and 6 vary significantly, and that this character is not sufficient to define *Corallichirus* Manning, 1992 as a monophyletic clade.

In the collection of the Zoological Museum, University of Copenhagen, nine species of the genus *Glypturus* were found: *G. xuthus* (Manning, 1988) from Clipperton Island; *G. articulatus* (Rathbun, 1906) from Hawaii; *G. lanceolatus* (Edmondson, 1944) from Hawaii; *G. winslowi* (Edmondson, 1944) from Hawaii; *G. haswelli* (Poore & Griffin, 1979) from Whitsunday Group, Queensland, Australia; *G. martensi* (Miers, 1884) from Mauritius; *G. assimilis* (De Man, 1928b) from Ambon, Indonesia; *G. armatus* (A. Milne-Edwards, 1870) from Fiji; *G. coutierei* (Nobili, 1904) from Djibouti.

Type species. — *Glypturus acanthochirus* Stimpson, 1866: 46, by monotypy. Gender of generic name, *Glypturus*, masculine.

Species included. — East Atlantic species: *Glypturus intesi* (De Saint Laurent & Le Loeuff, 1979). West Atlantic species: *G. acanthochirus* Stimpson, 1866; *G. hartmeyeri* (Schmitt, 1935); *G. longiventris* (A. Milne-Edwards, 1870); *G. rabalaisae* sp. nov. East Pacific species: *G. xuthus* (Manning, 1899). Indo-West Pacific species: *G. armatus* (A. Milne-Edwards, 1870); *G. articulatus* (Rathbun, 1906); *G. assimilis* (De Man, 1928); *G. collaroy* (Poore & Griffin, 1979); *G. coutierei* (Nobili, 1904); *G. haswelli* (Poore & Griffin, 1979); *G. lanceolatus* (Edmondson, 1944); *G. martensi* (Miers, 1884); *G. winslowi* (Miers, 1884).

East Atlantic species

Glypturus intesi (De Saint Laurent & Le Loeuff, 1979)

Callichirus intesi De Saint Laurent & Le Loeuff, 1979: 69, figs. 14g, 16c, 17b, 18b, 19b, 21a-c,
   23 j-m; Tudge et al., 2000: 144.
Glypturus intesi; Sakai, 1999c: 73.

Diagnosis. — Mxp3 ischium-merus operculiform, merus rounded on distomesial margin, male Plp1 blade-shaped, uniramous, two-segmented, chelate
distally, male Plp2 biramous, endopod with appendix interna, telson concave
on posterior margin, medially convex.

Type locality. — Port Dakar, Senegal.

Distribution. — Goree and Dakar, Senegal.

West Atlantic species

Glypturus acanthochirus Stimpson, 1866

Glypturus acanthochirus Stimpson, 1866: 46; Stimpson, 1871: 121; Kingsley, 1899: 821 (foot-
   note); De Man, 1928b: 19, 25, 180; Manning, 1987 (partim): 390, 398, fig. 3 (not figs. 4, 5 =
   G. armata; fig. 4 after A. Milne-Edwards, 1870; fig. 5 after Kensley, 1975); Poore &
   Suchanek, 1988: 201, fig. 4d; Manning & Felder, 1991: 778, figs. 2, 4; Dworschak, 1992:
   209; Dworschak & Ott, 1993: 282; Sakai, 1999c: 73, fig. 14i; Tudge et al., 2000: 144.
? *Glypturus acanthochirus*; Schmitt, 1924: 93.
Callianassa (Callichirus) acanthochirus; Schmitt, 1935b: 4, 20, pl. 1 fig. 6, pl. 2 fig. 5, pl. 3 fig.
   4, pl. 4 fig. 6; Heard, 1979: 52.
Callianassa acanthochirus; Gurney, 1944: 84; Biffar, 1971a: 655, figs. 3, 4; Heard, 1979: 52,
   fig. 1.
Callichirus acanthochirus; De Saint Laurent & Le Loeuff, 1979: 96.

Not *Callianassa acanthochirus*; Rabalais et al., 1981: 103, fig. 3 [= *Glypturus rabalaisae* sp. nov.]

Material examined. — SMF 23508, 1 female (Tl/Cl 89.0/21.5), Key Largo, Florida Bay,
Florida, U.S.A., 16.v.1971, leg. J. Doerjes.

Diagnosis. — Mxp3 ischium-merus operculiform, merus oblique and
rounded on distomesial margin, male Plp1 blade-shaped, uniramous, two-segmented, chelate distally, male Plp2 biramous, endopod with appendix interna
and appendix masculina (Biffar, 1971a: 655, fig. 3h, i), telson rounded on posterior margin.

Remarks. — P4 subchelate. P5 chelate. Plp3-5 with narrow, triangular appendices internae.

Type locality. — Florida Keys.

Distribution. — Atlantic coast of Florida, Gulf of Mexico, West Indies (Dry Tortugas; Puerto Rico; Jamaica; Barbados; Antigua); Caribbean coast of Colombia and Venezuela.

Glypturus hartmeyeri (Schmitt, 1935)

Glypturus grandimanus; Balss, 1924: 179, figs. 3, 4; Schmitt, 1935b: 4; Biffar, 1971a: 640, 649; Manning, 1987: 399. [Not *Callianassa grandimana* Gibbes, 1850.]
Callianassa (Callichirus) hartmeyeri Schmitt, 1935b: 3, 4. [Replacement name for *Glypturus grandimanus* sensu Balss, 1924.]
Callianassa hartmeyeri; Biffar, 1971a: 640, 641, 649, 651, 653; Manning, 1987: 388, 399.
Callichirus sp. aff. *placidus*; De Saint Laurent & Le Loeuff, 1979: 97.
Corallianassa hartmeyeri; Manning, 1988: 884, figs. 1, 2; Manning & Chace, 1990: 34, figs. 18, 19; Manning & Felder, 1991: 777.
Corallichirus hartmeyeri; Manning, 1992: 571; Tudge et al., 2000: 144.
Glypturus hartmeyeri; Sakai, 1999c: 74.

Diagnosis. — Mxp3 ischium-merus subsquare, merus convex on distomesial margin, male Plp1 blade-shaped, uniramous, male Plp2 biramous, endopod with appendix interna (Manning & Felder, 1991: 777), telson trapezoid, convex, with a rounded median projection on posterior margin.

Type locality. — Jamaica Is., Kingston [Kingston Harbor, 17°57'N 76°47'W].

Distribution. — Jamaica, Caribbean Sea, and Ascension, South Atlantic.

Glypturus longiventris (A. Milne-Edwards, 1870)

Callianassa longiventris A. Milne-Edwards, 1870: 92, 101; De Man, 1928a: 24, figs. 12a-h; Gurney, 1944: 85, figs. 1, 2; Biffar, 1971a: 651, 653, 685, figs. 13-14; Chace et al., 1986: 334, pl. 110.
Callianassa (Callichirus) longiventris; Borradaile, 1903: 547; De Man, 1928b: 19, 29 (list), 94, 108 (key); Schmitt, 1935b: 4, pl. 1 fig. 4, pl. 2 fig. 3, pl. 3 fig. 3, pl. 4 fig. 3.
Callichirus longiventris; De Saint Laurent & Le Loeuff, 1979: 97.
Corallianassa longiventris; Manning, 1987: 392, fig. 6; Manning & Felder, 1991: 777; Dworschak, 1992: 214; Dworschak & Ott, 1993: 281; Tudge et al., 2000: 144.
Glypturus longiventris; Sakai, 1999c: 74.

Not ? *Callianassa (Callichirus) longiventris*; Borradaile, 1904: 752, pl. 58 fig. 2 (= *Glypturus coutierei* (Nobili, 1904)).

Material examined. — SMF 23509, 1 ovig. female (Tl/Cl 11.0/2.4), Bahia Gairaca, ca. 20 km N. E. of Santa Marta, Magdalena, Colombia, 16.iv.1980.

Diagnosis. — Mxp3 ischium-merus operculiform, merus oblique and convex on distomesial margin, male Plp1 blade-shaped, uniramous, two-segmented, chelate distally, male Plp2 biramous, endopod with appendix interna, telson trapezoid, convex, with a rounded median projection on posterior margin (Biffar, 1971a: 686, 688, figs. 13e, g, 14a, e).

Remarks. — The following observations were made: P1 unequal, dissimilar in shape; P4 subchelate; P5 chelate; Plp3-5 with triangular appendices internae.

Type locality. — Martinique.

Distribution. — Bermuda; Carrie Bow Cay; southeastern Florida; Belize; Caribbean Sea (Jamaica, Martinique, Virgin Islands).

Glypturus rabalaisae sp. nov.

Callianassa acanthochirus; Rabalais et al., 1981: 103, fig. 3.

Diagnosis. — Mxp3 ischium-merus subsquare, merus short and straight on distomesial margin, male Plp1 blade-shaped, uniramous, two-segmented, chelate distally, male Plp2 biramous, endopod with appendix interna and appendix masculina (Biffar, 1971a: 655, fig. 3h, i), telson hexagonal, rounded on posterior margin, bearing an acute median posterior spine.

Type locality. — Off Galveston, Texas, 15-65 to 90 m.

Remarks. — The present new species, *G. rabalaisae*, is separated from *G. acanthochirus*, because in *G. rabalaisae* the antennular peduncle is slightly longer than the antennal one, and the telson bears a posteromedian spine, whereas in *G. acanthochirus* the distal segment of the antennular peduncle overreaches the antennal peduncle, and the telson bears no posteromedian spine.

East Pacific species

Glypturus xuthus (Manning, 1988)
(fig. 28A, B)

Callianassa hartmeyeri; Hult, 1938: 7, figs. 1-4, pl. 1; Schmitt, 1939: 15. [Not *Callianassa hartmeyeri* Schmitt, 1935b.]
Callianassa (Callichirus) placida; Chace, 1962: 617 (partim). [Not *Callianassa placida* De Man, 1905.]

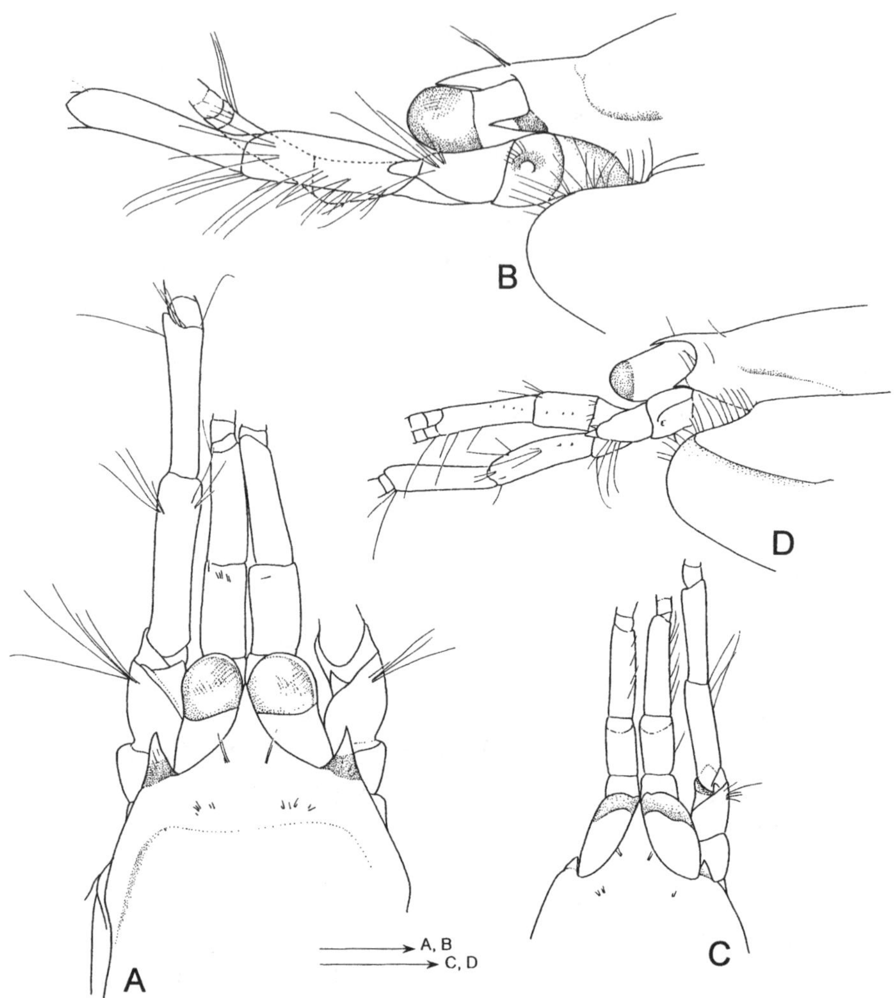

Fig. 28. A-B, *Glypturus xuthus* (Manning, 1988), SMNH 13883, 1 male (Tl/Cl, 44.0/10.0), Santa Cruz, Galapagos, 1934, leg. Rolf Blomberg; C-D, *Glypturus coutierei* (Nobili, 1904), SMNH 16239, 1 male (lacking posterior part to abdominal somite 1), Viti Levu, Namuka, Fiji Is., barrier reef, 18.vi.1917, leg. S. Bock. A, anterior part of carapace, dorsal view; B, same, lateral view in *Glypturus xuthus*; C, anterior part of carapace, dorsal view; D, same, lateral view in *Glypturus coutierei*. Scales 1 mm.

Callianassa placida; Hernández-Aguilera et al., 1986: 206. [Not *Callianassa placida* De Man, 1905.]
Corallianassa xutha Manning, 1988: 885, fig. 3; Manning & Felder, 1991: 777, figs. 1, 2, 5; Lemaitre & León, 1992: 44; Lemaitre & Ramos, 1992: 347; Hendrickx, 1995: 390.
Corallichirus xuthus; Manning, 1992: 571, figs. 1-2; Hernández-Aguilera, 1998: 304; Tudge et al., 2000: 144.
Glypturus xuthus; Sakai, 1999c: 75.

Material examined. — SMNH 13883, 1 male (Tl/Cl 44.0/10.0), Santa Cruz, Galapagos, leg. Rolf Blomberg, 1934.

Diagnosis. — Mxp3 ischium-merus subsquare, merus rounded on distomesial margin, male Plp1 blade-shaped, uniramous, two-segmented, chelate distally, male Plp2 biramous, endopod with appendix interna and slender appendix masculina, telson trapezoid, straight on posterior margin, lacking a rounded median projection. See fig. 28A, B.

Type locality. — Clipperton Is. (10°18'N 106°33'W).

Distribution. — East Pacific: Maria Madre Island, Tres Marias Islands, Baja California, 7-18 m; Clipperton Island; Port Utria, Colombia; Socorro Island, Clarión Island, Mexico; and Gorgona Island, Galapagos Islands [Ecuador], lagoon.

Indo-West Pacific species

Glypturus armatus (A. Milne-Edwards, 1870)

Callianassa armata A. Milne-Edwards, 1870: 90, 101, pl. 1; De Man, 1902: 754; Kensley, 1975: 48, fig. 1A-H; Manning, 1987: 392 (partim).
Callianassa (Callichirus) armata; Borradaile, 1903: 547; De Man, 1928b: 28, 93, 109.
Callichirus laurae De Saint Laurent (in Vaugelas & De Saint Laurent), 1984: 147, pl. 1 figs. A-D; Abu-Hilal et al., 1988: 233, 4 figs. [Type locality: Gulf of Aqaba, Red Sea.]
Glypturus laurae; Poore & Suchanek, 1988: 201, fig. 4c; Dworschak, 1992: 209; Tudge et al., 2000: 144.
Glypturus armatus; Poore & Suchanek, 1988: 201; Sakai, 1999c: 76; Tudge et al., 2000: 144.

Material examined. — ZMUC CRU-3857, 1 male (Tl/Cl 93.0/20.5), no locality.

Diagnosis. — Mxp3 ischium-merus subsquare, merus rounded on distomesial margin, male Plp1 blade-shaped, uniramous, two-segmented, male Plp2 biramous, endopod with appendix interna and appendix masculina, telson rounded on posterior margin, lacking a rounded median projection.

The branchial formula is given in table VIII.

Remarks. — The branchial formula shows that P5 is provided with an arthrobranch.

Type locality. — Fiji.

Distribution. — Mataiva, Tuamotu Arch.; Mauritius; Ternate, Indonesia; Fiji; Djibouti, Gulf of Aden; Aqaba, Red Sea; 5 to 45 m.

TABLE VIII
Branchial formula of *Glypturus armatus* (A. Milne-Edwards)

	Maxillipeds			Pereiopods				
	1	2	3	1	2	3	4	5
Exopods	1	1	–	–	–	–	–	
Epipods	1	–	–	–	–	–	–	–
Podobranchs	–	r	–	–	–	–	–	–
Arthrobranchs	–	–	2	2	2	2	2	1
Pleurobranchs	–	–	–	–	–	–	–	–

(r = rudimentary)

Glypturus articulatus (Rathbun, 1906)

Callianassa articulata Rathbun, 1906: 892, fig. 47; Chilton, 1911: 511.
Callianassa (*Callichirus*) *articulata*; De Man, 1928b: 28, 94, 108; Edmondson, 1944: 54, fig. 9a-j.
Corallianassa articulata; Dworschak, 1992: 210, fig. 14a-e; Tudge et al., 2000: 144.
Glypturus articulatus; Sakai, 1999c: 76, fig. 15a-f.

Diagnosis. — Mxp3 ischium-merus subsquare, merus rounded on distomesial margin, male Plp1 blade-shaped, uniramous, two-segmented, male Plp2 unknown, telson subsquare, roundish notch on posterior margin, lacking a median projection.

Type locality. — Vicinity of Modu Manu, Hawaii, 23-33 fathoms (c. 34-50 m).

Distribution. — Honolulu, Oahu, Hawaii; Kermadec Islands; Gilbert Island.

Glypturus assimilis (De Man, 1928)

Callianassa Martensi; De Man, 1888: 482-483, pl. 21 fig. 1. [Not *Callianassa Martensi* Miers, 1884.]
Callianassa (*Callichirus*) *assimilis* De Man, 1928b: 28, 109.
Glypturus assimilis; Sakai, 1999c: 78, fig. 16a-f.
Corallichirus bayeri Kensley, 2001: 328, figs. 1, 2. [Type locality: Agat Bay, north of Alutom Island, Guam, among rocks, 2.5-6 m.]
Callianassa assimilis; Tudge et al., 2000: 143.

Diagnosis. — Mxp3 ischium-merus subsquare, merus obliquely curved distally, rounded on mesial margin, male Plp1-2 undescribed, telson rounded on posterior margin, lacking a median projection.

Remarks. — *Corallichirus bayeri* Kensley, 2001 is a synonym of *G. assimilis* De Man, 1928 and should be included without doubt in the genus *Glypturus*, as explained above in the generic remarks. *C. bayeri* has the same characteristic denticulations of the P1 ischium, merus, and fingers on both sides, and the same shape of Mxp3 and uropods. Kensley (2001: 328) mentioned that the telson has a broadly rounded posterior margin, and that the mid-dorsal length is about 0.7 times the basal width. This description agrees with the telson described for *G. assimilis* (cf. Sakai, 1999c, fig. 16b). The curvature of the posterior margin of the telson would seem to be different between the two species, but this may just be due to the angle at which they have been drawn.

Type locality. — Ambon (= Amboina), Indonesia.

Distribution. — Ambon (= Amboina), Indonesia; Gilbert Is.; Agat Bay, north of Alutom Island, Guam.

Glypturus collaroy (Poore & Griffin, 1979)
(fig. 29)

Callianassa collaroy Poore & Griffin, 1979: 260, figs. 24, 25.
Glypturus collaroy; Sakai, 1988: 61.
Corallianassa collaroy; Sakai, 1992b: 212, fig. 1; Tudge et al., 2000: 144; Davie, 2002: 460.
Neocallichirus collaroy; Sakai, 1999c: 94 (key), 98.

Material examined. — NMW Cr. 9278, 1 ovig. female (Tl ca. 11.2, cephalothorax broken), Flax Bush Bay, New Zealand, gravely-sand bottom, in vertical mud-lined tube, leg. R. V. Grace, 10.xii.1976; NMW Cr. 9279, 1 ovig. female (Tl 11.2, cephalothorax broken), Flax Bush Bay, gravely-sand bottom, in vertical mud-lined tube, leg. R. V. Grace, 10.xii.1976.

Colour. — Chelae dark pink, body pale pink (note made by Mr. R. V. Grace, who made the collections by SCUBA).

Diagnosis. — Mxp3 ischium-merus oval, merus rounded with a spine on distomesial margin, male Plp1 blade-shaped, uniramous, two-segmented, bilobed distally, male Plp2 biramous, endopod with appendix interna and appendix masculina, telson subsquare, concave on posterior margin, lacking a median projection.

Remarks. — Sakai (1999) erroneously placed *Callianassa collaroy* Poore & Griffin, 1979 in the genus *Neocallichirus*, but it should be transferred to Glypturus, because subsequent examination of the above specimens has revealed that the anteroateral spines of the carapace are separated from the front

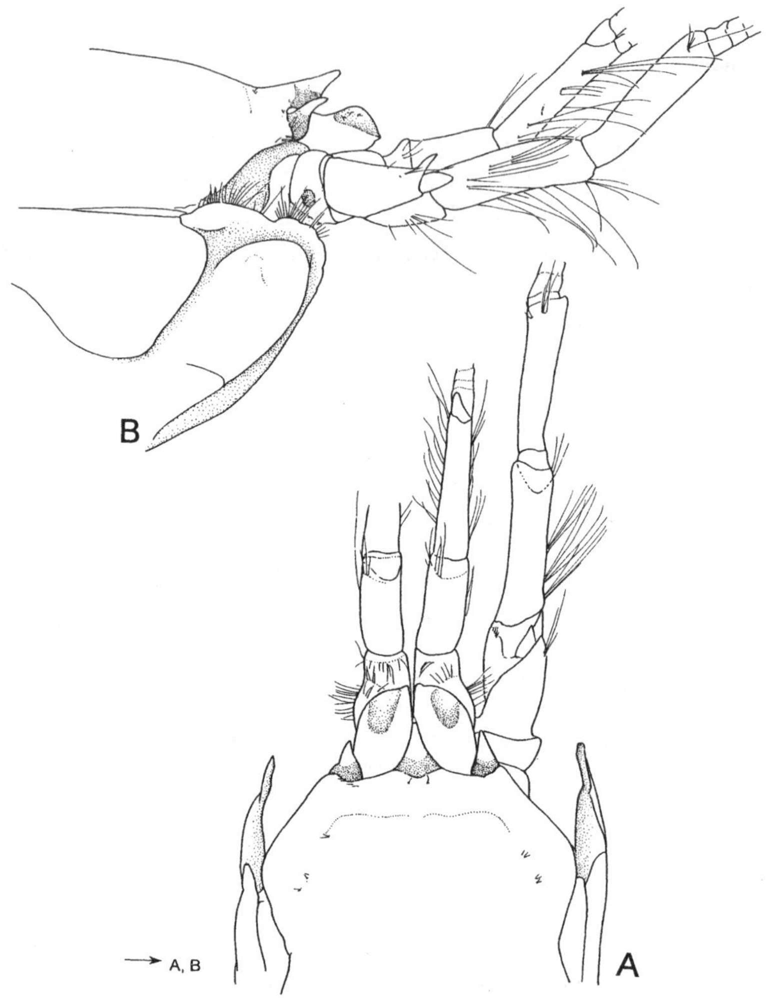

Fig. 29. *Glypturus collaroy* (Poore & Griffin, 1979). NMW Cr. 9278, 1 ovig. female (Tl ca. 11.2 mm), Flax Bush Bay, New Zealand, leg. R. V. Grace. A, anterior part of carapace, dorsal view; B, same, lateral view. Scale 1 mm.

by a non-calcified membrane (fig. 29A, B) as in *Glypturus*. The A1 peduncle is not longer or stouter than the A2 peduncle, but this character has turned out to be not always good enough to characterize *Neocallichirus*, because it has recently been found that in *N. angelikae* the A1 peduncle is slightly longer than the A2 peduncle. This species, *G. collaroy*, is characterized by the Mxp3 merus bearing a small spine on its anterolateral margin.

Type locality. — Collaroy, Long Reef, New South Wales, Australia.

Distribution. — Collaroy, New South Wales; Society Islands to Maharepa, Moorea, French Polynesia; coral reef, sand bottom, low intertidal among boulders.

Glypturus coutierei (Nobili, 1904)
(fig. 28C, D)

Callianassa (Callichirus) Coutierei Nobili, 1904: 237; Nobili, 1906a: 60; Nobili, 1906b: 110, pl. 7 fig. 1; De Man, 1928b: 28, 109, 174, 179.
? *Callianassa (Callichirus) longiventris*; Borradaile, 1904: 752, pl. 58 fig. 2. [Type locality: Hulue, Male Atoll, Maldives.] [Not *Callianassa longiventris* A. Milne-Edwards, 1870.]
Callianassa placida De Man, 1905: 612; Hernández-Aguilera et al., 1986: 206. [Type locality: off Laiwui, coast of Obi Major, Indonesia.]
Callianassa (Challichirus) longiventris var. *Borradailei* De Man, 1928a: 27; De Man, 1928b: 29, 108. [Type locality: Goidu, Goifurfehendu Atoll, Maldive Archipelago.]
Callianassa (Callichirus) placida; De Man, 1928b: 29, 93, 108, 171, pl. 18 fig. 29-29b, pl. 19 fig. 29c-e; Chace, 1962: 617.
Callianassa (Callichirus) borradailei; Ward, 1942: 62.
Callianassa (Callichirus) oahuensis Edmondson, 1944: 56, fig. 10a-h. [Type locality: Hanauma Bay, Oahu.]
Corallianassa borradailei; Manning, 1987: 394, figs. 7, 8; Manning, 1992, fig. 1a; Tudge et al., 2000: 144.
Corallichirus placidus; Manning, 1992: 571; Tudge et al., 2000: 144.
Glypturus coutierei; Sakai, 1999c: 78, figs. 17a-f, 18a-d.
Callianassa coutierei; Tudge et al., 2000: 143.

Material examined. — ZMUC CRU-3858, 1 female (Tl/Cl 60.0/12.0), 18°09'S 178°24'E, reef flat, Suva Harbour, Fiji Is. at low tide, living corals, leg. T. Wolff, 17.v.1965; SMNH 16239, 1 female (lacking posterior part to abdominal somite 2), Viti Levu, Namuka, Fiji Is., barrier reef, leg. S. Bock, 18.vi.1917.

Diagnosis. — Mxp3 ischium-merus subsquare, merus rounded on distomesial margin, male Plp1 blade-shaped, uniramous, two-segmented, male Plp2 biramous, endopod with appendix interna and appendix masculina (De Man, 1928b: 172), telson trapezoid, straight on posterior margin, lacking a rounded median projection.

Remarks. — De Man (1928b: 172) described the male pleopod of the present species as *Callianassa (Callichirus) placida* by the female specimen from Sta. 142, anchorage off Laiwui, coast of Obi Major.

Type locality. — Djibouti.

Distribution. — Hawaii (Hanauma Bay, Oahu); Mindanao, Philippines; Tahiti; Fiji; Goidu, Goifurfehendu Atoll, Maldive Archipelago; Indonesia (off

Seba, Savu; off Laiwui, coast of Obi Major); Gulf of Aden (Perim; Djibouti; Aden); Tuléar, S. W. Madagascar.

Glypturus haswelli (Poore & Griffin, 1979)

Callianassa haswelli Poore & Griffin, 1979: 263, figs. 26, 27.
Glypturus haswelli; Sakai, 1999c: 82; Davie, 2002: 460.

Diagnosis. — Mxp3 ischium-merus subsquare, merus rounded on distomesial margin, male Plp1 blade-shaped, uniramous, two-segmented, pointed distally, male Plp2 biramous, endopod without appendix interna and appendix masculina (Poore & Griffin, 1979, fig. 27m, n), telson trapezoid, straight with a rounded median projection on posterior margin.

Type locality. — Queensland, Whitsunday Group, Australia.

Distribution. — Torres Strait; Queensland, Islands off north and central Queensland coast.

Glypturus lanceolatus (Edmondson, 1944)

Callianassa (Callichirus) lanceolata Edmondson, 1944: 52, fig. 8a-i.
Glypturus lanceolatus; Sakai, 1999c: 83.

Diagnosis. — Mxp3 ischium-merus subsquare, merus rounded on distomesial margin, male Plp1-2 unknown, telson trapezoid, straight with a rounded median projection on posterior margin.

Type locality. — Hanauma Bay, Oahu, in shallow water on the reef.
Distribution. — Oahu, Hawaii.

Glypturus martensi (Miers, 1884)

Callianassa Martensi Miers, 1884: 13-15, pl. 1 fig. 1; Lanchester, 1900: 261, pl. 12 fig. 4, 4a; Nobili, 1906b: 111, fig. 7.
Callianassa (Callichirus) Martensi; Borradaile, 1903: 547; De Man, 1928b: 29, 109, 171.
Callianassa (Callichirus) nakasonei Sakai, 1967: 46, fig. 3. [Type locality: East coast of Tonaki Island, Okinawa-group, Ryukyu Is.]
Callianassa (Callichirus) martensi; Tirmizi, 1974: 286, figs. 1-4.
Callichirus martensi; De Saint Laurent & Le Loeuff, 1979: 97.
Callianassa martensi; Sakai, 1984: 99, fig. 3; Dworschak, 1992: 200, fig. 8a-e.
Callianassa nakasonei; Sakai, 1987a: 306; Tudge et al., 2000: 143.
Glypturus martensi; Sakai, 1988: 61; Tudge et al., 2000: 144; Davie, 2002: 460.

Diagnosis. — Mxp3 ischium-merus subsquare, merus rounded on distomesial margin, male Plp1 blade-shaped, uniramous, two-segmented, chelate distally, male Plp2 biramous, endopod with appendix interna and appendix masculina, telson trapezoid, straight with a rounded median projection on posterior margin (Tirmizi, 1974, figs. 1, 3, 4).

Type locality. — Mauritius.

Distribution. — East coast of Tonaki Island, Okinawa-group, Ryukyu Is.; Mauritius; Ambon (= Amboina), Indonesia; north and central Queensland, Australia; Sri Lanka; Belligom, West Pakistan.

Glypturus winslowi (Edmondson, 1944)

Callianassa (*Callichirus*) *winslowi* Edmondson, 1944: 59, fig. 11a-g.
Glypturus winslowi; Sakai, 1999c: 84.
Callianassa winslowi; Tudge et al., 2000: 143.

Diagnosis. — Mxp3 ischium-merus subsquare, merus rounded on distomesial margin, male Plp1 blade-shaped, uniramous, two-segmented, chelate distally, male Plp2 undescribed, telson trapezoid, convex with a pointed median projection on posterior margin (Edmondson, 1944, fig. 11f).

Type locality. — Maui, Hawaii.

Distribution. — Maui, Hawaii.

Genus **Lepidophthalmus** Holmes, 1904

Lepidophthalmus Holmes, 1904: 310; Manning & Felder, 1991: 778, figs. 5, 13; Poore, 1994: 102; Sakai, 1999c: 64; Sakai & Apel, 2002: 278.

Definition. — [Revised from Sakai, 2002.] Carapace with dorsal oval, lacking rostral carina, hepatic sulcus, and prominence. Rostrum spinous, sharply triangular; frontal margin of carapace with or without a pair of anterolateral spines on the front. A1 peduncle longer and stouter than A2 peduncle. Mxp3 ischium-merus broadened, subpediform, propodus ovate, dactylus narrow, digitiform, exopod rudimentary. P1 unequal, larger cheliped with a meral hook. P2 chelate. P3 propodus broadened. P4 subchelate. P5 chelate. Abdominal somites 3-5 without dorsal ornament. Male Plp1 uniramous, biarticulate, distal segment usually chelate or exceptionally simple; male Plp2 biramous, usually with appendix interna or with appendices interna and masculina, or exceptionally simple. Female Plp1 uniramous, biarticulate, female Plp2 biramous and with appendix interna. Plp3-5 foliaceous with appendices internae in both sexes. Telson broader than long, posterior margin usually with a medial, broadly convex lobe. Uropodal endopod rhombic or leaf-like, exopod broadened, distal half tapering or ovoid distally.

Remarks. — The genus *Lepidophthalmus* was defined as "Rostrum sharply triangular; without a pair of anterolateral spines (or projections) in front (Sakai, 1999c: 64)", but this should be amended as: "Rostrum spinous, carapace with or without anterolateral projections", because, as in *L. rosae*, a distinct anterolateral projection is present on the front of the carapace.

Type species. — *Callianassa bocourti* A. Milne-Edwards, 1870, by monotypy. Gender of the generic name, *Lepidophthalmus*, masculine.

Species included. — East Atlantic-Mediterranean species: *Lepidophthalmus turneranus* (White, 1861). West Atlantic species: *L. jamaicense* (Schmitt, 1935); *L. louisianensis* (Schmitt, 1935); *L. manningi* Felder & Staton, 2000; *L. richardi* Felder & Manning, 1997; *L. sinuensis* Lemaitre & Rodrigues, 1991; *L. siriboia* Felder & Rodrigues, 1993. East Pacific species: *L. bocourti* (A. Milne-Edwards, 1870); *L. rafai* Felder & Manning, 1998. Indo-West Pacific species: *L. grandidieri* (Coutière, 1899); *L. rosae* (Nobili, 1904); *L. socotrensis* Sakai & Apel, 2002; *L. tridentatus* (Von Martens, 1869).

East Atlantic and Mediterranean species

Lepidophthalmus turneranus (White, 1861)

Callianassa turnerana White, 1861a: 42, pl. 6; White, 1861b: 479; A. Milne-Edwards, 1870: 89, 101; Nobili, 1900: 3; Rathbun, 1900a: 308; Lenz, 1911: 316, figs. 1-11; Vanhöffen, 1911: 105, fig. a, d; Balss, 1916: 33; Monod, 1927: 595; Holthuis, 1991: 250, 246, fig. 455; Clark & Presswell, 2001: 155.
Callianassa Krukenbergi Neumann, 1878: 34; Borradaile, 1903: 548. [Type locality: ?Central America.]
Callianassa diademata Ortmann, 1891: 56, pl. 1 fig. 11. [Type locality: Africa.]
Callianassa (Callichirus) diademata; Borradaile, 1903: 547.
Callichirus Turnerana; Borradaile, 1903: 547; De Man, 1928b: 30, 114.
Callichirus diademata; Borradaile, 1903: 547.
Callianassa (Callichirus) Krukenbergi; De Man, 1928a: 51.
Callianassa (Callichirus) Turnerana; De Man, 1928a, pl. 12 fig. 21-21d (not f); De Man, 1928b: 94.
Callichirus turneranus; Le Loeuff & Intès, 1974: 40, fig. 10a-s; De Saint Laurent & LeLoeuff, 1979: 64, figs. 14e, 19e, 20a-d, 23a-e.
Callianassa ?turnerana; Dworschak, 1992: 205.
Lepidophthalmus turneranus; Sakai, 1999c: 65; Tudge et al., 2000: 144.

Diagnosis. — Mxp3 ischium-merus subsquare, merus rounded on distomesial margin, male Plp1 blade-shaped, uniramous, two-segmented, pointed distally, male Plp2 biramous, without appendix interna and appendix masculina, telson oval and W-shaped, with a distinct median projection on posterior margin (Le Loeuff & Intès, 1974: 40, fig. 10).

Type locality. — Cameroon, fresh water.

Distribution. — Togo, Cameroon to Congo, W. Africa. Lagoons and estuaries to practically fresh water.

West Atlantic species

Lepidophthalmus jamaicense (Schmitt, 1935)

Callianassa (Callichirus) jamaicense Schmitt, 1935b: 1, 4, 9-12, pl. 1, fig. 1 pl. 2 figs. 6, 8, pl. 4 fig. 1.

Callianassa jamaicense; Biffar, 1971a: 650, 654; Coelho & Ramos, 1973: 162 (partim); Abele & Kim, 1986: 27, 295, 296, 302-303, figs. j, k, l (partim); Manning, 1987: 397; Dworschak, 1992: 196 (partim) (not fig. 4a-d = *L. siriboia*).

Callianassa jamaicensis; Holthuis, 1974: 231.

Callichirus jamaicensis; De Saint Laurent & Le Loeuff, 1979: 67, 69 (only species name); Coelho & Ramos-Porto, 1987: 30.

Lepidophthalmus jamaicense; Felder et al., 1991: 101A (partim); Manning & Felder, 1991: 778, fig. 13a-e (not f, which is defined as *L. jamaicense* var. *Louisianensis*); Felder & Rodrigues, 1993: 357, 358, 367, 373; Sakai, 1999c: 66; Tudge et al., 2000: 144.

Diagnosis. — Mxp3 ischium-merus subsquare, merus concave on distal margin, male Plp1-2 unknown, telson subrectangular and trilobate, shallow with a smooth median lobe on posterior margin (Schmitt, 1935b: 9-12, pl. 3 fig. 1).

Type locality. — Montego Bay, Jamaica.

Distribution. — Caribbean Sea: Dangriga and Sao Luiz-MA, Rio Anil, Belize; Honduras, sandy beach between stones; Montego Bay, Jamaica, blackish pond.

Lepidophthalmus louisianensis (Schmitt, 1935)

Callianassa (Callichirus) jamaicense var. *louisianensis* Schmitt, 1935b: 1, 12-15, pl. 1 fig. 2, pl. 2 figs. 2, 7, pl. 4 fig. 4; Hedgpeth, 1950: 113-114; Biffar, 1971a: 641-642, 650; Phillips, 1971: 165-196, figs. 1-3a, c, e, 6, 7b, c, d, 8b, c, d, tabs. 1-3, 5-7; Felder et al., 1986: 91; Manning & Felder, 1989: 9.

Callianassa stimpsoni; Reed, 1941: 42, 47 (partim). [Not *Callianassa stimpsoni* Gabb, 1864.]

Callianassa jamaicense louisianense; Anonymous, 1941: 5; Willis, 1942: 1, 2, 4, 5; Behre, 1950: 21; Hedgpeth, 1950: 114, tab. 1; Darnell, 1958: 369, 400; Pounds, 1961: 26, pl. 1 fig. 1; Leary, 1964: 26-27; Dawson, 1967: 224; Felder, 1973: 3, 24; Fotheringham & Brunenmeister, 1975: 114-116, 166, figs. 6, 12; Fotheringham, 1980: 63, 106, figs. 7, 12; Fotheringham & Brunenmeister, 1989: 62, 118, figs. 7, 12.

Callianassa jamaicense; Hedgpeth, 1950: 114; Menzel, 1971: 78. Phillips, 1971: 166; Felder, 1973: 3, 24, pl. 2 figs. 6-8; Felder, 1978: 409-427, figs. 2, 3, 5-10, tabs. I, II; Felder, 1979: 125-136, figs. 1-6; Rabalais et al., 1981: 96, 105, 112; Heard, 1982: 47; Felder et al., 1984: 67A; Lovett & Felder, 1984: 74A; Abele & Kim, 1986: 27, 295 (partim, not 302, 303, figs. j, k, l); Felder et al., 1986: 91-104, figs. 1-8.

Callianassa jamaicensis louisianensis; Humm, 1953: 6; Tiefenbacher, 1976: 314-316 (partim), fig. 1a, b (not fig. 1c, d = *L. siriboia*).

Callianassa jamaicense louisianensis; Wass, 1955: 46, 148; Menzel, 1956: 43.

Callianassa (Callichirus) jamaicense; Rodrigues, 1971: 198-204 (partim), figs. 21-40, tab. 2.

Callianassa jamaicensis; Coelho & Ramos, 1973: 162 (partim); Manning & Felder, 1986: 439; Britton & Morton, 1989, tab. 1-1; Williams et al., 1989: 28.

Callichirus jamaicense; Felder, 1975: i-x, (Part I) 1-63, tab. 1, figs. 1, 4, 6, 8, 10-15, (Part II) 74-110, tabs. 1, 2, figs. 1-6; Felder, 1979: 125; Abele & Kim, 1986: 296.

Callianassa jamaicense?; Shipp, 1977: 48-60, figs. 32-37.

Callianassa jamaicense var.; Felgenhauer & Felder, 1986: 34A.

Callichirus jamaicensis; Coelho & Ramos-Porto, 1987: 30 (partim).

Callianassa louisianensis; Manning, 1987: 397; Staton et al., 1988: 125A; Britton & Morton, 1989: 6, 121, 193, 195, 209, figs. 6, 7e, j, 7-9T; Felder & Lovett, 1989: 540-552, figs. 1-6, tabs. 1-3; Lovett & Felder, 1989: 530, figs. 1, 2, tabs. 1, 2; Manning & Felder, 1989: 9; Rabalais et al., 1989: 32-34, tab. 3; Williams et al., 1989: 28, fig. 4; Griffis & Suchanek, 1991, tab. 2; Dworschak, 1992: 198, fig. 7a-f.

Lepidophthalmus louisianensis; Felder & Staton, 1990: 137A; Felder et al., 1991: 101A (partim); Lemaitre & Rodrigues, 1991: 629; Manning & Felder, 1991: 778 (partim); Manning & Felder, 1992: 560; Felder & Felgenhauer, 1993: 263-276, figs.; Dworschak, 2000a: 99; Sakai, 1999c: 67, fig. 14a-b; Tudge et al., 2000: 144.

Lepidophthalmus jamaicense; Manning & Felder, 1991, fig. 13f (not *L. jamaicense*).

Diagnosis. — Mxp3 ischium-merus subsquare, merus concave on distal margin, male Plp1 blade-shaped, uniramous, two-segmented, chelate distally, male Plp2 biramous, with appendix interna and appendix masculina (Sakai, 1999c, fig. 14b), telson subrectangular, shallowly trilobate with a median projection on posterior margin (Rodrigues, 1993: 202, figs. 27, 36, 37, 40).

Type locality. — Grand Isle, Louisiana.

Distribution. — Florida (Perdido Key, Big Lagoon, 20-50 cm); Alabama (Nobile Bay and Dauphin Is. near airport, intertidal); Mississippi (Bay St. Louis, intertidal), and Louisiana.

Lepidophthalmus manningi Felder & Staton, 2000

Lepidophthalmus sp. "a"; Staton et al., 2000: 161, figs. 1-3, tabs. 3-5.
Lepidophthalmus manningi Felder & Staton, 2000: 170, figs. 1-2.

Diagnosis. — Mxp3 ischium-merus subsquare, merus obliquely curved and rounded on distomesial margin, male Plp1 blade-shaped, uniramous, two-segmented, subspatulate distally, male Plp2 biramous, with appendix interna and appendix masculina, telson subrectangular, shallowly trilobate with a median projection on posterior margin (Felder & Staton, 2000: 170, figs. 1j, 2f, k, q, r).

Remarks. — It seems that this species is closely similar to *L. louisianensis* (Schmitt, 1935) and *L. richardi* Felder & Manning, 1997, but it is very difficult to differentiate *L. manningi* from *L. louisianensis*: this can be done by specialized techniques for species discrimination, involving allozymic analysis of the sclerites. However, *L. manningi* is separated from *L. louisianensis* by the absence of sclerites on the abdominal somites, and from *L. richardi* "by the deeper proximal meral notch on the superior margin of the major cheliped, the larger lobe on the proximal article of the female first pleopod, the larger appendix interna on the male second pleopodal endopod, the narrower distal lobe on the female second pleopodal endopod, and the less strongly developed carination on mesial margin of the lateral ventral plates on the second abdominal somite of mature males" (Felder & Staton, 2000: 180).

Type locality. — Laguna San Augustin, near village of Palma Sola, 19°55.23'N 96°31.85'W, Veracruz, Mexico.

Distribution. — Veracruz, Tabasco, and western Campeche, Mexico, intertidal and shallow subtidal muddy sand and sandy mud substrates in coastal lagoons and river mouths.

Lepidophthalmus richardi Felder & Manning, 1997

Lepidophthalmus jamaicense complex; Felder et al., 1991: 101A (partim).
Lepidophthalmus richardi Felder & Manning, 1997: 320-329, figs. 4a-j, 5a-f, 6a-i, 7a; Sakai, 1999c: 68; Tudge et al., 2000: 144.

Diagnosis. – Mxp3 ischium-merus subsquare, merus obliquely curved and rounded on distomesial margin, male Plp1 blade-shaped, uniramous, two-segmented, subspatulate distally, male Plp2 biramous, with appendix interna and appendix masculina, telson subrectangular, shallowly trilobate with an indistinct median projection on posterior margin (Felder & Manning, 1997, figs. 4a-j, 5a-f, 6a-i, 7a).

Remarks. — Felder & Manning (1997: 328) separated the present species from the Gulf of Mexico endemic, *L. louisianensis*, and the Brazilian species, *L. siriboia*, both of which lack elaborate cuticular armour or plating on the ventral surfaces of the abdomen. However, this feature of distinction is questionable as a means to separate those three species.

Type locality. — Intertidal shoreline at Pelican Beach Hotel, near Dangriga, Belize.

Distribution. — Dangriga, Belize.

Lepidophthalmus sinuensis Lemaitre & Rodrigues, 1991

Lepidophthalmus sinuensis Lemaitre & Rodrigues, 1991: 623, figs. 1-4; Manning & Felder, 1991: 778; Nates et al., 1999: 526-541; Sakai, 1999c: 68; Tudge et al., 2000: 144.

Diagnosis. — Mxp3 ischium-merus subsquare, merus obliquely curved and rounded on distomesial margin, male Plp1 blade-shaped, uniramous, two-segmented, subspatulate distally, male Plp2 biramous, with appendix interna but without appendix masculina, telson subrectangular, shallowly trilobate with a median projection on posterior margin (Lemaitre & Rodrigues, 1991: 623, figs. 1b, c, d, 2f).

Type locality. — Colombia, mouth of Rio Sin (9°07'N 75°0'W).

Distribution. — Caribbean coast of Colombia. Intertidal, 1.5 m.

Lepidophthalmus siriboia Felder & Rodrigues, 1993

Callianassa jamaicense; Biffar, 1971a: 650, 654 (partim); Abele & Kim, 1986: 27 (partim); Griffis & Suchanek, 1991, tab. 2; Dworschak, 1992: 196 (partim), fig. 4a-d. [Not *L. jamaicense* (Schmitt, 1935b).]
Callianassa (*Callichirus*) *jamaicensis*; Rodrigues, 1971: 198 (partim).
Callianassa jamaicensis; Rodrigues, 1971: 202-204, figs. 21-40, tab. 2 (partim); Coelho & Ramos, 1973: 162 (partim); Tiefenbacher, 1976: 314 (partim), fig. 1c, d (not fig. 1a, b = *L. louisianensis* (Schmitt, 1935b)); Griffis & Suchaneck, 1991, tab. 2.
Callichirus jamaicensis; De Saint Laurent & Le Loeuff, 1979: 67, 96 (partim); Coelho & Ramos-Porto, 1987: 30 (partim).
Lepidophthalmus siriboia Felder & Rodrigues, 1993: 367, figs. 2e-h, 4a-f, 6a-l; Sakai, 1999c: 69.
Lepidophthalmus siriboius; Tudge et al., 2000: 144.

Diagnosis. — Mxp3 ischium-merus subsquare, merus obliquely curved and rounded on distomesial margin, male Plp1 blade-shaped, uniramous, two-segmented, subspatulate distally, male Plp2 biramous, with appendix interna and appendix masculina, telson subrectangular, convex with an indistinct median projection on posterior margin (Felder & Rodrigues, 1993, figs. 4a-f, 6f, g, h, j).

Type locality. — Brazil: Maranhão, São Luís, mouth of Rio Anil.

Distribution. — Brazil: mouth of Rio Anil, São Luís, Maranhão; Marapanin, Pará; mouth of Rio Gramame, Joáo Pessoa, Paraiba; mouth of Rio Caravelas, beach.

East Pacific species

Lepidophthalmus bocourti (A. Milne-Edwards, 1870)

Callianassa bocourti A. Milne-Edwards, 1870: 95, 101.
Callianassa (Callichirus) Bocourti; Borradaile, 1903: 547; De Man, 1928b: 28, 94, 115.
Lepidophthalmus Eiseni Holmes, 1904: 311, pl. 35 figs. 6-13. [Type locality: Lower California.]
Callianassa (Callichirus) Eiseni; De Man, 1928b: 28.
Callianassa Eiseni; De Man, 1928b: 110.
Callianassa (Callichirus) eiseni; Schmitt, 1935b: 9.
Callianassa eiseni; Holthuis, 1954a: 12-15, fig. 3; Holthuis, 1954b: 160; Bott, 1955: 47, fig. 6a-g.
Callianassa bocorti; De Saint Laurent & Le Loeuff, 1979: 96.
Lepidophthalmus bocourti; Manning & Felder, 1991: 778; ?Lemaitre & León, 1992: 44; ?Lemaitre & Ramos, 1992: 349, fig. 4; Felder & Rodrigues, 1993: 373; Hendrickx, 1995: 390; Felder & Manning, 1997: 319; Felder & Manning, 1998: 398, 406; Sakai, 1999c: 70, fig. 14c-d; Felder & Staton, 2000: 171, 179; Staton et al., 2000: 167; Tudge et al., 2000: 144; Felder, 2003: 429, figs. 1-19.
Lepidophthalmus bocourti aff. Staton et al., 2000: 158, 167, fig. 2 [designated as *Lepidophthalmus* nr. [near] *bocourti*].
Lepidophthalmus eiseni; Felder & Rodrigues, 1993: 373; Tudge et al., 2000: 144; Felder, 2003: 429, figs. 20-29.

[Not *Lepidophthalmus bocourti*; Biffar, 1972, excluded as published lit. by ICZN, Article 8·4.]

Material examined. — SMF 2186, 1 male, 1 female, Estero near La Playa de las Flores near La Libertad, El Salvador, leg. O. Schuster, 18.ix.1952.

Diagnosis. — Mxp3 ischium-merus subsquare, merus obliquely curved and truncate on mesial margin, male Plp1 blade-shaped, uniramous, two-segmented, subchelate distally, male Plp2 biramous, with appendix masculina and appendix interna, the latter bearing minute, curved hooks (cf. Holthuis, 1954a, fig. 3h, i; Sakai, 1999c, fig. 14c, d), telson subrectangular and trilobate, with a broad median projection on posterior margin.

Remarks. — The following observations are reiterated: P4 subchelate, propodus forming a triangle ventrodistally; P5 chelate. Male Plp1 uniramous, 2-segmented, distal segment tapering distally; Plp2 biramous, appendix interna and appendix masculina present (Sakai, 1999c, fig. 14c, d); Plp3-5 bearing triangular appendices internae. Female Plp1 uniramous, 3-segmented, distal segment forming a flagellum; Plp2 biramous.

Felder (2003) separated *Lepidophthalmus bocourti* and *L. eiseni* by the morphological differences of the ventral abdominal sclerite. However, it is difficult to separate those two species by the shape of the sclerite, due to the variability in this formation, which is affected by the locality, and by its unknown function, as that author mentions that small, immature specimens lack the structure in either *L. bocourti* or *L. eiseni* (cf. Felder, 2003: 434).

The records from Malaga Bay and Isla Gorgona, Colombia, of Lemaitre & León (1992: 44) and Lemaitre & Ramos (1992: 349, fig. 4) are not assigned to the present species, *L. bocourti* (cf. Felder, 2003: 431).

Type locality. — San José del Cabo, Lower California.

Distribution. — East Pacific from San José del Cabo, Baja California, Mexico, and El Salvador to Colombia, intertidal (Lemaitre & Ramos, 1992).

Lepidophthalmus rafai Felder & Manning, 1998

Lepidophthalmus rafai Felder & Manning, 1998: 398, figs. 1-3; Tudge et al., 2000: 144.

Diagnosis. — Mxp3 ischium-merus subovoid, merus concave distally and rounded on distomesial angle of merus, male Plp1 blade-shaped, uniramous, two-segmented, subchelate distally, male Plp2 biramous, with appendix masculina, and with appendix interna bearing minute, curved hooks, telson subrectangular, weakly trilobate on posterior margin (Felder & Manning, 1998: 398, figs. 1g, 3a, c, f).

Remarks. — The present species is similar to *L. bocourti* in the shape of the telson, but Felder & Manning (1998: 406) touched upon an unpublished Ph. D. Dissertation for their reference to the related species, mentioning that "*Lepidophthalmus rafai* differs from known populations of congeneric eastern Pacific species (? = *L. bocourti*) in lacking a strongly trilobate posterior margin on the telson, such as was figured by Bott (1955: fig. 6g) (not 7g) and Biffar (1972: fig 17a). …" Such citation of Biffar's unpublished paper is not admitted, as it is an unpublished work in the sense of zoological nomenclature, and it should be excluded from further references to the proposed new species.

Type locality. — Beach at Playa Basura, Bahia de Buenaventura, Pacific coast of Colombia.

Distribution. — Known only from the type locality.

Indo-West Pacific species

Lepidophthalmus grandidieri (Coutière, 1899)

Callianassa Grandidieri Coutière, 1899: 285, figs. 1-5; Tudge et al., 2000: 143.
Callianassa (Callichirus) Grandidieri; Borradaile, 1903: 547; De Man, 1928b: 28, 92, 110.
Lepidophthalmus grandidieri; Sakai, 1999c: 71.

Diagnosis. — Mxp3 and male Plp1-2 unknown; telson subrectangular, rounded on posterior margin.

Type locality. — Northeastern coast of Madagascar, River Mahanara.

Distribution. — River Mahanara, northeastern coast of Madagascar.

Lepidophthalmus rosae (Nobili, 1904)
(figs. 30-32)

Callianassa (Callichirus) Rosae Nobili, 1904: 237; Nobili, 1906b: 108, pl. 7 fig. 2; De Man, 1928b: 29, 110; Balss, 1933: 88.
Lepidophthalmus rosae; Sakai, 1999c: 71, fig. 14g-h.
Callianassa rosae; Tudge et al., 2000: 143.

Material examined. — ZMUC CRU-3859, 1 female (Tl/Cl 61.0/13.1), 1 female (46.7/12.7), 1 female (Cl 12.0, carapace without abdomen), 1 female (ABl 45.0, abdomen without carapace), 1 male (chelipeds only, without carapace and abdomen), Puerto Galera, Mindoro, Philippines, sandy coast, 9 m depth, Th. Mortensen's Pacific Expedition, leg. Th. Mortensen, 03.ii.1914. ZMUC CRU-3860, 1 female (53.0/10.5), 03°S 128°E, Saparaea Bay, Kei Is., Danish Exped. to Kei Islands, 1922, leg. Th. Mortensen, 12.iii.1922.

Diagnosis. — Rostrum spinous and slightly carinate dorsally (fig. 30A, B). Anterolateral projections present. A1 peduncle (fig. 30A-B) distinctly longer than A2 peduncle. Mandible with denticulate incisor edge (fig. 30C). Mxp3 ischium-merus subrectangular, merus largely convex on ventral margin. Male P1 unequal and dissimilar. Male larger cheliped (fig. 31A) with rod-like ischium, dorsal margin sinuous and unarmed, ventral margin with a few denti-

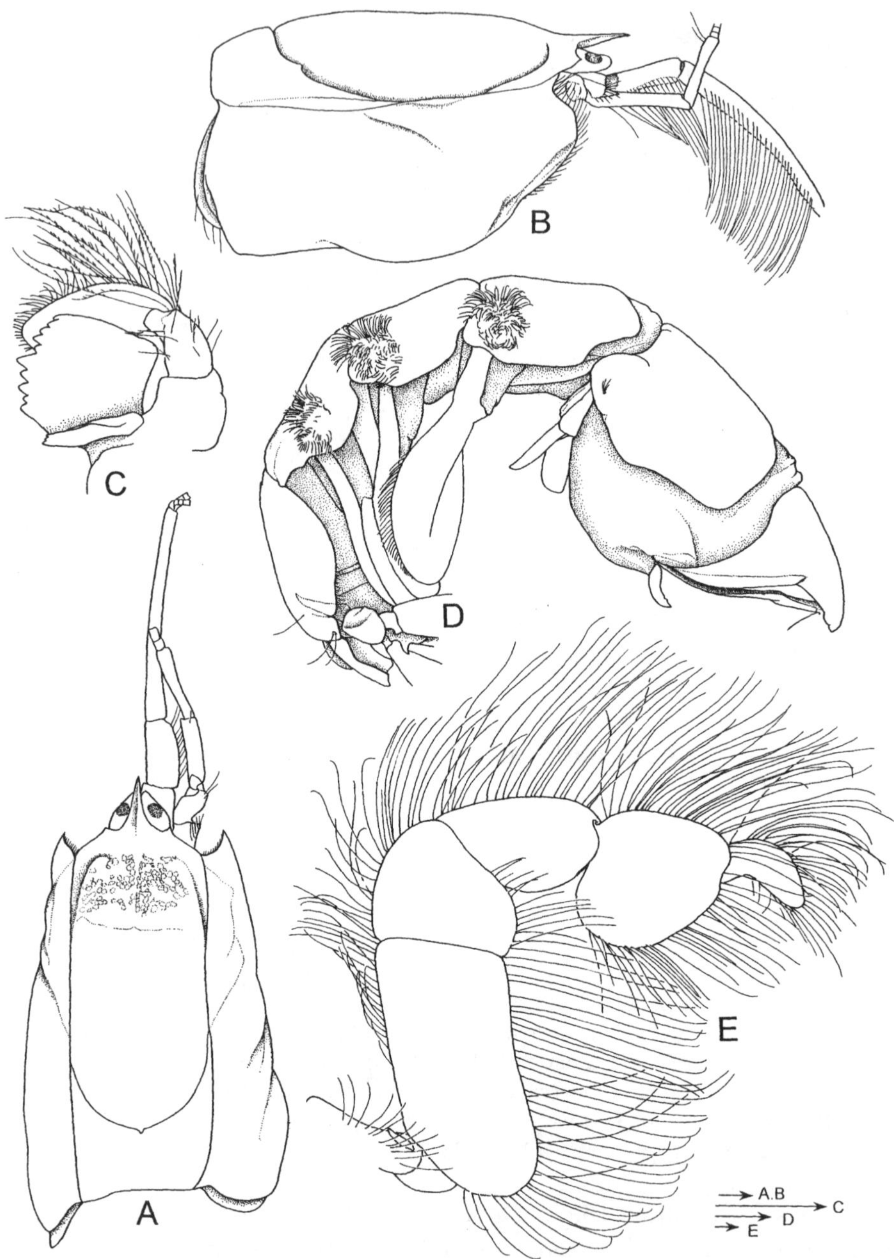

Fig. 30. *Lepidophthalmus rosae* (Nobili, 1904). A, carapace, dorsal view; B, same, lateral view; C, mandible, mesial view; D, abdomen, lateral view; E, maxilliped 3, right side, lateral view. A-E, ZMUC 91, female, Puerto Galera, Mindanao, Philippines, sand coast. Scales 1 mm.

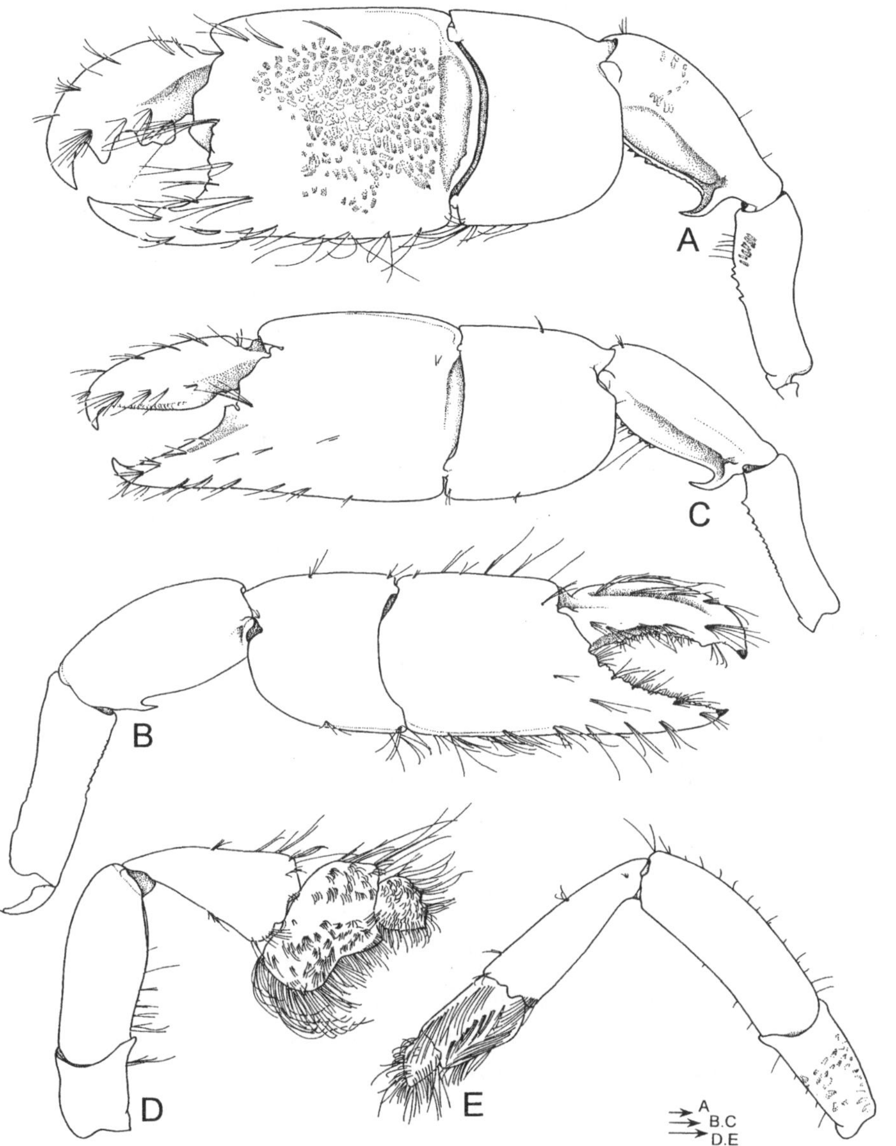

Fig. 31. *Lepidophthalmus rosae* (Nobili, 1904). A, male larger cheliped, lateral view; B, male smaller cheliped, lateral view. C, female larger cheliped, lateral view; D, pereopod 3, male, right lateral view; E, pereopod 4, left lateral view. A-B, D-E, ZMUC 91, male, Puerto Galera, Mindanao, Philippines, sand coast; C, ZMUC 91, female, same locality. Scales 1 mm.

cles medially; merus 1.2 times as long as ischium, 2.0 times as long as high, dorsal margin slightly arcuate and smooth, ventral margin armed with a strong proximal tooth and distal to it armed with a row of denticles, outer surface depressed in ventral half. Carpus 1.5 times as high as long, 0.7 times as long as merus, posterior margin broadened and truncate. Chela three times as long as carpus; palm 2.0 times as long as carpus, about 1.2 times as long as high, smooth on dorsal and ventral margins; distal gap armed with a truncate, flat swelling, and below it remarkably descending downward to fixed finger, extending distally to tip; prehensile margin denticulate only at proximal part and distally smooth. Dactylus stout, dorsal margin entirely roundish incurved, prehensile margin armed with a truncate proximal tooth at proximal fourth, a triangular median tooth, and a sharp distal tooth. Smaller cheliped with ischium slender, dorsal margin slightly sinuous and roughly denticulate proximally, ventral margin with minute denticles in distal half; merus about as long as ischium, slightly shorter than 2.0 times as long as high, dorsal margin slightly arcuate and smooth, ventral margin armed with a simple proximal tooth and distally smooth. Carpus broadened, about as long as high and 0.8 times as long as merus, posterior margin largely rounded. Chela 2.2 times as long as carpus; palm 1.3 times as long as carpus, about 1.1 times as long as high, dorsal and ventral margins smooth and distal gap diverging distally to fixed finger; fixed finger 0.8 times as long as palm, prehensile margin entirely concave, with a row of denticles; dactylus distally incurved on dorsal margin, prehensile margin denticulate in proximal two-thirds and armed with a triangular subdistal tooth. Smaller cheliped (fig. 31B) different from larger one in shape; merus oblong, ventral margin with subproximal tooth; propodus about as long as broad, anterior margin curved downward to cutting margin of fixed finger; dactylus slightly longer than fixed finger, cutting margin with subdistal lower tooth.

Female larger cheliped (fig. 31C) with ischium slender, dorsal margin slightly sinuous and unarmed, ventral margin with minute denticles increasing in size distally; merus, carpus, and chela similar to those of male cheliped, but prehensile margins of dactylus and fixed finger unarmed. P3 propodus broadened, bilobed on ventral margin; dactylus trilobed on distal margin (fig. 31D). P4 subchelate (fig. 31E).

Abdominal somite 1 dorsally smooth, without anterior dome. Abdominal somite 6 convex posteriorly on lateral margins (fig. 30D). Male Plp1 (fig. 32A) of a single segment, chelate distally; Plp2 biramous, endopod bearing on its distomesial margin an appendix masculina with a few long terminal setae; attached to margin mesial of appendix masculina is an appendix interna with terminal hooks (cf. Sakai, 1999c: 71, fig. 14g-h). Female Plp1 (fig. 32B) 3-

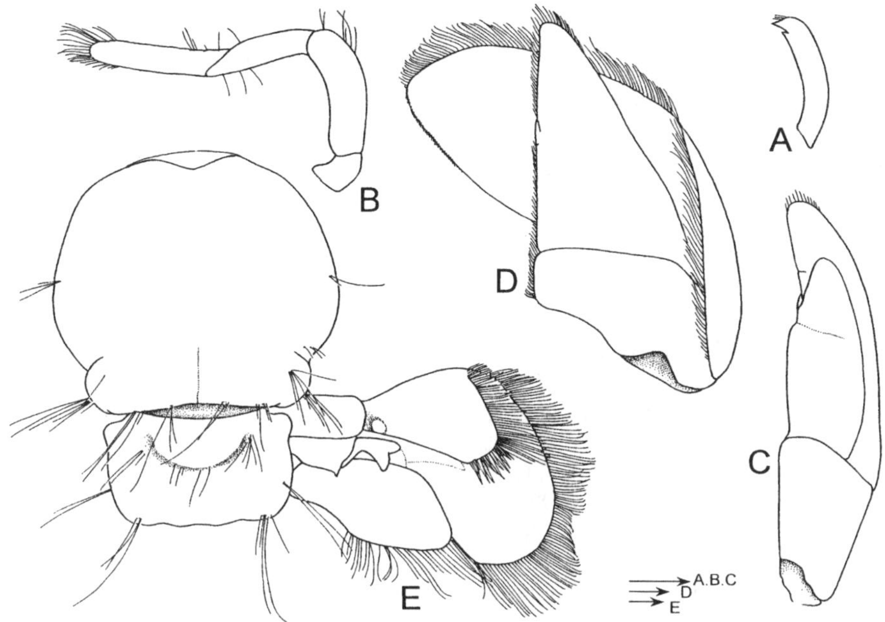

Fig. 32. *Lepidophthalmus rosae* (Nobili, 1904). A, male left Plp1, lateral view; B, female Plp1, mesial view; C, female Plp2, posterior view; D, female Plp3, posterior view; E, abdominal somite 6 and telson, with uropod. A, ZMUC 91, male, Puerto Galera, Mindanao, Philippines, sand coast; B-E, ZMUC 91, female, same locality. Scales 1 mm.

segmented; Plp2 (fig. 32C) biramous, foliaceous, endopod with an appendix interna. Plp3 (figs. 30D, 32D) with a fused appendix interna. Telson (fig. 32E) rectangular, 1.7 times as broad as long, lateral margins protruded proximally, posterior margin slightly convex medially and not armed with a projection midway. Uropodal endopod characteristically lanceolate in shape.

Remarks. — This is the first record of the species from the Philippines.

Type locality. — Red Sea.

Distribution. — Madagascar; Red Sea; Lombok, Indonesia; Mindanao, Philippines.

Lepidophthalmus socotrensis Sakai & Apel, 2002

Lepidophthalmus socotrensis Sakai & Apel, 2002: 278, figs. 3-7.

Diagnosis. — Mxp3 ischium-merus subsquare, merus rounded on distomesial angle, male Plp1 blade-shaped, uniramous, two-segmented, subchelate distally, male Plp2 biramous, with appendix masculina, and appendix interna fused, telson subrectangular, weakly trilobate on posterior margin.

Type locality. — Khawr Girmah, Socotra Island, Republic of Yemen, intertidal mudflat.

Distribution. — Khawr Girmah; Khawr Qariyah; Qalansiyah Lagoon; Ras Kharmah, Socotra Island, Socotra Archipelago, Yemen, intertidal mud, sand/ mud in eulittoral, open lagoon, eulittoral.

Lepidophthalmus tridentatus (Von Martens, 1869)

Callianassa tridentata Von Martens, 1869: 614; A. Milne-Edwards, 1870: 94, 101.
Callianassa (Callichirus) tridentate; Borradaile, 1903: 547; De Man, 1928a: 27, pl. 7 fig. 13-13h; De Man, 1928b: 30, 93, 110, 171, 175; Sakai, 1970b: 393, figs. 1-3.
Lepidophthalmus tridentatus; Sakai, 1999c: 71, fig. 14e-f.
Corallichirus tridentatus; Tudge et al., 2000: 144.

Diagnosis. — Mxp3 ischium-merus subsquare, merus straight distally and rounded on distomesial margin, male Plp1 blade-shaped, uniramous, two-segmented, subchelate distally, male Plp2 biramous, with appendix interna and appendix masculina, telson subrectangular, broadly rounded on posterior margin.

Type locality. — Java, Indonesia.

Distribution. — Java, Indonesia; Sri Lanka.

Genus **Michaelcallianassa** Sakai, 2002

Michaelcallianassa Sakai, 2002: 480.

Definition. — [Revised from Sakai, 2002: 480.] Rostrum spinous. Carapace with post-dorsal oval posterior to cervical groove, linea thalassinica entire. Eyestalks flattened, contiguous; cornea dorsal. A1 peduncle slightly longer than A2 peduncle. Maxilla 2 scaphognathite without a posterior elongate seta. Mxp3 ischium-merus subpediform, propodus broadened and dactylus digitiform; no exopod. P1 unequal, chelate. P2 chelate. P3 propodus distinctly protruded on posterior; dactylus trilobed on dorsal margin. P4 subchelate. P5 chelate. Abdominal somite 1 with an anterodorsal dome and laterally with a row of specialized setae and pits. Abdominal terga 3-5 with a pair of longitudinal, anteriorly convergent grooves, laterally with a strong pattern of rounded, integumental patches, terga 4-5 with a transverse row of setae. Abdominal somite 6 distinctly concave posteriorly on lateral margin. Male Plp1 uniramous, two-segmented; Plp2 uniramous, two-segmented, distal segment bilobed distally, and without appendix interna or appendix masculina. Female Plp1 uniramous, of three segments; Plp2 uniramous, of three segments, with a small appendix interna. Plp3-5 in males and females biramous, foliaceous, and appendices internae fused with the respective endopods. Telson wider than long, conspicuously concave posteriorly on lateral margins, posterior margin entirely concave and without marginal setae. Uropodal endopod broadly triangular.

Remarks. — This genus is different from *Callichirus*, *Glypturus*, and *Lepidophthalmus* by the form of the male Plp2, because in *Michaelcallianassa* the male Plp2 is uniramous and bilobed distally, whereas in *Callichirus*, *Glypturus* and *Lepidophthalmus* it is biramous, usually bearing appendix interna and appendix masculina. The female Plp2 (fig. 33G) consists of two segments in the present specimen, though previous observation (Sakai, 2002: 477, fig. 14H) shows that it consists of three segments.

Type species. — *Michaelcallianassa indica* Sakai, 2002, by original designation and monotypy. The gender of the generic name, *Michaelcallianassa*, is feminine.

Species included. — *Michaelcallianassa indica* Sakai, 2002.

Indo-West Pacific species

Michaelcallianassa indica Sakai, 2002
(fig. 33)

Michaelcallianassa indica Sakai, 2002: 481, figs. 11A-C, 12A-D, 13A-G, 14A-J.

Material examined. — ZMUC CRU-3861, 1 male (Tl/Cl 16.0/3.8, P1-3, 4 absent), 7°00'N 99°22'E, W. Malay Peninsula, 27 m depth, Thai/Danish Exped. 1966, Station 1052, 10.ii.1966; ZMUC CRU-3862, paratype, 1 male (18.0/3.5), 11°06'N 80°05'E, off Tranquebar, S. E. India, 28 m depth, "Galathea" Exped., 1950-1952, Station 291, leg. R/V "Galathea", 21.iv.1951; ZMUC CRU-3863, 1 male (16.0/3.0, cheliped on left side absent, P2 on right side absent, P3 broken), 1 male (10.0/2.2, cheliped on right side absent), 11°10'N 079°59'E, off Tranquebar, coarse sand and mud, 50 m depth, "Galathea" Exped., 1950-1952, Station 294, leg. R/V "Galathea", 22.iv.1951; ZMUC CRU-3864, 1 female (27.0/5.3), 20°37'N 087°33'E, Bay of Bengal, muddy sand with shells, 50 m depth, "Galathea" Exped., 1950-1952, Station 304, leg. R/V "Galathea", 26.iv.1951.

Diagnosis. — Carapace with a white calcified dorsal semi-oval posterior to cervical groove. Eyestalks triangular distally, longer than broad in dorsal view.

Antennular peduncle reaching middle of distal antennal peduncular segment. Mxp3 (fig. 33A-B) ischium-merus subpediform, ischium with crista dentata with a row of strong denticles; merus declined and rounded on distomesial margin, propodus broadened; no exopod. P1 unequal in size and dissimilar in shape. Female larger cheliped with ischium rod-like, 3.0 times as long as broad, dorsal margin straight and unarmed, ventral margin almost straight and armed with a row of denticles; merus almost as long as ischium, 1.8 times as long as high, dorsal margin slightly arched and smooth, ventral margin almost straight and entirely denticulate, no proximal lobe. Carpus broadened, slightly shorter than merus and 1.2 times as high as long, ventral margin entirely declined. Chela slender, about 2.0 times as long as carpus; palm slightly longer than high, dorsal margin entirely convex and ventral margin straight and smooth, distal gap convex and minutely denticulate, declined to fixed finger, fixed finger armed with a distinct triangular tooth at proximal third and entirely denticulate on prehensile margin, mesial surface with a distinct carina medially (fig. 33B). Dactylus narrow and distally incurved downward, prehensile margin denticulate (fig. 33D). Smaller cheliped (fig. 33C) with ischium slightly more slender than that of larger cheliped; merus about as long as ischium and slightly less than 2.0 times as long as high, dorsal margin arched and unarmed, ventral margin slightly convex and unarmed. Carpus 1.2 times as long as high and about as long as merus, ventral margin entirely declined. Chela 1.8 times as long as carpus; palm slightly higher than long, unarmed on dorsal and ventral margins, distal gap sloping down to fixed finger; fixed finger entirely concave, roughly denticulate in the middle and slightly incurved distally. Dactylus slender, indistinctly denticulate medially on prehensile margin. P3 propodus kidney-shaped, posterior angle entirely swollen in oval form, ventral margin straight and setose; dactylus trilobed on dorsal margin. P4 subchelate. P5 chelate.

Abdominal somite 6 longer than broad, convergent laterally in posterior fourth. Male Plp1 blade-shaped, uniramous, two-segmented, subchelate distally, male Plp2 uniramous, two-segmented, distal segment one-third length of proximal one, and bilobed distally, the outer distal lobe shorter and smaller than the inner one, smooth distally, and the inner distal lobe distally with some long setae. Female Plp1 (fig. 33F) 3-segmented; Plp2 (fig. 33G) 2-segmented, distal segment with a small appendix interna apically. Plp3-5 narrow, biramous, appendices internae fused with mesial margin of endopod. Telson trapezoid, proximally concave on both sides, posterior margin entirely concave

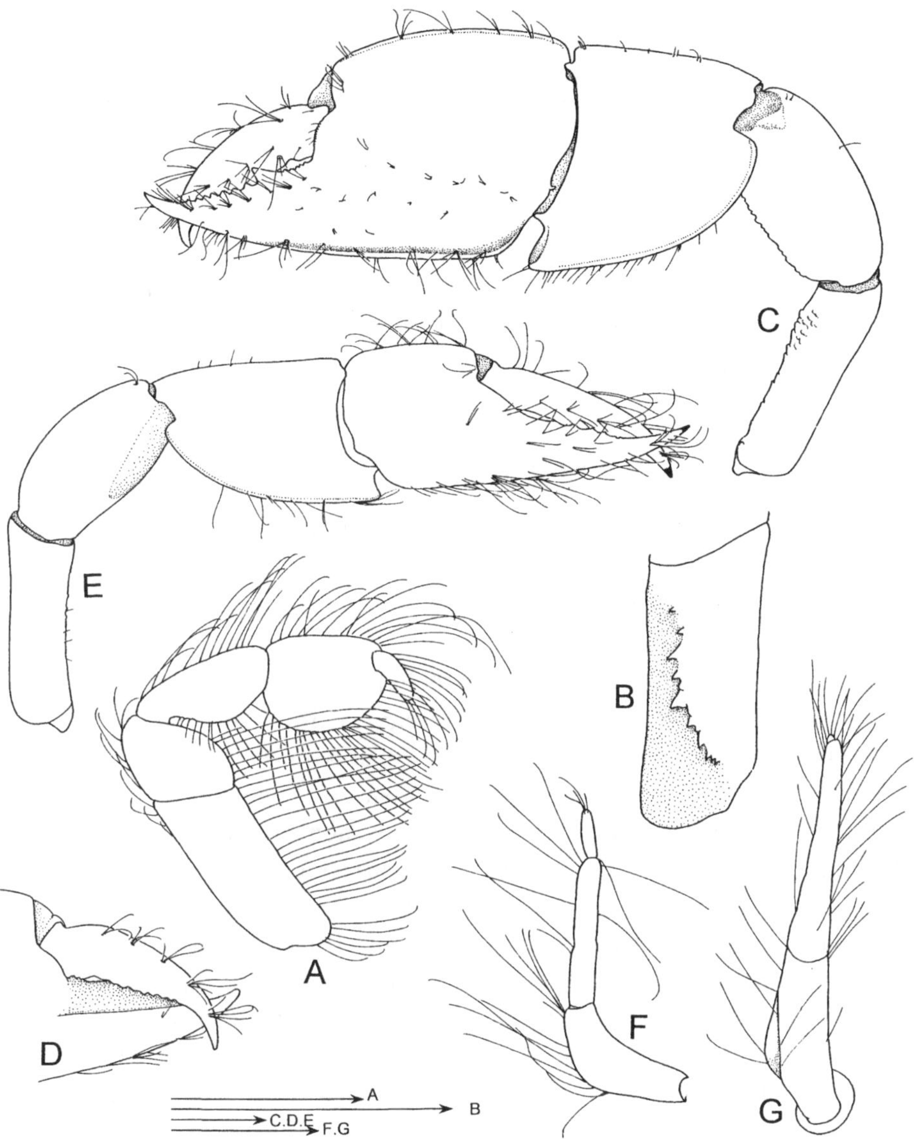

Fig. 33. *Michaelcallianassa indica* Sakai, 2002. A, Mxp3, lateral view; B, ischium of Mxp3, mesial view; C, larger cheliped, lateral view; D, distal part of larger cheliped, mesial view; E, smaller cheliped, lateral view; F, female Plp1; G, female Plp2 with appendix interna. A-B, ZMUC 22, male, "Galathea" Exped., Sta. 294, 11°10'N 79°59'E, off Tranquebar, coarse sand and mud, 50 m depth; C-G, ZMUC 172, female, "Galathea" Exped., Sta. 304, 20°51'N 87°58'E, Bay of Bengal, muddy sand with shells, 50 m depth. Scales 1 mm.

without a median spine; uropodal endopod subsquare and without a median carina.

Remarks. — The small male specimen (ZMUC CRU-3862, Tl measuring 16 mm) has distally triangular eyestalks that lack a sharp, distomedial tip; P1 similar to that of female; while in the larger male holotype, measuring 43.0 mm in total length (Sakai, 2002), it bears a sharp distomedian tip.

Type locality. — Persian (= Arabian) Gulf, 29°29.897'N 49°54.108'E, 32 m depth.

Distribution. — Persian Gulf (29°29.897'N 49°54.108'E, 32 m); Andaman Sea: 7°15.087'N 99°02.918'E; 7°00.082'N 99°15.660'E-9°00.062'N 97°53.366'E; 9°30.991'N 97°57.706'E; 17.0-65.4 m, mud, muddy sand and sand with shell fragments.

Genus **Neocallichirus** Sakai, 1988

Neocallichirus Sakai, 1988: 61; Manning & Felder, 1991: 779, figs. 1, 3, 4; Poore, 1994: 102; Sakai, 1999c: 84; Sakai, 2000: 92; Davie, 2002: 461.
Sergio Manning & Lemaitre, 1994: 40, fig. 1; Poore, 1994: 102.
Corallianassa Manning, 1987 (partim): 392; Manning & Felder, 1991: 776.

Definition. — [Revised from Sakai, 1999c.] Carapace with dorsal oval; rostral spine present or absent and, when present, calcified proximally with frontal margin; rostral carina, cardiac prominence, and transverse cardiac sulci absent. A1 peduncle not longer or stouter than A2 peduncle. Mxp3 without exopod, ischium-merus subquadrate, propodus subquadrate, and dactylus narrow, digitiform. P1 unequal, chelate, male larger cheliped with or without meral hook. P2 chelate. P3 simple. P4 subchelate. P5 chelate. Male Plp1 uniramous, biarticulate, usually distal segment chelate distally, or exceptionally simple with a pointed tip or with an obtuse tip; male Plp2 usually biramous, with appendix interna, or with appendix interna and masculina, or exceptionally uniramous, unsegmented; female Plp1 uniramous, 2- or 3-articulate; female Plp2 biramous and with or without appendix interna; Plp3-5 foliaceous, appendices internae present in both sexes. Telson broader than long. Uropodal endopod subquadrate, broadened distally, or slender, tapering distally.

Remarks. — In the species of *Neocallichirus*, the shape of the uropodal endopod is variable: it shows diverse forms:

1. A **subquadrate** form, broader than long, distal margin truncate: *N. angelikae* Sakai, 2000; *N. cacahuate* Felder & Manning, 1995 (broad, subrectangular, slightly broader than long, posterior margin truncate, nearly straight; Felder & Manning, 1995: 484); *N. grandimanus* (Gibbes, 1850); *N. horneri*

Sakai, 1988; *N. indicus* (= *N. variabilis* (Edmondson, 1944)); *N. jousseaumei* (Nobili, 1904); *N. lemaitrei* Manning, 1993 (endopod broader than long, widening posteriorly, posterior margin flattened; Manning, 1993: 109); *N. manningi* Kazmi & Kazmi, 1992); *N. moluccensis* (De Man, 1905); *N. nickellae* Manning, 1993; *N. taiaro* Ngoc-Ho, 1995.

2. A **subsquare** form with a convex distal margin, little longer than wide, greatest width at distal third: *N. limosus* (Poore, 1975) (subtriangular, greatest width in distal third, little longer than wide; Poore, 1975: 204); *N. pachydactylus* (A. Milne-Edwards, 1870).

3. A **rhomboid** form, longer than wide, greatest width in the middle part: *N. calmani* (Nobili, 1904); *N. darwinensis* Sakai, 1988; *N. mauritianus* (Miers, 1882); *N. mucronatus* (Strahl, 1862) (= *N. brevicaudata* (A. Milne-Edwards, 1870)); *N. ranongensis* (Sakai, 1983); *N. rathbunae* (Schmitt, 1935b) (rhomboid, rounded distolaterally; Biffar, 1971b); *N. sulfureus* (Lemaitre & Felder, 1996) (longer than broad; Lemaitre & Felder, 1996: 461).

4. A **triangular** shape, greatest width proximally: *N. audax* (De Man, 1911); *N. caechabitator* Sakai, 1988; *N. calmani* (Nobili, 1904); *N. guara* Rodrigues, 1971 (longer than wide and almost triangular in outline; Rodrigues, 1971: 212); *N. sassandrensis* (Le Loeuff & Intès, 1974); *N. vigilax* (De Man, 1916).

5. An **elongate triangular** form, longer than wide: *N. denticulatus* Ngoc-Ho, 1994 (lanceolate, widest proximally; Ngoc-Ho, 1994: 59); *N. guaiqueri* (Blanco Rambla & Liñero Arana, 1994); *N. guassutinga* Rodrigues, 1971 (= *Sergio mericeae* Manning & Felder, 1995; *Sergio* sp. Staton & Felder, 1955) (broadly triangular with rounded extremities; Rodrigues, 1971: 207); *N. karumba* (Poore & Griffin, 1979) (lanceolate, greatest width at proximal third, width half [of] length; Poore & Griffin, 1979: 268); *N. mirim* (nearly 2.0 times as long as wide, and oval in outline; Rodrigues, 1971: 218); *N. monodi* (De Saint Laurent & Le Loeuff, 1979); *N. motupore* (Poore & Suchanek, 1988) (ovoid-triangular, 2.0 times as long as wide, widest at about proximal one-third and converging to rounded apex; Poore & Suchanek, 1988: 198); *N. pentagonocephalus* (Rossignol, 1962); *N. trilobatus* (Biffar, 1970).

6. A **narrow triangle**, posterior margin straight: *N. kempi* Sakai, 1999.

In the collection of the Zoologisk Museum, Copenhagen, there are also two new species, *N. auchenorhynchus* sp. nov. and *N. mortenseni* sp. nov. This is the first record of the species of the genus *Neocallichirus* from the East Pacific region.

Type species. — *Neocallichirus horneri* Sakai, 1988, by original designation. Gender of the generic name, *Neocallichirus*, masculine.

Species included. — East Atlantic species: *Neocallichirus monodi* (De Saint Laurent & Le Loeuff, 1979); *N. pachydactylus* (A. Milne-Edwards, 1870); *N. pentagonocephalus* (Rossignol, 1926); *N. sassandrensis* (Le Loeuff & Intès, 1974). West Atlantic species: *N. cacahuate* Felder & Manning, 1995; *N. grandimanus* (Gibbes, 1850); *N. guaiqueri* (Blanco Rambla & Liñero Arana, 1994); *N. guara* (Rodrigues, 1971); *N. guassutinga* (Rodrigues, 1971); *N. lemaitrei* Manning, 1993; *N. mirim* (Rodrigues, 1971); *N. nickellae* Manning, 1993; *N. rathbunae* (Schmitt, 1935); *N. sulfureus* (Lemaitre & Felder, 1996); *N. trilobatus* (Biffar, 1970); *N.* sp. Blanco Rambla & Liñero Arana, 1994. East Pacific species: *N. mortenseni* sp. nov. Indo-West Pacific species: *N. angelikae* Sakai, 2000; *N. audax* (De Man, 1911); *N. caechabitator* Sakai, 1988; *N. calmani* (Nobili, 1904); *N. darwinensis* Sakai, 1988; *N. denticulatus* Ngoc-Ho, 1994; *N. horneri* Sakai, 1988; *N. indicus* (De Man, 1905); *N. jousseaumei* (Nobili, 1904); *N. karumba* (Poore & Griffin, 1979); *N. kempi* Sakai, 1999; *N. manningi* Kazmi & Kazmi, 1992; *N. mauritianus* (Miers, 1882); *N. moluccensis* (De Man, 1905); *N. motupore* (Poore & Suchanek, 1988); *N. mucronatus* (Strahl, 1862); *N. ranongensis* (Sakai, 1983); *N. vigilax* (De Man, 1916); *N.* sp. (Rathbun, 1906). Unknown locality: *N. auchenorhynchus* sp. nov.

East Atlantic species

Neocallichirus monodi (De Saint Laurent & Le Loeuff, 1979)

Callichirus monodi De Saint Laurent & Le Loeuff, 1979: 71, figs. 14h, 16f, 17c, 18a, 19h, 22a-b, 23n-r; Tudge et al., 2000: 144.
Neocallichirus monodi; Sakai, 1999c: 86.

Diagnosis. — Mxp3 ischium-merus subsquare, merus obliquely curved distally and rounded on distomesial margin. Male Plp1 blade-shaped, uniramous, two-segmented, subchelate distally, male Plp2 biramous, endopod bilobed distally, with an appendix masculina (appendix interna uncertain, but presumably present). Female Plp1 uniramous, three-segmented, female Plp2 bilobed, endopod with bilobed appendix interna. Telson subovoid, broadly concave medially with a pointed median protrusion on posterior margin (De Saint Laurent & Le Loeuff, 1979: 71, figs. 17c, 18a, 19h, 23n-r).

Type locality. — Senegal.

Distribution. — Senegal, intertidal.

Neocallichirus pachydactylus (A. Milne-Edwards, 1870)

Callianassa pachydactyla A. Milne-Edwards, 1870: 86, pl. 2 fig. 1a-d.
Callianassa (*Cheramus*) *pachydactyla*; Borradaile, 1903: 545; De Man, 1928b: 19, 26, 94, 100, 121, 160-164.
Callichirus pachydactyla; De Saint Laurent & Le Loeuff, 1979: 76, figs. 14j, 16d-e, 18e, 19k, 22c-d, 23s-v.
Neocallichirus pachydactylus; Sakai, 1999c: 86, fig. 20a-c; Tudge et al., 2000: 144.

Not *Callianassa pachydactyla*; Longhurst, 1958: 42 [= *Podocallichirus foresti* (Le Loeuff & Intès, 1974)].

Diagnosis. — Mxp3 ischium-merus subrectangular, merus obliquely curved distally and rounded on distomesial margin, male Plp1 blade-shaped, uniramous, two-segmented, subchelate distally, male Plp2 trilobed, median lobe of appendix masculina with long setae, laterally with appendix interna bearing minute, curved hook, telson trapezoid, narrow, and concave on posterior margin, lacking a median spine (De Saint Laurent & Le Loeuff, 1979: 76, figs. 18e, 19k, 23s-v; Sakai, 1999c: 86).

Type locality. — Cape Verde Is.

Distribution. — Cape Verde Is.; Senegal; Ghana; Ile Príncipe, intertidal.

Neocallichirus pentagonocephalus (Rossignol, 1962)

Callianassa pentagonocephala Rossignol, 1962: 139, fig. 1a-c.
Callichirus pentagonocephala; De Saint Laurent & Le Loeuff, 1979: 78, figs. 14k, 16b, 18d, 19i, 21d, 23w-x; Tudge et al., 2000: 144.
Neocallichirus pentagonocephala; Sakai, 1999c: 87.

Diagnosis. — Mxp3 ischium-merus subrectangular, merus obliquely curved distally and rounded on distomesial margin, male Plp1 blade-shaped, uniramous, two-segmented, shallowly bilobed distally, male Plp2 biramous, endopod with distally separated lobe of appendix masculina bearing long setae, appendix interna uncertain, telson pentagonal, shallowly concave on posterior margin, lacking a median spine (De Saint Laurent & Le Loeuff, 1979: 78, figs. 18d, 19i, 23w-x).

Type locality. — Bay of Pointe Noire, Congo.

Distribution. — Congo; Cameroon; 6-7 m.

Neocallichirus sassandrensis (Le Loeuff & Intès, 1974)

Callichirus sassandrensis Le Loeuff & Intès, 1974: 43, fig. 11a-t; De Saint Laurent & Le Loeuff, 1979: 71, figs. 14i, 18c, 19i.
Neocallichirus sassandrensis; Sakai, 1999c: 87; Tudge et al., 2000: 144.

Diagnosis. — Mxp3 ischium-merus subrectangular, merus obliquely curved distally and straight on mesial margin, male Plp1-2 unknown, telson pentagonal, straight on posterior margin, bearing a median spine (Le Loeuff & Intès, 1974: 43, fig. 11k, q-t).

Type locality. — Ivory Coast, Sassandra (4°58.8'N 6°01'W), 10 m.

Distribution. — Ivory Coast.

West Atlantic species

Neocallichirus cacahuate Felder & Manning, 1995

Neocallichirus cacahuate Felder & Manning, 1995: 478, figs. 1a-c, 2, 3a-e, 4a-c, 5; Sakai, 1999c: 88; Tudge et al., 2000: 144.

Diagnosis. — Mxp3 ischium-merus subrectangular, merus obliquely declined on distomesial margin, male Plp1 blade-shaped, uniramous, two-segmented, chelate distally, male Plp2 biramous, endopod distally with separated lateral lobe of appendix interna bearing long distal setae, laterally bearing a minute appendix interna, telson trapezoid, concave on posterior margin, lacking a median spine (Felder & Manning, 1995: 478, figs. 1c, 2g, 4a, c).

Type locality. — Florida, West Palm Beach County, Lake Worth, north side of Peanut Island, sparsely vegetated sandy to shelly sand, intertidal flats (26°46.7'N 80°2.9'W).

Distribution. — North side of Peanut Island, Lake Worth, West Palm Beach County, Florida.

Neocallichirus grandimanus (Gibbes, 1850)

Callianassa grandimana Gibbes, 1850: 194; Stimpson, 1866: 47; Stimpson, 1871: 122; Kingsley, 1899: 823; Schmitt, 1935b: 2; Biffar, 1971a: 649, 671-674; Manning, 1987: 388, 397, fig. 2; Dworschak, 1992: 196.
Glypturus branneri Rathbun, 1900b: 150, pl. 8 figs. 5-8; Rathbun, 1901: 93; Rathbun, 1920: 328, fig. 3; Verrill, 1922: 33, pl. 1 fig. 2, pl. 8 fig. 1a-e; Schmitt, 1924: 93; Schmitt, 1935a: 194, fig. 55; Manning, 1987: 397. [Type locality: Mamanguape Stone Reef, Brazil.]
Glypturus grandimanus; Rathbun, 1900b: 151.

Glypturus grandimana; Borradaile, 1903: 548; De Man, 1928b: 25.
Glypturus siguanensis Boone, 1927: 85, fig. 17; Manning, 1987: 397. [Type locality: Siguana Bay, Isle of Pines near Cuba, Gulf of Mexico.]
Callianassa branneri; Schmitt, 1935b: 4; Gurney, 1944: 82, figs. 16, 17; Weimer & Hoyt, 1964: 764; Biffar, 1971a: 652, 654, 661, figs. 5, 6; Coelho & Ramos, 1973: 161; Manning, 1987: 398.
Callianassa siguanensis; Biffar, 1971a: 649.
Neocallichirus grandimana; Sakai, 1988: 61; Manning & Felder, 1991: 779, figs. 3, 4; Lemaitre & León, 1992: 44; Lemaitre & Ramos, 1992: 349, fig. 5; Dworschak & Ott, 1993: 281; Felder & Manning, 1995: 478, figs. 1d-f, 3f, 4, d-h.
Neocallichirus grandimanus; Manning, 1993: 113; Sakai, 1999c: 89; Tudge et al., 2000: 144.

Not *Glypturus grandimanus*; Balss, 1924: 179, figs. 3, 4; Manning, 1987: 399. (= *C. hartmeyeri* Schmitt.)

Diagnosis. — Mxp3 ischium-merus subrectangular, merus obliquely declined on distomesial margin, male Plp1 blade-shaped, uniramous, two-segmented, bilobed distally, male Plp2 biramous, endopod with distally demarcated appendix interna with long setae, laterally with appendix interna, telson trapezoid, concave on posterior margin, lacking a median spine (Felder & Manning, 1995: 478, figs. 1f, 4d, h).

Type locality. — Key West, Florida.

Distribution. — Atlantic side: Bermuda; southeast of Florida; west coast of Florida, including Keys and Dry Tortugas; Bimini and Little San Salvador, Bahamas; Cuba; Puerto Rico; Barbados; Tobago; Curaçao; South Water Cay (lagoonside intertidal to 0.5 m) and Belize; Mamanguape, Brazil. Pacific side: Panama; Ecuador; Gorgona Island, Colombia.

Neocallichirus guaiqueri Blanco Rambla & Liñero Arana, 1994

Neocallichirus guaiqueri Blanco Rambla & Liñero Arana, 1994: 20-22, figs. 4-5.
Sergio guaiqueri; Blanco Rambla et al., 1995: 102, text-figs. 1-3.
Sergio guaiqueri; Lemaitre & Felder, 1996: 453; Tudge et al., 2000: 144.
Neocallichirus guaiqueri; Sakai, 1999c: 90.

Diagnosis. — Mxp3 ischium-merus subrectangular, merus obliquely declined on distomesial margin, male Plp1 blade-shaped, uniramous, two-segmented, tapered distally, male Plp2 biramous, endopod bilobed distally, appendix masculina and appendix interna uncertain, telson trapezoid, concave on posterior margin, lacking a median spine (Blanco Rambla et al., 1995: 102, text-figs. 1g, 2f, 3a, b).

Type locality. — Venezuela, Anzoategui State, north of Jose (10°08.40'N 64°50.10'W), Petersen grab.

Distribution. — Venezuela.

Neocallichirus **guara** (Rodrigues, 1971)

Callianassa (*Callichirus*) *guara* Rodrigues, 1971: 210, figs. 61-76; Biffar, 1971a: 654; Coelho & Ramos, 1973: 162; Manning, 1987: 397.
Callianassa guara; Biffar, 1971a: 652, 654.
Neocallichirus guarus; Manning & Felder, 1991: 779.
Sergio guara; Manning & Lemaitre, 1994: 41; Lemaitre & Felder, 1996: 453; Tudge et al., 2000: 144.
Neocallichirus guarus; Sakai, 1999c: 90.

Diagnosis. — Mxp3 ischium-merus subrectangular, merus obliquely declined on distomesial margin, male Plp1 blade-shaped, uniramous, two-segmented, distally chelate, male Plp2 biramous, endopod trilobed distally, appendix interna and appendix masculina uncertain, telson trapezoid, concave on posterior margin, lacking a median spine (Rodrigues, 1971: 210, figs. 67, 74, 75, 76).

Type locality. — Brazil, São Sebastião, São Paulo, beach.

Distribution. — Tampa Bay, Miami; Lemon Bay, Florida; Brazil.

Neocallichirus **guassutinga** (Rodrigues, 1971)

Callianassa (*Callichirus*) *guassutinga* Rodrigues, 1971: 204, figs. 41-60.
Callianassa guassutinga; Biffar, 1971a: 651, 653, 674, figs. 9, 10; Coelho & Ramos, 1973: 162; Abele & Kim, 1986: vii, 26, 296, 298, 299, figs. a-c; Manning, 1987: 397; Williams et al., 1989: 28, 61.
Callianassa; Rabalais et al., 1989: 35.
Neocallichirus guassutinga; Manning & Felder, 1991: 779.
Sergio guassutinga; Manning & Lemaitre, 1994: 40; Tudge et al., 2000: 144.
Sergio mericeae Manning & Felder, 1995: 267, figs. 1a-f, 2a-f, 3a-f, 4a-f, 5a-g; Tudge et al., 2000: 144. [Type locality: 27°28.2'N 80°18.8'W, north side of Fort Pierce Inlet, Florida, Indian River Lagoon, intertidal sandflat.]
Sergio sp. Staton & Felder, 1995: 505.
Neocallichirus guassutingus; Sakai, 1999c: 90.

Diagnosis. — Mxp3 ischium-merus subrectangular, merus obliquely declined on distomesial margin, male Plp1 blade-shaped, uniramous, two-segmented, bilobed distally, male Plp2 biramous, endopod demarcated distal'ly with a small distal lobe, appendix interna and appendix masculina uncertain, telson pentagonal, triangularly concave on posterior margin, lacking a median spine (Rodrigues, 1971: 204, figs. 48, 56, 57, 60; Biffar, 1971a, figs. 9d, f, h, 10f).

Type locality. — Brazil, São Sebastião.

Distribution. — North side of Fort Pierce Inlet, Florida; Indian River Lagoon, St. Lucie County, Florida; Louisiana, 12-13 m; Texas; Barra del Tordo, Tamaulipas, Mexico; Gulf of Panama; Brazil.

Neocallichirus lemaitrei Manning, 1993

Neocallichirus lemaitrei Manning, 1993: 107, figs. 1-3; Felder & Manning, 1995: 488, fig. 6; Sakai, 1999c: 91; Tudge et al., 2000: 144.

Diagnosis. — Mxp3 ischium-merus subrectangular, merus obliquely declined on distomesial margin, male Plp1 blade-shaped, uniramous, two-segmented, chelate distally, male Plp2 biramous, endopod with distally demarcated inner lobe of appendix interna and laterally with appendix masculina, telson pentagonal, straight on posterior margin, lacking a median spine (Manning, 1993: 107, figs. 1h, 3a, c; Felder & Manning, 1995, fig. 6b, c).

Type locality. — Isla del Rosario (10°10'N 75°46'W), Colombia, beach on south side.

Distribution. — Colombia, Atlantic side.

Neocallichirus mirim (Rodrigues, 1971)

Callianassa (Callichirus) mirim Rodrigues, 1971: 214, figs. 77-98.
Callianassa mirim; Biffar, 1971a: 654; Coelho & Ramos, 1973: 162; Manning, 1987: 397; Dworschak, 1992: 202.
Callichirus mirim; Ferrari, 1981: 12, fig. 1; Rodrigues, 1983: 31, figs. 53-60; Rodrigues, 1984a: 239, figs. 1-39 [larval stage]; Rodrigues, 1984b: 914 [larval stage]; Rodrigues & Hödl, 1990: 50, fig. 1.
Neocallichirus mirim; Manning & Felder, 1991: 779.
Sergio mirim; Manning & Lemaitre, 1994: 41; Tudge et al., 2000: 144.
Neocallichirus mirim; Sakai, 1999c: 91, fig. 20d.

Diagnosis. — Mxp3 ischium-merus subrectangular, merus obliquely declined on distomesial margin, male Plp1 blade-shaped, uniramous, two-segmented, bilobed distally, male Plp2 biramous, endopod distally trilobed, with median lobe of appendix masculina distinct and distally armed with long setae, and with laterally attached a small lobe of the appendix interna, telson pentagonal, posterior margin concave, medially with a sharp spine (Rodrigues, 1971: 214, figs. 84, 94, 95, 98).

Type locality. — São Paulo, São Sebastião, Brazil.

Distribution. — Santos, State of São Paulo, Brazil, to Argentina. Common in the lower intertidal and shallow subtidal of sandy beaches.

Neocallichirus nickellae Manning, 1993

Neocallichirus nickellae Manning, 1993: 110, figs. 4-6; Sakai, 1999c: 92; Tudge et al., 2000: 144.

Diagnosis. — Mxp3 ischium-merus subrectangular, merus concave distally, convex on distomesial margin, male Plp1 blade-shaped, uniramous, two-segmented, chelate distally, male Plp2 biramous, endopod distally trilobed, with distinct median lobe of appendix masculina with long setae and lateral lobe of appendix interna, telson pentagonal, posterior margin convex, without a median spine (Manning, 1993: 110, figs. 4c, f, 6a, b).

Type locality. — Republic of Trinidad and Tobago, Tobago, Buccoo Reef, Coral Garden (11°11'N 60°49'W).

Distribution. — Republic of Trinidad and Tobago.

Neocallichirus rathbunae (Schmitt, 1935)

Callianassa (Callichirus) rathbunae Schmitt, 1935b: 4, 15, pl. 1 fig. 5, pl. 2 fig. 2, pl. 3 fig. 1, pl. 4 fig. 2.
Callianassa rathbunae; Biffar, 1971a: 651, 654, 699, figs. 19, 20; Manning & Heard, 1986: 347-349, fig. 1; Manning, 1987: 397; Dworschak, 1992: 202, fig. 10.
Callichirus rathbunae; De Saint Laurent & Le Loeuff, 1979: 97.
Neocallichirus rathbunae; Manning & Felder, 1991: 779; Manning, 1993: 113; Sakai, 1999c: 92; Tudge et al., 2000: 144.
Neocallichirus raymanningi Blanco Rambla & Lemaitre, 1999: 768, 4 figs. [Type locality: Playa Cristal, south coast of Gulf of Cariaco, Sucre State, Venezuela.]

Diagnosis. — Mxp3 ischium-merus subrectangular, merus convex on distomesial margin, male Plp1 blade-shaped, uniramous, two-segmented, chelate distally, male Plp2 biramous, endopod simple, articulated with appendix masculina distally and laterally with appendix interna, telson trapezoid, posterior margin convex, without a median spine (Biffar, 1971a: 699, figs. 19g, h; 20e, f).

Remarks. — *Neocallichirus raymanningi* from Venezuela was distinguished from *N. rathbunae* by such differences as the rostrum, the merus of the male major cheliped, and male Plp2 (Blanco Rambla et al., 1999: 775). However, those characters are often variable, so the species of Blanco Rambla & Lemaitre (1999) can be synonymized with the present species.

Type locality. — Bluefields, Jamaica.

Distribution. — Indian River, Miami, Florida; west coast of Florida; Bimini, Bahamas; Jamaica; St. Croix, Virgin islands; Twin Cays, Cassiopeia Cove, Belize; N. of Santa Marta, Colombia; and Venezuela, which is included by the distribution of *N. raymanningi*.

Neocallichirus sulfureus (Lemaitre & Felder, 1996)

Sergio sulfureus Lemaitre & Felder, 1996: 453-463, text-figs. 1-6; Tudge et al., 2000: 145.
Neocallichirus sulfureus; Sakai, 1999c: 92.

Diagnosis. — Mxp3 ischium-merus subrectangular, merus obliquely declined distally and convex on distomesial margin, male Plp1 blade-shaped, uniramous, two-segmented, distal segment longitudinally subdivided into two lobes, male Plp2 biramous, endopod trilobed distally, distal median lobe of appendix masculina weakly separated from its base, sparsely setose distally, appendix interna uncertain, telson trapezoid, posterior margin concave, without a median spine (Lemaitre & Felder, 1996: 453-463, text-figs. 2h, 4d, 6b, d, e, j).

Type locality. — Caribbean coast of Colombia, S. W. shoreline of Barú, beach facing Rosario Islands.

Distribution. — Caribbean coast of Colombia.

Neocallichirus trilobatus (Biffar, 1970)

Callianassa trilobata Biffar, 1970: 36, fig. 1; Biffar, 1971a: 653, 654, 704, figs. 21, 22; Manning, 1987: 397.
Neocallichirus trilobata; Manning & Felder, 1991: 779.
Sergio trilobatus; Manning & Lemaitre, 1994: 41; Tudge et al., 2000: 145.
Neocallichirus trilobatus; Sakai, 1999c: 93, fig. 20e-f.

Diagnosis. — Mxp3 ischium-merus subrectangular, merus obliquely declined distally and convex on distomesial margin, male Plp1 blade-shaped, uniramous, two-segmented, distal segment chelate distally, male Plp2 with a distal lobe of appendix masculina separated from endopod, laterally with a minute appendix interna, telson subrectangular, posterior margin concave medially, without a median spine (Biffar, 1970: 36, fig. 1f, l, m, o).

Type locality. — Off Pinellas Point, Tampa Bay, Florida, 2-3 m.

Distribution. — Tampa Bay, Miami; Lemon Bay, Florida.

Neocallichirus sp. Blanco Rambla & Liñero Arana, 1994

Neocallichirus sp. Blanco Rambla & Liñero Arana, 1994: 20, figs. 4, 5.

Diagnosis. — Mxp3 ischium-merus subrectangular, merus obliquely declined on distomesial margin, male Plp1 blade-shaped, uniramous, two-segmented, distal segment chelate distally, male Plp2 bilobed, endopod simple distally, telson subrectangular, posterior margin concave medially, without a median spine (Blanco Rambla & Liñero Arana, 1994: 20, figs. 4e, 5a, b, d).

Locality. — North [of] Jose, 10°08'40"N 64°50'10"W, Venezuela; 24 m depth, clay-silt.

East Pacific species

Neocallichirus mortenseni sp. nov.
(figs. 34-36)

Material examined. — ZMUC CRU-3865, 1 female (holotype) (Tl/Cl 21.0/5.0), shore collection on a small island off N. coast of Taboga, Panama, Th. Mortensen's Pacific Expedition, leg. Th. Mortensen, 03.ii.1916.

Diagnosis. — Rostrum triangular and pointed distally; frontal margin of carapace with distinct anterolateral projections. A1 peduncle slightly shorter than A2 peduncle. Mxp3 pediform, subrectangular, propodus broadened and rounded on ventral margin. Male Plp1-2 unknown. Telson trapezoid, lateral margins convex in proximal half, convergent in distal half, posterior margin slightly concave medially, bearing a median spine. Uropodal endopod leaf-like and armed with a spinule on lateral margin. Uropodal exopod broadly rounded distally.

Description of female holotype (fig. 34A). — Rostrum (fig. 34B) triangular and pointed at tip; frontal margin of carapace armed with triangular anterolateral projections; dorsal oval conspicuous; cervical groove located approx.

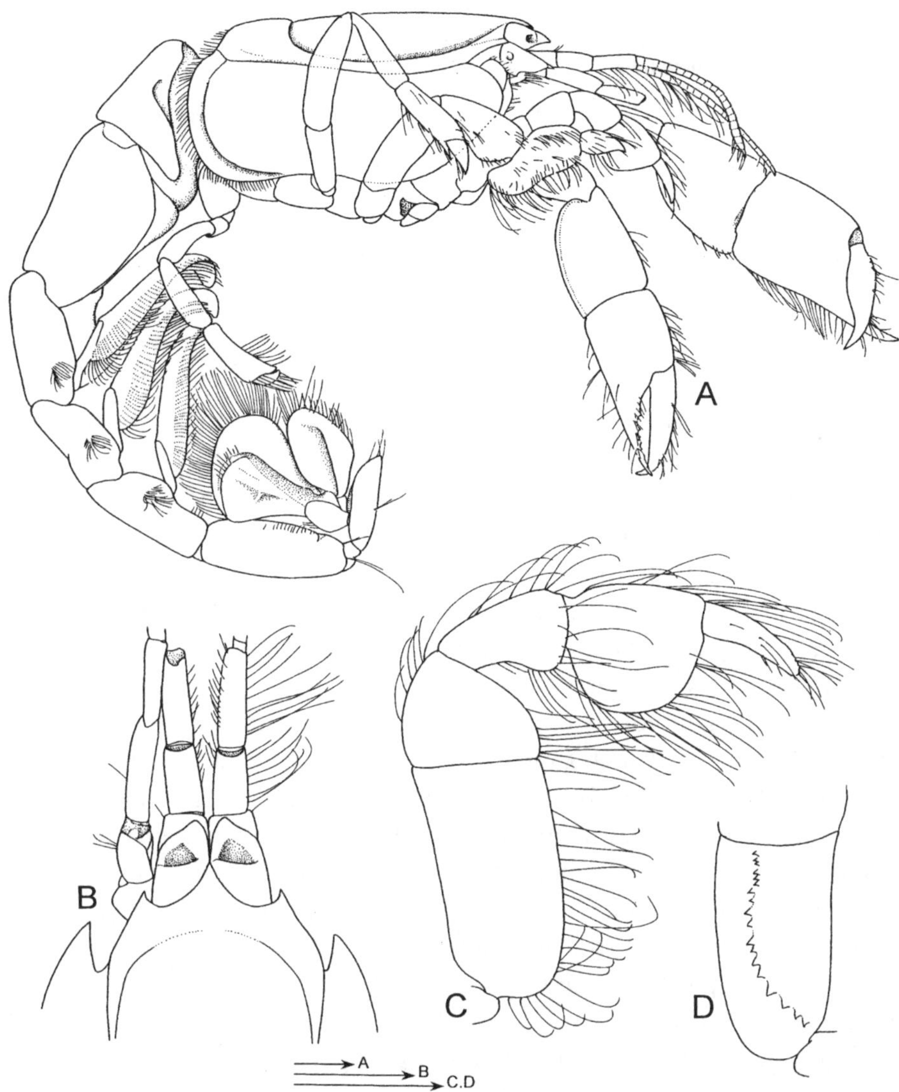

Fig. 34. *Neocallichirus mortenseni* sp. nov. A, whole body, lateral view; B, anterior part of carapace, dorsal view; C, Mxp3, lateral view; D, ischium of Mxp3, mesial view. A-D, ZMUC 210, holotype female, Taboga, Panama, shore on small island off the northern coast. Scales 1 mm.

in posterior third of carapace; linea thalassinica entire. Eyestalks convergent in distal part, longer than broad, shorter than A1 basal segment, and convex on dorsal surface; cornea large and rounded, located medially, pigmented yellow in alcohol specimens. A1 peduncle slightly longer than A2 peduncle, terminal

TABLE IX

Branchial formula of *Neocallichirus mortenseni* sp. nov.

	Maxillipeds			Pereiopods				
	1	2	3	1	2	3	4	5
Exopods	1	1	–	–	–	–	–	
Epipods	1	–	–	–	–	–	–	–
Podobranchs	–	r	–	–	–	–	–	–
Arthrobranchs	–	–	2	2	2	2	2	–
Pleurobranchs	–	–	–	–	–	–	–	–

(r = rudimentary)

segment about 1.2 times as long as penultimate. A2 terminal segment 0.8 times as long as penultimate, scaphocerite oval and vestigial; A2 flagellum more than four times as long as A1 flagella. Mxp3 merus-ischium (fig. 34C) subrectangular, elongate; ischium subrectangular, 2.0 times as long as broad, crista dentata (fig. 34D) with row of 18 denticles; merus subtriangular, slightly shorter than broad, convergent distally on lateral and mesial margins; carpus triangular, slightly longer than merus, mesial margin entirely convergent in its distal half; propodus broadened, about as long as carpus and about as long as broad, ventral margin broadly convex; dactylus digitiform, 0.8 times as long as propodus; no exopod present. Branchial formula as shown in table IX.

P1 subequal and similar. In larger cheliped (fig. 35A) on left side, ischium slender, dorsal margin almost straight and unarmed, ventral margin with a row of minute denticles; merus spindle-shaped, about as long as ischium, about 2.0 times as long as high, dorsal margin slightly convex and smooth, ventral margin slightly arched, with a row of minute denticles, lacking a distinct, enlarged proximal lobe; carpus broadened, about as high as long and about 0.8 times as long as merus, entirely diverging distally on posteroventral margin. Chela heavy, 2.0 times as long as carpus; palm as long as carpus, about as long as high, dorsal and ventral margins smooth, distal gap declining to tip of fixed finger; fixed finger 0.6 times as long as palm, prehensile margin entirely concave and smooth; dactylus incurved downwards distally, prehensile margin inclined distally with a subproximal denticle. Smaller cheliped on right side (fig. 35B) more slender than larger cheliped but similar in form; ischium narrow, dorsal margin unarmed, ventral margin minutely denticulate; merus weakly spindle-shaped, about as long as ischium, dorsal and ventral margins slightly convex and unarmed; carpus broadened, about as long as merus, 1.5 times as long as high and entirely arched on proximoventral margin. Chela 1.5 times as

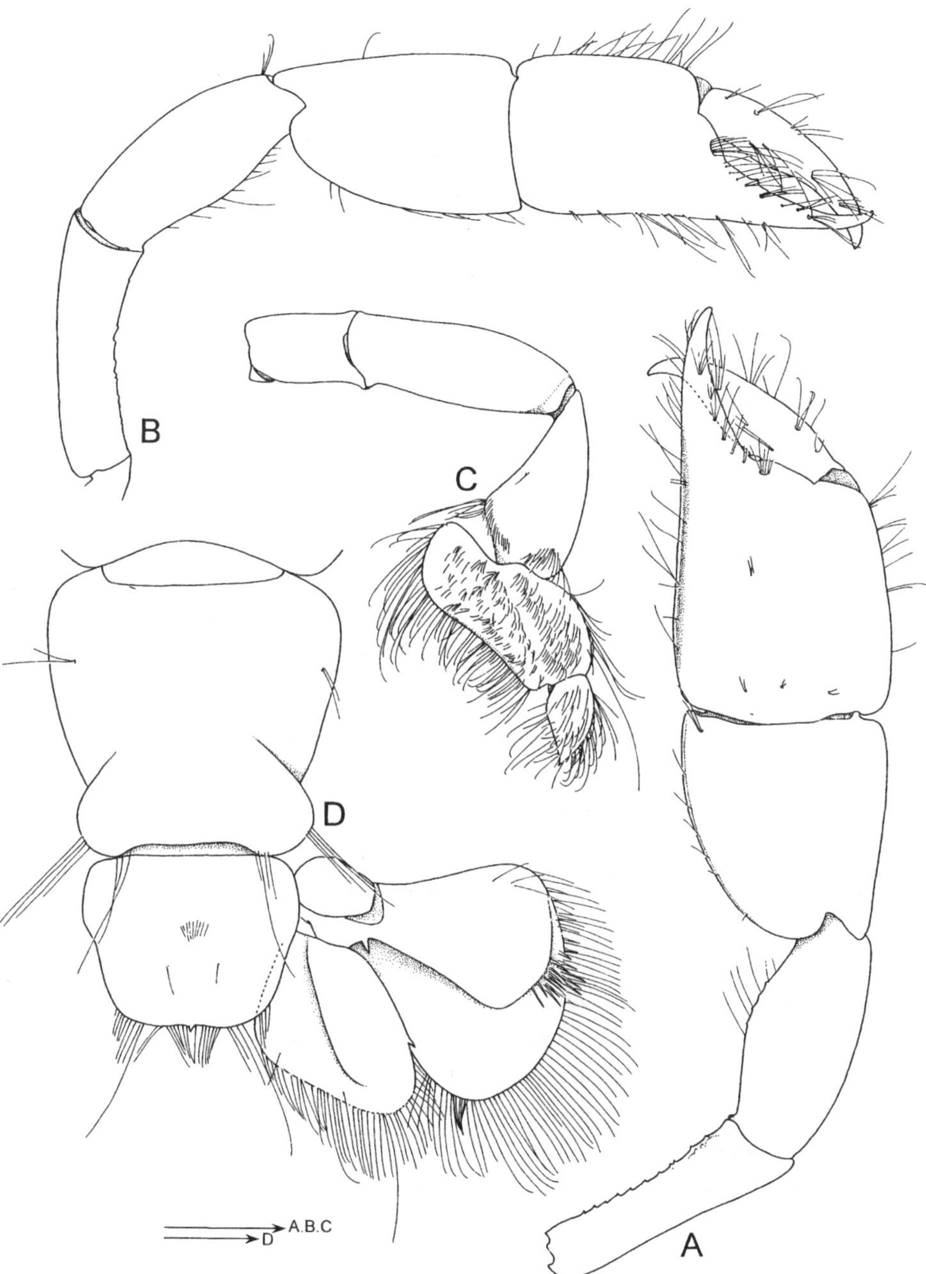

Fig. 35. *Neocallichirus mortenseni* sp. nov. A, larger cheliped, lateral view; B, smaller cheliped, lateral view; C, pereiopod 3, lateral view; D, abdominal somite 6 and telson, with uropod on right side. A-D, ZMUC 210, holotype female, Taboga, Panama, shore on small island off the northern coast. Scales 1 mm.

long as carpus; palm subsquare, about 1.1 times as long as high; distal gap convex and unarmed, incurved downward to tip of fixed finger, fixed finger half as long as palm, prehensile margin entirely concave and armed with a triangular denticle at proximal fourth; dactylus slender, slightly longer than palm, slightly longer than fixed finger, prehensile margin entirely concave in distal half and unarmed. P2 chelate, ischium about as long as broad; merus 3.0 times as long as high, and 3.5 times as long as ischium; carpus 2.0 times as long as high and 0.7 times as long as merus, chela slightly shorter than carpus, dactylus about 2.0 times as long as palm; missing on right side. P2 chelate, missing on right side. P3 (fig. 35C) ischium 1.3 times as long as broad; merus 2.5 times as long as high and 2.3 times as long as ischium; carpus 1.8 times as long as broad and 0.8 times as long as merus, divergent on dorsal and ventral margins; propodus kidney-shaped, 1.2 times as long as high, posteroventral margin entirely protruded posteriorly in triangular form, lateral surface setose, ventral margin broadened, slightly concave, and setose; dactylus triangular in shape, slightly longer than palm, convergent distally on dorsal and ventral margins. P4 (fig. 34A) subchelate, ischium 2.5 times as long as broad; merus 3.5 times as long as high and 1.5 times as long as ischium; carpus about 0.8 times as long as merus; subchela about as long as carpus, palm 0.7 times as long as carpus and 2.0 times as long as high, distoventral corner triangularly protruded, forming a subchela with dactylus; dactylus triangular, 0.5 times as long as propodus. P5 chelate; ischium as long as broad; merus 5.0 times as long as broad, carpus 0.7 times as long as merus; propodus slightly shorter than carpus, protruded on distoventral margin, forming a chela with dactylus. Dactylus deflected.

Abdominal somites smooth, glabrous dorsally; pleura 2-5 each with a tuft of setae laterally. Abdominal somite 6 (fig. 35D) about as long as broad, convergent posteriorly to posterior fourth on lateral margins. Plp1 (fig. 36A) uniramous and narrowly 2-segmented. Plp2 (fig. 36B) biramous, endopod with an appendix interna. Plp3 (fig. 36C) to Plp5 biramous, slender, foliaceous, each bearing an appendix interna. Telson (fig. 35D) trapezoid, lateral margins slightly convex proximally and then convergent posteriorly to posterolateral angle, posterior margin entirely convex, setose, and with a median spine; dorsal surface with a transverse row of setae in the middle. Uropodal endopod broadened in trapezoid form; dorsal surface with a median longitudinal carina. Uropodal exopod broadly rounded distally, almost as long as broad, and larger than endopod; dorsal surface with a longitudinal ridge running to secondary setose lobe in anterodistal half.

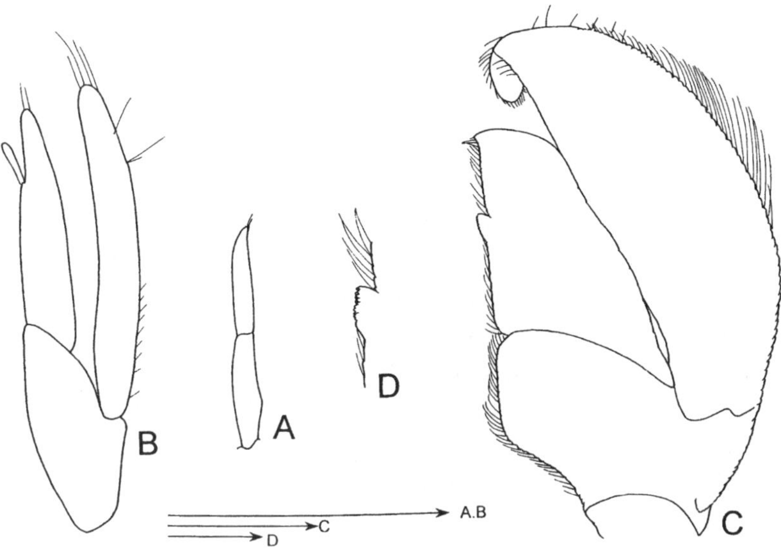

Fig. 36. *Neocallichirus mortenseni* sp. nov. A, male Plp1; B, male Plp2; C, Plp3; D, fused appendix interna on Plp3. A-D, ZMUC 210, holotype female, Taboga, Panama, shore on small island off the northern coast. Scales 1 mm.

Etymology. — The species name, *mortenseni*, is dedicated to the famous Danish scientist, Theodor Mortensen, in memory of his exploration of the eastern Pacific coasts of Panama and of many other locations besides South Africa, Thailand, Japan, and Hawaii, to mention only some remarkable trips and collecting for the era around the turn of the 19th to the 20th century. The specific name thus is a noun in the genitive singular.

Remarks. — *Neocallichirus mortenseni* is the first species of the genus *Neocallichirus* reported from the East Pacific region and it is similar to the Indian Ocean species, *N. manningi* Kazmi & Kazmi, 1992 from Pakistan. Sakai (1999: 99) erroneously included *N. manningi* in *Callianassa indica* De Man, 1905. *N. manningi* is different from *N. indicus* in the form of the telson, as Kazmi & Kazmi (1992) correctly reported. *N. mortenseni* is also different from *N. manningi*, collected from a different locality, i.e., Karachi, in the eye bearing no denticles distally.

Indo-West Pacific species

Neocallichirus angelikae Sakai, 2000

Neocallichirus angelikae Sakai, 2000: 91, figs. 1-3; Davie, 2002: 461.

Diagnosis. — Mxp3 ischium-merus subrectangular, merus obliquely declined distally and rounded on distomesial margin, male Plp1 blade-shaped, uniramous, two-segmented, distal segment slightly bilobed distally, male Plp2 with distal lobe of appendix masculina clearly separated and laterally with minute appendix interna with minute hook, telson subrectangular, posterior margin narrow and straight, with a median spine (Sakai, 2000: 91, figs. 1E, 2A, 3G, H, I).

Type locality. — Murat Bay, Great Australian Bight, South Australia, stony flat (32°07'S 133°40'E).

Distribution. — Murat Bay, Great Australian Bight, South Australia (32°07'S 133°40'E).

Neocallichirus audax (De Man, 1911)

Callianassa audax De Man, 1911: 223; Dworschak, 1992: 190, fig. 1a-d; Tudge et al., 2000: 143.
Callianassa (Callichirus) audax; De Man, 1928b: 1, 18, 28, 113, 179, pl. 20 fig. 31-31i; Rao & Kartha, 1967: 279, figs. 1-2; ?Tirmizi, 1967: 151-154, figs. 1-2.
Callichirus audax; De Saint Laurent & Le Loeuff, 1979: 97.
Neocallichirus audax; Sakai, 1999c: 95, fig. 21d-f.

Diagnosis. — Mxp3 ischium-merus subrectangular, merus obliquely declined distally and rounded on distomesial margin, male Plp1 blade-shaped, uniramous, two-segmented, distal segment bilobed distally, separated longitudinally by indistinct suture, male Plp2 distally with a subrectangular lobe of appendix masculina bearing long setae distally, appendix interna uncertain, telson subrectangular, posterior margin weakly trilobed, lacking a median spine (Tirmizi, 1967: 151-154, figs. 1B, F, 2D, E).

Type locality. — Straits of Malacca.

Distribution. — Straits of Malacca, Malay Peninsula; West Pakistan; Ratnagiri, India; east coast of India.

Neocallichirus caechabitator Sakai, 1988

Neocallichirus caechabitator Sakai, 1988: 67, figs. 9, 10; Sakai, 1999c: 96; Tudge et al., 2000: 144; Davie, 2002: 461.

Diagnosis. — Mxp3 ischium-merus subrectangular, merus obliquely declined distally and rounded on distomesial margin, male Plp1-2 unknown, telson subovoid, posterior margin narrow, straight and medially concave, lacking a median spine (Sakai, 1988: 67, figs. 9E, 10A).

Type locality. — Shoal Bay, False Creek Point, Darwin, Northern Territory, Australia, 1.0 m.

Distribution. — Shoal Bay, False Creek Point, Darwin, Northern Territory, Australia.

Neocallichirus calmani (Nobili, 1904)

Callianassa (*Cheramus*) *Calmani* Nobili, 1904: 236; Nobili, 1906b: 104, pl. 5 fig. 2; De Man, 1928b: 26, 100.
Callichirus calmani; De Saint Laurent & Le Loeuff, 1979: 97.
Callianassa calmani; Dworschak, 1992: 192, fig. 3a-f; Tudge et al., 2000: 143.
Neocallichirus calmani; Sakai, 1999c: 96, fig. 22a-d.

Diagnosis. — Mxp3 ischium-merus subrectangular, merus obliquely declined distally and rounded on distomesial margin, male Plp1 uniramous, two-segmented, bilobed distally, male Plp2 biramous, endopod trilobed distally, median lobe of appendix masculina large, with long setae and laterally fused with appendix interna bearing a few hooks, telson trapezoid, posterior margin convex, lacking a median spine (Dworschak, 1992, fig. 3c; Sakai, 1999c: 97, fig. 22b, c, d).

Type locality. — Obock, Djibouti.

Distribution. — Obock, Djibouti, Bay of Aden, in the sandy intertidal and shallow subtidal; Aqaba, Red Sea.

Neocallichirus darwinensis Sakai, 1988

Neocallichirus darwinensis Sakai, 1988: 62, figs. 5, 6; Sakai, 1999c: 98; Tudge et al., 2000: 144; Davie, 2002: 461.

Diagnosis. — Mxp3 ischium-merus subrectangular, merus obliquely declined distally and convex on distomesial margin, male Plp1-2 unknown, telson trapezoid, posterior margin convex, lacking a median spine (Sakai, 1988, figs. 5G, 6G).

Type locality. — Mindil Beach, Darwin, Northern Territory, Australia.

Distribution. — Mindil Beach, Darwin, Northern Territory, Australia.

Neocallichirus denticulatus Ngoc-Ho, 1994

Neocallichirus denticulatus Ngoc-Ho, 1994: 56, fig. 4; Sakai, 1999c: 98; Tudge et al., 2000: 144; Davie, 2002: 461.

Diagnosis. — Mxp3 ischium-merus subrectangular, merus obliquely declined distally and convex on distomesial margin, male Plp1-2 unknown, telson trapezoid, posterior margin convex, lacking a median spine (Ngoc-Ho, 1994: 56, fig. 4h, m).

Type locality. — Queensland, N. W. of Townsville (18°56'S 146°50'E), 24 m, muddy sand.

Distribution. — N. W. of Townsville, Queensland, Australia.

Neocallichirus horneri Sakai, 1988

Neocallichirus horneri Sakai, 1988: 65, figs. 7, 8; Manning & Felder, 1991: 779; Sakai, 1999c: 99; Tudge et al., 2000: 144; Davie, 2002: 461.

Diagnosis. — Mxp3 ischium-merus subrectangular, merus obliquely declined distally and shortened on distomesial margin, male Plp1 uniramous, two-segmented, bilobed distally, male Plp2 biramous, endopod with distally demarcated appendix masculina bearing distal setae, appendix interna uncertain, telson trapezoid, posterior margin convex, lacking a median spine (Sakai, 1988: 65, figs. 7F, 8A, H, I).

Type locality. — Australia, Northern Territory, Darwin, Nightcliff.

Distribution. — Nightcliff and West Shoal Bay, Darwin, Northern Territory, Australia; intertidal.

Neocallichirus indicus (De Man, 1905)

Callianassa (*Cheramus*) *indica* De Man, 1905: 605; De Man, 1928b: 26, 93, 100, 159, 160, pl. 17 fig. 26-26g.

Callianassa (*Cheramus*) *variabilis* Edmondson, 1944: 47, figs. 1, 6a-i, 7a-j, l, p. [Type locality; Hanauma Bay, Oahu, in a gravel bed of the intertidal zone.]

Callianassa natalensis Barnard, 1946: 379; Barnard, 1950: 506, 511, fig. 95f-h; Kensley, 1974: 277. [Type locality: Zululand, South Africa.]
Callianassa indica; Kensley, 1975: 50, fig. 2A-E; Sakai, 1987a: 302, 306.
Neocallichirus taiaro Ngoc-Ho, 1995: 212, figs. 1-2. [Type locality: Taiaro atoll, Tuamotu Isle, French Polynesia.]
Neocallichirus indicus; Sakai, 1999c: 99, fig. 23a-e; Sakai & Apel, 2002: 277, fig. 2; Tudge et al., 2000: 144.
Callianassa variabilis; Tudge et al., 2000: 143.
Neocallichirus natalensis; Tudge et al., 2000: 144.
Neocallichirus taiaro; Tudge et al., 2000: 144.

Diagnosis. — Mxp3 ischium-merus subrectangular, merus obliquely declined distally and convex on distomesial margin, male Plp1 uniramous, two-segmented, bilobed distally, male Plp2 biramous, endopod with distally demarcated appendix masculina bearing distal setae and laterally with appendix interna, telson subovoid, posterior margin convex, lacking a median spine (De Man, 1928b, pl. 17 fig. 26c, d, f; Sakai, 1999c: 99, fig. 23b, d, e).

Type locality. — Indonesia, south coast of Kangeang, Bay of Kankamaran (6°59'S 115°24.7'E).

Distribution. — Hanauma Bay, Oahu, Hawaii; Tuamotu Isle, French Polynesia; Tonaki Is., Ryukyu Is.; Flores, Indonesia; Mauritius; Kangeang Reef, Bay of Kankamaran, Indonesia; Sandspit, Karachi; Djibouti, Gulf of Aden; Red Sea; Zululand, South Africa.

Neocallichirus jousseaumei (Nobili, 1904)

Callianassa (Cheramus) Jousseaumei Nobili, 1904: 236; Nobili, 1906b: 101, pl. 6 fig. 2; De Man, 1928b: 26 (list), 100 (key), pl. 18 fig. 27-27a.
Callichirus jousseaumei; De Saint Laurent & Le Loeuff, 1979: 97.
Callianassa jousseaumei; Dworschak, 1992: 198, figs. 5a-d, 6a-c.
Neocallichirus jousseaumei; Sakai, 1999c: 100, fig. 22e-g; Tudge et al., 2000: 144.

Diagnosis. — Mxp3 ischium-merus subrectangular, merus obliquely declined distally and convex on distomesial margin, male Plp1 uniramous, two-segmented, chelate distally, male Plp2 biramous, endopod distally simple with appendix masculina bearing distal setae and with not-demarcated appendix interna, telson trapezoid, posterior margin convex, lacking a median spine (De Man, 1928b, pl. 18 fig. 27a, c; Sakai, 1999c: 100, fig. 22f, g).

Type locality. — Djibouti, Perim, Gulf of Tadjourah, Red Sea.

Distribution. — Djibouti and Perim, Gulf of Aden; Safaga, Tubaya Al-Kabir, Red Sea; Aqaba, Gulf of Aqaba, 9 m.

Neocallichirus karumba (Poore & Griffin, 1979)

Callianassa karumba Poore & Griffin, 1979: 266, figs. 30-31.
Glypturus karumba; Sakai, 1988: 61; Tudge et al., 2000: 144; Davie, 2002: 460.
Neocallichirus karumba; Sakai, 1999c: 101.

Diagnosis. — Mxp3 ischium-merus subrectangular, merus obliquely declined distally and rounded on distomesial margin, male Plp1 uniramous, two-segmented, chelate distally, male Plp2 biramous, endopod with distally demarcated appendix masculina and laterally attached a narrow appendix interna, telson subrectangular, posterior margin convex, lacking a median spine (Poore & Griffin, 1979: 266, figs. 30c, 31f, g, h).

Type locality. — Norman River, Karumba, Queensland, Australia.

Distribution. — Norman River, Karumba, Queensland, Australia.

Neocallichirus kempi Sakai, 1999

Callianassa (*Callichirus*) *maxima*; Kemp, 1915: 252, pl. 13 figs. 1-5; De Man, 1928b: 29, 92, 112. [Not *Callianassa maxima* A. Milne-Edwards, 1870.]
Callianassa maxima; Pillai, 1954: 23-26; Tudge et al., 2000: 143.
Neocallichirus kempi Sakai, 1999c: 101, fig. 24a-e.

Diagnosis. — Mxp3 ischium-merus subrectangular, merus obliquely declined distally and rounded on distomesial margin, male Plp1-2 uncertain, telson subovoid, posterior margin laterally convergent toward median portion, lacking a median spine (Kemp, 1915: 252, pl. 13 fig. 5; Sakai, 1999c: 101, fig. 24b, c).

Type locality. — Bangka, Indonesia.

Distribution. — Bangka, Indonesia; Nalbano Island and Barhampur Island, Chilka Lake near Madras, India; Kayamkulam Lake, central Travancore, India.

Neocallichirus manningi Kazmi & Kazmi, 1992

Neocallichirus manningi Kazmi & Kazmi, 1992: 296, fig. 1; Tudge et al., 2000: 144.
Neocallichirus indicus; Sakai, 1999c: 99; Sakai, 2000: 277.

Diagnosis. — Mxp3 ischium-merus subrectangular, merus obliquely declined distally and rounded on distomesial margin, male Plp1-2 unknown, telson trapezoid, posterior margin slightly convex, lacking a median spine (Kazmi & Kazmi, 1992: 296, fig. 1D, C).

Remarks. — This species was erroneousely included in *N. indicus* by Sakai (1999c), but it is different from *N. indicus* in the form of the telson. In *N. manningi*, the telson is trapezoid, whereas in *N. indicus* the lateral margins are convergent posteriorly, to a narrow posterior margin. Unfortunately, the type specimen of *N. manningi* is a female, so it is difficult to compare the male chelipeds of both species until further material will become available.

Type locality. — Sandspit, Karachi, 24°48'N 66°58'E, lower tidal region.

Distribution. — Known only from the type locality.

Neocallichirus mauritianus (Miers, 1882)

Callianassa mauritiana Miers, 1882: 341; Miers, 1884: 15, pl. 1 fig 2; Nobili, 1906b: 106, figs. 5, 6; Michel, 1974: 256; Kensley, 1975: 51, fig. 3A-H; Tudge et al., 2000: 143.
Callianassa (Trypaea) mauritiana; Borradaile, 1903: 546.
Callianassa (Cheramus) mauritiana; De Man, 1928a: 10, pl. 2 fig. 4; De Man, 1928b: 26, 99, 160.
Neocallichirus mauritianus; Sakai, 1999c: 103, fig. 21a-c.

Diagnosis. — Mxp3 ischium-merus subrectangular, merus obliquely declined distally and rounded on distomesial margin, male Plp1 uniramous, two-segmented, and with distally demarcated chela, male Plp2 biramous, endopod with distally demarcated appendix masculina, laterally fused with an appendix interna bearing minute hooks, telson subovoid, posterior margin convex, with a median spine (Kensley, 1975: 51, fig. 3B, C, H; Sakai, 1999c: 103, fig. 21b, c).

Type locality. — Mauritius.

Distribution. — Mauritius; Red Sea.

Neocallichirus moluccensis (De Man, 1905)

Callianassa (Cheramus) moluccensis De Man, 1905: 606; De Man, 1928b: 26, 93, 99, 159, pl. 16 fig. 25-25a, pl. 17 fig. 25b-c.
Neocallichirus moluccensis; Sakai, 1999c: 104, fig. 25a-f; Tudge et al., 2000: 144.

Diagnosis. — Mxp3 ischium-merus subrectangular, merus obliquely declined distally and convex on distomesial margin, male Plp1 uniramous, two-segmented, chelate distally, male Plp2 biramous, endopod with distally demarcated appendix masculina, and laterally attached, protruded appendix interna with minute hooks, telson trapezoid, posterior margin convex, lacking a median spine (De Man, 1928b, pl. 17 fig. 25b-c; Sakai, 1999c: 104, fig. 25c, d, e).

Type locality. — Amboina, reef.

Distribution. — Amboina, Indonesia.

Neocallichirus motupore (Poore & Suchanek, 1988)

Glypturus motupore Poore & Suchanek, 1988: 198, figs. 1-3, 4a, tab. 1; Tudge et al., 2000: 144.
Neocallichirus motupore; Sakai, 1999c: 105.

Diagnosis. — Mxp3 ischium-merus subrectangular, merus obliquely declined distally and rounded on distomesial margin, male Plp1 uniramous, two-
segmented, chelate distally, male Plp2 biramous, endopod distally trilobed,
median lobe of appendix masculina and laterally attached, narrow appendix interna, telson subrectangular, posterior margin rounded, lacking a median spine
(Poore & Suchanek, 1988: 198, figs. 1e, 3f, g, h, i).

Type locality. — Bootless Inlet (9°32'S 147°16'E), Motupore Is., Papua
New Guinea; intertidal.

Distribution. — Bootless Inlet, Motupore Is., Papua New Guinea.

Neocallichirus mucronatus (Strahl, 1862)

Callianassa mucronata Strahl, 1862a: 1056; Strahl, 1862b: 383; A. Milne-Edwards, 1870: 94;
    De Man, 1888: 484, pl. 21 fig. 2; Ortmann, 1891: 57; Ortmann, 1894: 23; Estampador, 1937:
    499; Tirmizi, 1977: 21 (partim), figs. 1-3 (not fig. 1b = *Callianassa gruneri* Sakai, 1999c);
    Poore & Griffin, 1979: 273, figs. 34, 35; Dworschak, 1992: 202, fig. 9a-f.
Callianassa brevicaudata A. Milne-Edwards, 1870: 91, 101, pl. 2 fig. 2, 2a, 2b; Tudge et al.,
    2000: 143. [Type locality: Zanzibar.]
Callianassa (*Cheramus*) *novaeguineae* Thallwitz, 1891: 31, pl. 1 fig. 9. [Type locality: Papua
    New Guinea.]
Callianassa novaeguineae; De Man, 1902: 757.
Callianassa (*Callichirus*) *novae-guineae*; Borradaile, 1903: 547.
Callianassa (*Callichirus*) *brevicaudata*; Borradaile, 1903: 547; De Man, 1928b: 28, 115.
Callianassa (*Callichirus*) *mucronata*; Borradaile, 1903: 547; Nobili, 1906b: 108; De Man,
    1928b: 112, 175, pl. 19 fig. 30-30e.
Glypturus mucronatus; Sakai, 1988: 61; Tudge et al., 2000: 144.
Neocallichirus mucronatus; Sakai, 1999c: 105, fig. 26a-i; Sakai & Apel, 2002: 278.
Glypturus novaeguineae; Davie, 2002: 461.

Diagnosis. — Mxp3 ischium-merus subrectangular, merus obliquely declined distally and convex on distomesial margin, male Plp1 uniramous, two-
segmented, chelate distally, male Plp2 biramous, endopod with distally demarcated appendix masculina, appendix interna uncertain, telson subrectangular,

posterior margin convex, lacking a median spine (Poore & Griffin, 1979: 273, fig. 34c, e, f, g).

Type locality. — Luzon, Philippines.

Distribution. — Papua New Guinea; Queensland, Australia, central reef; Luzon, Philippines; Maldives; Amboina; N. Ceram; northwest coast of New Guinea; Siau Island and Karakelong Island, Indonesia; Djibouti and Perim, Gulf of Aden; Zanzibar; Madagascar, Baie de Pasandava.

Neocallichirus ranongensis (Sakai, 1983)

Callianassa ranongensis Sakai, 1983: 111-115, figs. 1-2; Sakai, 1987b: 45, figs. 1-2.
Neocallichirus ranongensis; Sakai, 1999c: 107.
Lepidophthalmus ranongensis; Tudge et al., 2000: 144.

Diagnosis. — Mxp3 ischium-merus subrectangular, merus obliquely declined distally and concave on distomesial margin, male Plp1 uniramous, two-segmented, chelate distally, male Plp2 biramous, endopod with distally demarcated appendix masculina and laterally attached, small appendix interna, telson subrectangular, posterior margin convex, lacking a median spine (Sakai, 1987b: 45, figs. 1E, F, J, 2A).

Type locality. — Hatsaikhao, Ranong Province, Thailand, muddy mangrove swamp.

Distribution. — Ranong Province, Thailand; Halmahera, Indonesia.

Neocallichirus vigilax (De Man, 1916)

Callianassa (Callichirus) vigilax De Man, 1916: 57, pl. 1 figs. 1-6; De Man, 1928b: 30, 93, 109.
Neocallichirus vigilax; Sakai, 1999c: 108.

Diagnosis. — Mxp3 ischium-merus subrectangular, merus obliquely declined distally and rounded on distomesial margin, male Plp1-2 unknown, telson subtrapezoid, posterior margin convex, lacking a median spine (De Man, 1916: 57, pl. 1 figs. 1-3, 4).

Type locality. — Ambon (= Amboina), Indonesia.

Distribution. — Ambon (= Amboina), Indonesia.

Neocallichirus sp. (Rathbun, 1906)

Callianassa sp. Rathbun, 1906: 893, fig. 48.
Callianassa (*Trypaea*) sp.; De Man, 1928b: 28.
Neocallichirus sp.; Sakai, 1999c: 108.

Diagnosis. — Mxp3 ischium-merus subrectangular, merus obliquely declined distally and rounded on distomesial margin, male Plp1-2 unknown, telson unknown (Rathbun, 1906: 893, fig. 48).

Distribution. — Honolulu, Hawaii.

Species of unknown locality

Neocallichirus auchenorhynchus sp. nov.
(fig. 37)

Material examined. — ZMUC CRU-3866, 1 female holotype (Tl/Cl 20.0/4.4), no data.

Diagnosis. — Frontal margin of carapace showing a neck-like form, with triangular rostrum and a pair of anterolateral projections. Eyestalks truncate distally, protruded distally as a flat distomedial lobe, cornea rounded, located medially. Mxp3 ischium-merus subpediform, merus obliquely declined on distomesial margin and rounded mesially, propodus broadened, rounded on ventral margin. Male Plp1-2 unknown. Telson subsquare, as long as broad, distal margin truncate, not armed with a median spine.

Description of female holotype. — Rostrum (fig. 37A-C) triangular; frontal margin of carapace with anterolateral projections; dorsal oval conspicuous; linea thalassinica present at full length; cervical groove located approx. in posterior fifth of carapace. Eyestalks (fig. 37A-C) truncate distally, longer than broad, touching dorsally with rostrum, and protruding distally as a flat distomedial lobe, almost reaching distal end of antennular basal segment; cornea rounded, located medially, pigments vanished in alcohol. A1 peduncle slightly shorter than A2 peduncle, terminal segment about as long as penultimate. A2 terminal segment slightly shorter than penultimate segment, scaphocerite small; antennal flagellum about 1.5 times as long as antennal peduncle. Mxp3 ischium-merus (fig. 37D-E) subpediform, three times as long as broad; ischium 2.2 times as long as broad; crista dentata with a sparse row of 13 denticles; merus obliquely declined on distomesial margin and rounded mesially; carpus triangular, 1.7 times as long as broad; propodus broadened, slightly longer than broad, rounded on ventral margin; dactylus digitiform, 0.8 times as long as propodus and with a strong seta distally; no exopod present.

P1 unequal and dissimilar. Larger cheliped absent. Smaller cheliped (fig. 37F) with ischium slender, dorsal margin slightly incurved distally and unarmed, ventral margin entirely with minute denticles; merus oblong, 0.8 times as long as ischium, 2.2 times as long as high, dorsal and ventral margins slightly arched and smooth. Carpus broadened, 1.5 times as long as high and 1.1 times as long as merus; posteroventral margin largely divergent. Chela 1.7 times as long as carpus; palm 0.7 times as long as carpus, about as long as high, dorsal and ventral margins smooth, distal gap unarmed, declined to tip of fixed finger; fixed finger about as long as palm, entirely concave on prehensile margin; dactylus narrow, prehensile margin armed with a triangular tooth at proximal fourth and two denticles about in the middle. P2 chelate, ischium slightly longer than broad; merus three times as long as high and 2.8 times as long as ischium; carpus 1.8 times as long as high and 0.6 times as long as carpus, chela slightly longer than carpus and dactylus about 2.0 times as long as palm. P3 absent on both sides. P4 subchelate, ischium 2.8 times as long as broad; merus three times as long as high and 1.5 times as long as ischium; carpus about 0.8 times as long as merus; propodus rectangular, 0.7 times as long as carpus and 2.3 times as long as high, distoventral corner densely setose and protruded; dactylus slightly less than half length of propodus. P5 chelate; propodus protruded on distodorsal margin, forming a chela with a fixed finger distoventally, ventral surface with dense setation; dactylus hooked towards external side of fixed finger, tip deflected.

Abdominal somite 1 with anterior tergum; abdominal pleura 2-5 each with a tuft of setae laterally; abdominal somite 6 about as long as broad, distinctly concave at posterior fourth on lateral margins. Plp1 uniramous and 2-segmented. Plp2 biramous, slender, with an appendix interna distally. Plp3-5 biramous, slenderly foliaceous, each bearing a small appendix interna fusing distally on mesial margin of endopod. Telson (fig. 37G) subsquare, as long as broad, lateral margins slightly convergent distally to rounded posterolateral corners, distal margin truncate, not armed with a median spine, dorsal surface with a row of setae medioproximally. Uropodal endopod broadened distally; uropodal exopod broadened, broadly rounded distally, anterior surface distinguished from posterior by longitudinal median carina, anterior margin slightly concave.

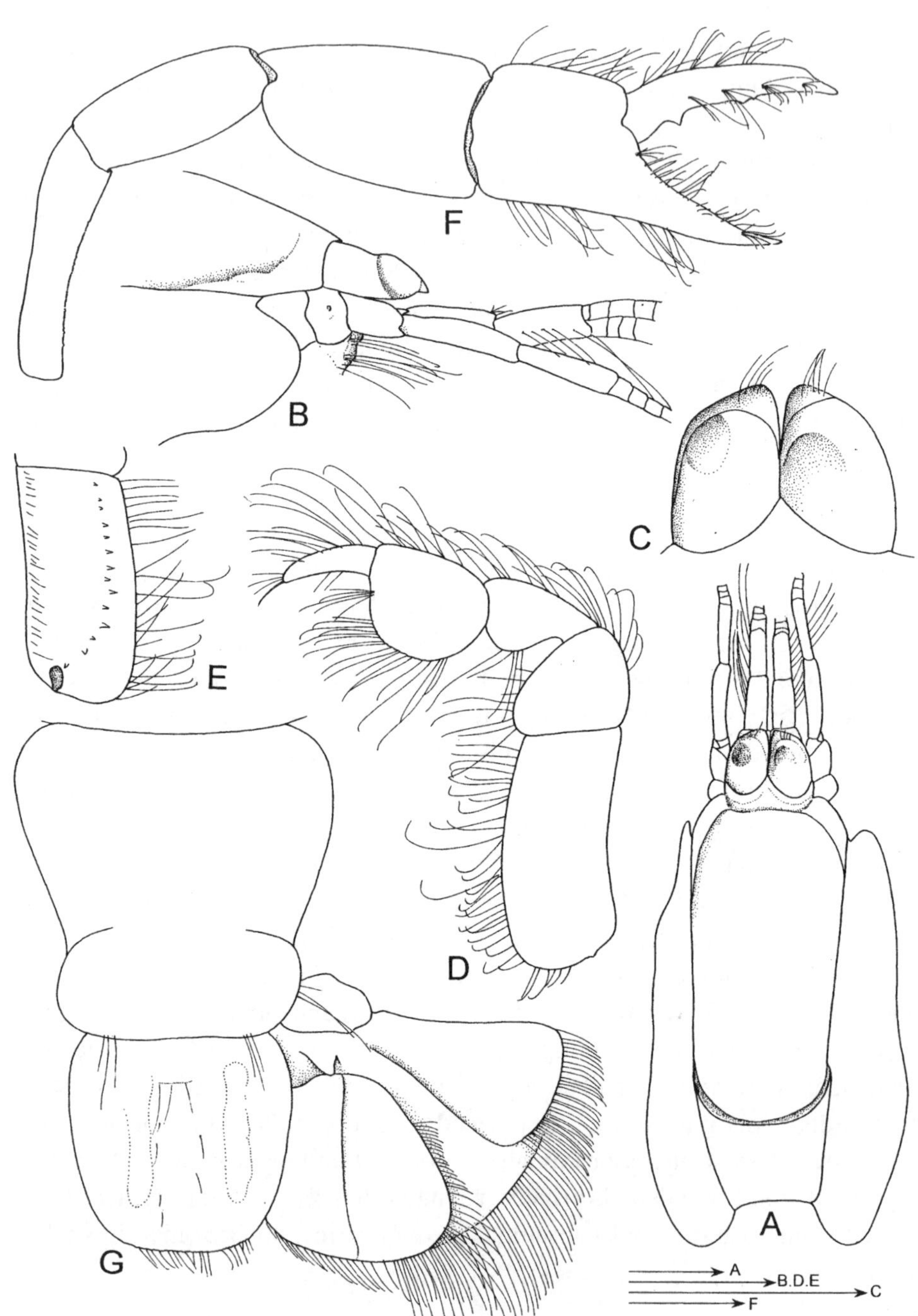

Remarks. — The present single specimen is characteristic in the form of the neck-like structure in the rostrum and a pair of anterolateral projections. This species is very similar to *Neocallichirus karumba* (Poore & Griffin, 1979) from Queensland, Australia, in that the eyes bear a rounded lobe mediodistally and the A1 peduncles are slightly shorter than those of A2. However, it differs as follows: in *N. karumba*, the anterolateral projections on the frontal margin of the carapace are not developed and the uropodal endopod is lanceolate, whereas in the present species a pair of anterolateral projections is remarkable, and the uropodal endopod is broadened.

Etymology. — The specific name derives from Greek "auchenos" (= neck) and "rhynchos" (= nose, snout, muzzle), and alludes to the neck-like structure of the rostrum. Consequently, it is a noun in apposition to the generic name.

Type locality. — Unknown.

Distribution. — Unknown.

Genus **Podocallichirus** Sakai, 1999

Podocallichirus Sakai, 1999c: 53.
Grynaminna Poore, 2000: 150 (new synonymy).

Definition. — Carapace with dorsal oval; rostral spine present or absent; rostral carina, hepatic sulcus, and prominence absent. A1 peduncle longer than A2 peduncle. Mxp3 ischium-merus pediform, keeping the same width from proximal to distal, propodus broadened, dactylus digitiform; exopod absent. P1 chelate, unequal, male larger cheliped with or without meral hook. P2 chelate. P3 simple. P4 subchelate. P5 chelate. Male Plp1 uniramous, uni- or biarticulate, distal segment chelate distally, or simple; male Plp2 uniramous, uni- or biarticulate, or biramous with appendix interna and masculina. Female Plp1 uni- or biramous, 3-segmented; female Plp2 uni- or biramous, with or without appendix interna. Uropodal endopod longer than wide and ovoid or rhombic in shape. Uropodal exopod with secondary setal lobe.

Remarks. — Poore (2000) established the genus *Grynaminna* and pointed out that it is different from *Callichirus* in possessing an elongate-oval uropodal

Fig. 37. *Neocallichirus auchenorhynchus* sp. nov. A, carapace with eyes, antennular and antennal peduncles, dorsal view; B, anterior part of carapace, lateral view; C, eyestalks, dorsal view; D, Mxp3, lateral view; E, ischium of Mxp3, mesial view; F, pereiopod 1, probably smaller, lateral view; G, abdominal somite 6 and telson, with uropod on right side, dorsal view. A-G, ZMUC 117, holotype female, no data. Scales 1 mm.

endopod and in having neither a meral hook on the major cheliped, nor symmetrical dorsal sculpture on abdominal somites 3-5, as in *Callichirus*. However, there is no justification in his references to establish the genus *Grynaminna*, because he neglected *Podocallichirus* Sakai, 1999, and paid no attention to the symmetrical dorsal sculpture on abdominal somites 3-5, which is found only in *Callichirus*, but not found in any other genera of the Callianassidae except in *Michaelcallianassa*. *Grynaminna* can be defined as a junior synonym of *Podocallichirus* Sakai, 1999, because Mxp3 is not fundamentally different between *Grynaminna* and *Podocallichirus*: in *Grynaminna*, the Mxp3 ischium-merus is oblong, the propodus is distally broadened, and the dactylus is digiform as in *Podocallichirus*. *Grynaminna* is, however, characteristic in having short antennular flagella (fig. 39), and in the Mxp3 merus bearing a produced mesiodistal angle (fig. 40A, B), but those characters are only sufficient to be considered as characters at the species level.

In the present reference, it has become clear that Poore (2000: 151) poorly defined the genus *Grynaminna* as "Rostral spine absent". However, the rostrum and the anterolateral spines of the carapace, though weakly developed, are definitely present as low, transparent projections (fig. 39A, B), as in *G. guineensis* (De Man, 1928), and in *G. foresti* (Le Loeuff & Intès, 1974). He further defined "Telson is wider than long", but it is, in fact, not so simply wide as seen in his figure (Poore, 2000: 155, fig. 1A), but rather hexagonal, with convex lateral margins, when observed right from above (fig. 40C). *Grynaminna* is more similar to *Lepidophthalmus* than to *Callichirus*, because the A1 peduncle is much longer than the A2 peduncle (fig. 39B), and the symmetrical dorsal sculpture on abdominal somites 3-5 is absent. However, in *Lepidophthalmus* the larger cheliped is provided with a meral hook, though in *Glypturus*, i.e., *G. acanthochirus* Stimpson, 1866 and *G. collaroy* (Poore & Griffin, 1979), the meral hook is absent as in *Grynaminna*. An elongate oval uropodal endopod (fig. 40C) is found in *Lepidophthalmus jamaicense* (Schmitt, 1935), *L. siriboia* Felder & Rodrigues, 1993, *L. rosae* (Nobili, 1904), and *Glypturus acanthochirus* Stimpson, 1866, as in *Grynaminna tamakii* Poore 2000. Regarding the form of the appendix interna in Plp2 in both sexes, *Grynaminna* resembles *Lepidophthalmus*: *L. jamaicense* (Schmitt, 1935), *L. bocourti* (A. Milne-Edwards, 1870), and *L. socotrensis* Sakai & Apel, 2002 bear the digitiform appendix interna as in *Grynaminna tamakii*.

Type species. — *Callianassa madagassa* Lenz & Richters, 1881. The gender of the generic name, *Podocallichirus*, is masculine.

Species included. — East Atlantic and Mediterranean species: *P. balssi* (Monod, 1933); *P. foresti* (Le Loeuff & Intès, 1974); *P. guineensis* (De Man, 1928); *P. tenuimanus* (De Saint Laurent & Le Loeuff, 1979). Indo-West Pacific species: *P. gilchristi* (Barnard, 1946); *P. madagassus* (Lenz & Richters, 1881); *P. masoomi* (Tirmizi, 1970); *P. tamakii* (Poore, 2000).

East Atlantic and Mediterranean species

Podocallichirus balssi (Monod, 1933)

Callianassa (*Callichirus*) *balssi* Monod, 1933: 13, fig. 2A-F.
Callianassa balssi; Longhurst, 1958: 45 (partim); Sakai, 1999c: 54.
Callichirus balssi; Le Loeuff & Intès, 1974: 46, fig. 13a-o; De Saint Laurent & Le Loeuff, 1979: 58, figs. 11c-e, 12d, 13a-b, 14a, 15a-d, 19a; Tudge et al., 2000: 144.

Diagnosis. — Mxp3 ischium-merus pediform, merus convex on distomesial angle, male Plp1 uniramous, two-segmented, minutely pointed distally, male Plp2 uniramous, unsegmented but slightly contracted medially, telson subrectangular, posterior margin convex, slightly concave medially, lacking a median spine (Le Loeuff & Intès, 1974: 46, fig. 13g, m; De Saint Laurent & Le Loeuff, 1979: 58, figs. 12d, 15a-d, 19a).

Type locality. — Bay of Repos, Nouadhibou, Mauritania.

Distribution. — Bay of Repos, Nouadhibou, Mauritania; Dakar, Senegal; Gambia; Pointe-Noire, Congo; 6-10 m.

Podocallichirus foresti (Le Loeuff & Intès, 1974)

Callianassa guineensis; Longhurst, 1958: 31 (partim). [Not *Callianassa* (*Callichirus*) *guineensis* De Man, 1928.]
Callianassa balssi; Longhurst, 1958: 31, 45, 47 (partim). [Not *Callianassa* (*Callichirus*) *balssi* Monod, 1933.]
Callianassa pachydactyla; Longhurst, 1958: 42, 44. [Not *Callianassa pachydactyla* A. Milne-Edwards, 1870.]
Callichirus foresti Le Loeuff & Intès, 1974: 46, fig. 14a-x; De Saint Laurent & Le Loeuff, 1979: 58, figs. 11f-g, 12c, 13c-d, 14b, 15i-l, 19b; ?Tudge et al., 2000: 144.
Podocallichirus foresti; Sakai, 1999c: 55.

Diagnosis. — Mxp3 ischium-merus pediform, merus angulate on distomesial angle, male Plp1 uniramous, two-segmented, chelate distally, male Plp2 uniramous, two-segmented, telson trapezoid, posterior margin straight, slightly concave medially, lacking a median spine (Le Loeuff & Intès, 1974: 46, fig. 14m, s, t, u; De Saint Laurent & Le Loeuff, 1979: 58, figs. 12c, 15i-l, 19b).

Type locality. — Grand Lahou (5°05'N 5°04.05'W), Ivory coast, 22 m.

Distribution. — Senegal; Guinea; Sierra Leone; Ivory Coast; Pointe-Noire, Congo; 5-30 m.

Podocallichirus guineensis (De Man, 1928)

Callianassa (*Callichirus*) *guineensis* De Man, 1928a: 45, pl. 10, fig. 19-19e, pl. 11 fig. 19f-19r; De Man, 1928b: 28, 94, 114.
Callichirus guineensis; Le Loeuff & Intès, 1974: 46, fig. 12a-o; De Saint Laurent & Le Loeuff, 1979: 64, figs. 11h, 12e, 13g-h, 14d, 15m-p, 19d; Tudge et al., 2000: 144.
Podocallichirus guineensis; Sakai, 1999c: 55.

Not *Callianassa guineensis*; Longhurst, 1958: 31 (partim) (= *Podocallichirus foresti*; *Callianassa diaphora*; *C. marchali*).

Diagnosis. — Mxp3 ischium-merus pediform, merus angulate on distomesial angle, male Plp1 uniramous, unsegmented, male Plp2 uniramous, unsegmented, telson subovoid, posterior margin rounded, lacking a median spine (De Man, 1928a: 45, pl. 10 fig. 19a, b; De Saint Laurent & Le Loeuff, 1979: 64, figs. 12e, 15m, n; 19d).

Type locality. — Prampram, Ghana, 9-10 m.

Distribution. — Ghana, 9-15 m; Nigeria.

Podocallichirus tenuimanus (De Saint Laurent & Le Loeuff, 1979)

Callianassa balssi; Longhurst, 1958: 31 (partim). [Not *Callianassa balssi* Monod, 1933.]
Callichirus tenuimanus De Saint Laurent & Le Loeuff, 1979: 61, figs. 11a-b, 12a-b, 13e-f, 14c, 15e-h, 19c; Tudge et al., 2000: 144.
Podocallichirus tenuimanus; Sakai, 1999c: 55.

Diagnosis. — Mxp3 ischium-merus pediform, merus angulate on distomesial angle, male Plp1 uniramous, two-segmented, chelate distally, male Plp2 uniramous, unsegmented, telson subrectangular, posterior margin concave, lacking a median spine (De Saint Laurent & Le Loeuff, 1979: 61, figs. 12b, 15e-f, 19c).

Type locality. — Sierra Leone, 7 m.

Distribution. — Sierra Leone.

Indo-West Pacific species

Podocallichirus gilchristi (Barnard, 1946)

Callianassa gilchristi Barnard, 1946: 379; Barnard, 1950: 509, fig. 95a-e; Kensley, 1974: 277;
Tudge et al., 2000: 143.
Podocallichirus gilchristi; Sakai, 1999c: 56, fig. 9a-d.

Diagnosis. — Mxp3 ischium-merus pediform, merus angulate on distomesial angle, male Plp1 uniramous, two-segmented, bilobed distally, male Plp2 unknown, telson subrectangular, posterior margin straight, lacking a median spine (Barnard, 1950, fig. 95a; Sakai, 1999c: 56, fig. 9b).

Type locality. — False Bay and Durban Bay, South Africa.

Distribution. — Saldanha Bay, False Bay to Port Elizabeth; Durban Bay and off Zululand coast, South Africa; 37 m.

Podocallichirus madagassus (Lenz & Richters, 1881)
(fig. 38)

Callianassa madagassa Lenz & Richters, 1881: 427, figs. 20-23; Tudge et al., 2000: 143.
Callianassa (*Callichirus*) *madagassa*; Borradaile, 1903: 547; De Man, 1928a: 42, fig. 18-18e;
De Man, 1928b: 29, 92, 113.
Podocallichirus madagassus; Sakai, 1999c: 56, fig. 10a-d.

Material examined. — SMF 7938, holotype, 1 female (Tl/Cl 52.0/11.1), Madagascar, leg C. Ebenau, 1880; SMF 4953, 1 male (49.0/10.7), paratype, Madagascar, leg. C. Ebenau, 1880.

Diagnosis. — Mxp3 ischium-merus subpediform, merus convex on distomesial margin, male Plp1 uniramous, two-segmented, chelate distally, male Plp2 biramous, endopod with distally demarcated appendix masculina, laterally with small appendix interna bearing minute hooks, telson subrectangular, posterior margin slightly convex medially, lacking a median spine.

Remarks. — The types were re-examined. The A1 flagella are slightly shorter than the last segment of the A1 peduncle. The female Plp1-2 (fig. 38A-B) were examined, and it was found that their structure, as that of the A1 flagella, is very similar to those of *P. tamakii*. The Plp3 (fig. 38C, D) to Plp5 appendices internae are fused with the mesial margins of the endopods.

Type locality. — Nossi Bé, N. W. Madagascar.

Distribution. — Nossi Bé, N. W. Madagascar.

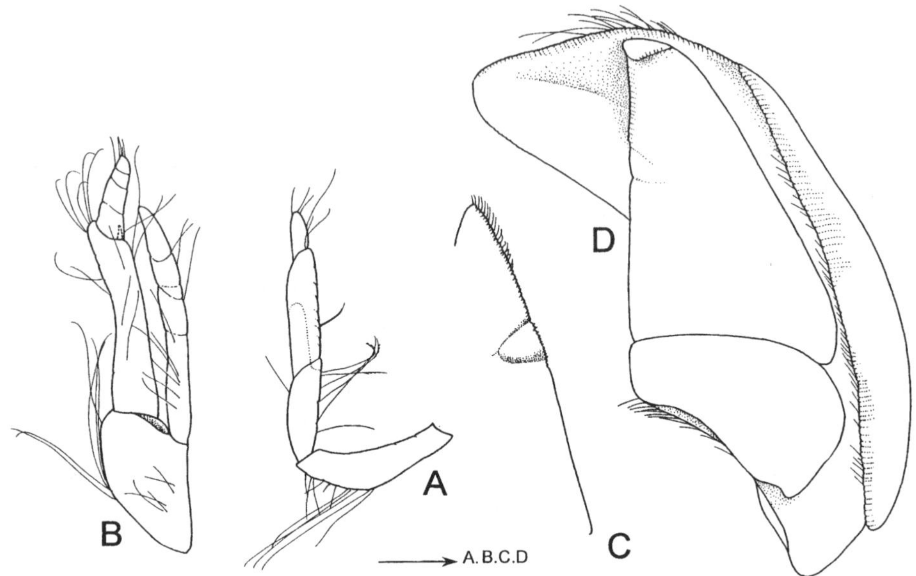

Fig. 38. *Podocallichirus madagassus* (Lenz & Richters, 1881). A, female Plp1; B, female Plp2; C, fused appendix interna on the mesial margin of Plp3; D, Plp3, posterior surface. A-B, SMF 7938, female holotype, Madagascar; C-D, SMF 4953, 1 male, paratype, Madagascar. Scale 1 mm.

Podocallichirus masoomi (Tirmizi, 1970)

Callianassa (*Callichirus*) *masoomi* Tirmizi, 1970: 245, figs. 1-3.
Callianassa (*Callichirus*) *kewalramanii* Sankolli, 1971: 94, figs. 5, 8. [Type locality: Ratnagiri, Bombay, India.]
Callichirus kewalramanii; Rodrigues, 1984a: 253.
Podocallichirus masoomi; Sakai, 1999c: 58.
Callianassa kewalramanii; Tudge et al., 2000: 143.
Callianassa masoomi; Tudge et al., 2000: 143.

Diagnosis. — Mxp3 ischium-merus pediform, merus angulate on distomesial angle, male Plp1 uniramous, two-segmented, simple distally, male Plp2 biramous, endopod simple distally, appendix interna and appendix masculina uncertain, telson subrectangular, posterior margin weakly trilobate, lacking a median spine (Sankolli, 1971: 94, figs. 5c, 6h, 8a, b).

Type locality. — Pakistan, Bholegi, W. of Karachi, intertidal zone of muddy sand beach with loose stones.

Distribution. — Bholegi, W. of Karachi, Pakistan; Ratnagiri, Bombay, India.

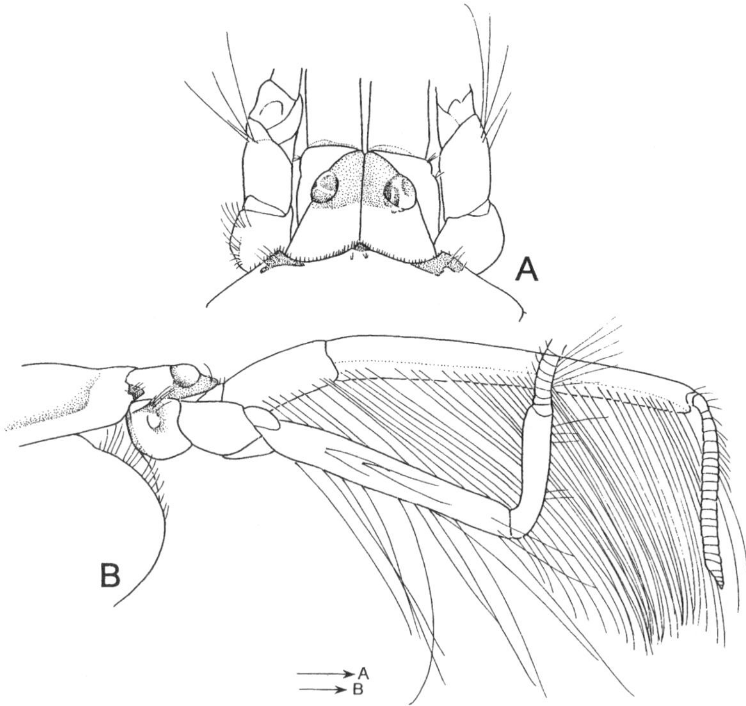

Fig. 39. *Podocallichirus tamakii* (Poore, 2000). NSMT-Cr. 12533, male (Tl/Cl 65.0/13.0), Mi-nami-Ariake-cho, S. of Shimabara Peninsula, Nagasaki Pref., 32°37'N 130°13'E, 8 July 1998, leg. A. Tamaki. A, anterior part of carapace, dorsal view; B, anterior part of carapace, and A1 and A2 peduncles, lateral view. Scales 1 mm.

Podocallichirus tamakii (Poore, 2000)
(figs. 39-40)

Grynaminna tamakii Poore, 2000: 150, figs. 1-4.

Material examined. — NSMT-Cr. 12532, paratype, male (Tl/Cl 66.0/13.5, chelipeds missing); NSMT-Cr. 12533, paratype, 1 male (65.0/13.0, chelipeds missing); NSMT-Cr. 12534, paratype, 1 male (68.0/15.3, chelipeds missing); NSMT-Cr.12526, paratype, 1 male (51.0/10.0, larger cheliped missing, smaller one detached); NSMT-Cr.12527, 1 male (62.0/13.5, larger and smaller chelipeds detached); NSMT-Cr. 12528, paratype, 1 female (43.0/8.2, right cheliped detached, left one missing); NSMT-Cr.12529, paratype, 1 male (38.0/7.9, with bopyrid, extra cheliped contained); NSMT-Cr. 12530, paratype, 1 female (38.0/7.9, left cheliped detached, right one missing); NSMT-Cr. 12531, paratype, 1 female (49.0/9.8, right and left chelipeds detached), S. end of Shimabara Peninsula, Nagasaki Pref., Kyushu, Japan, leg. A. Tamaki, 8.vii.1998.

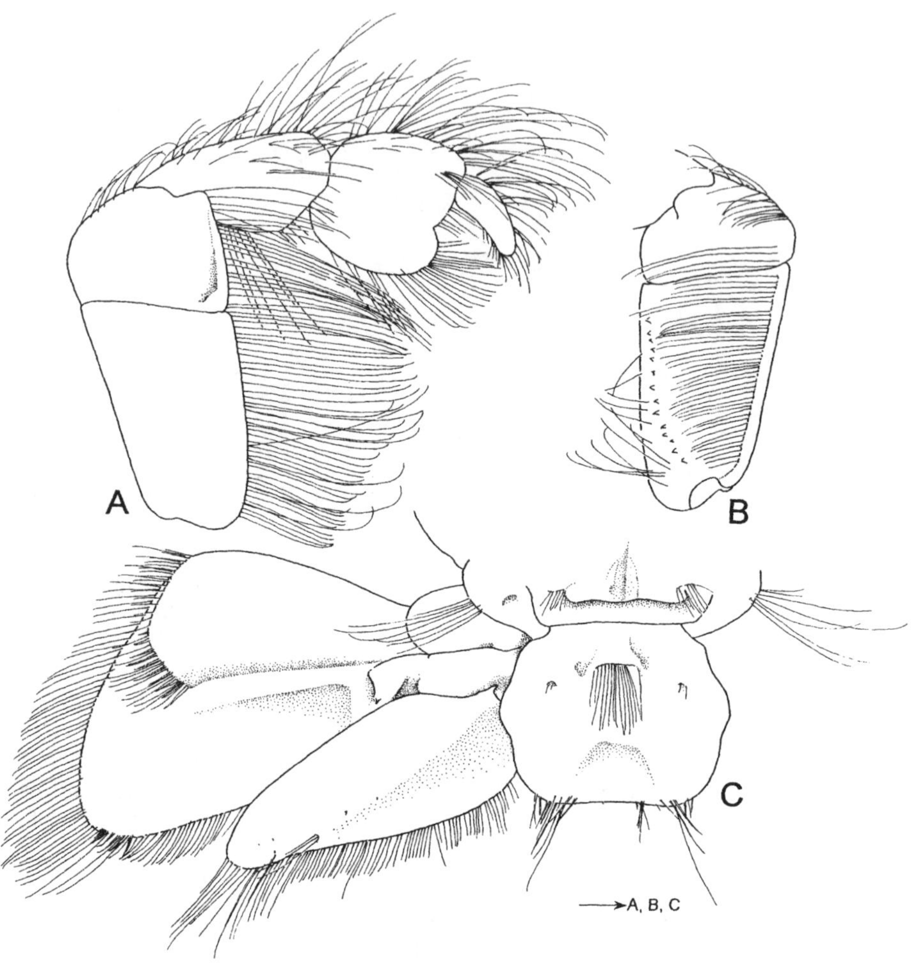

Fig. 40. *Podocallichirus tamakii* (Poore, 2000). NSMT-Cr. 12533, male (Tl/Cl 65.0/13.0), Minami-Ariake-cho, S. of Shimabara Peninsula, Nagasaki Pref., 32°37'N 130°13'E, 8 July 1998, leg. A. Tamaki. A, Mxp3, lateral view; B, ischium of Mxp3, mesial surface; C, telson and uropod on left side. Scale 1 mm.

Diagnosis. — Mxp3 ischium-merus subpediform, merus convex on distomesial angle, male Plp1 uniramous, two-segmented, chelate distally, male Plp2 biramous, endopod slightly contracted near tip, distally with appendix masculina, laterally with a protruded appendix interna bearing minute hooks distally, telson subrectangular, posterior margin slightly convex medially, lacking a median spine.

Remarks. — The relative length of the A1 flagella and the A1 last segment, the female appendix interna in Plp2, and the feature of the tail-fan, are all closely similar to the conditions found in the type species of the genus *Podocallichirus*, *P. madagassus*. Mxp3 seems characteristic for the angular distomesial angle, though in *P. madagassus* there is not a strong but yet a distinct swelling at the mesiodistal angle, like in the present species, *P. tamakii*.

Type locality. — Minami-Arima-cho, Shimabara Peninsula, Nagasaki Prefecture, Japan, 32°37'N 130°13'E.

Distribution. — Japan: Minami-Arima-cho, Shimabara Peninsula (Poore, 2000).

Subfamily EUCALLIACINAE Manning & Felder, 1991

Eucalliinae Manning & Felder, 1991: 781.
Eucalliacinae, Sakai, 1999c: 108; Ngoc-Ho, 2003.

Definition. — Rostrum developed or reduced, lacking rostral carina. Carapace without dorsal oval, cardiac prominence with or without a mid-pit; and transverse hepatic sulcus present or absent. Eyestalks dorsoventrally flattened and contiguous. Abdominal somite 6 lacking lateral projections. A2 scaphocerite developed as a small process. Mxp3 ischium-merus variform, from pediform to operculiform, without meral tooth; propodus broadened; dactylus ovate. P1 chelate, unequal or subequal, and similar or dissimilar, lacking meral hook in both chelipeds. P2 chelate. P3 propodus broadened. P4 simple. P5 chelate. Male Plp1 present or absent, when present, slender and uniramous. Female Plp1 uniramous. Male Plp2 blade-like and biramous, with appendix interna, or with appendix interna and appendix masculina. Female Plp2 blade-like and biramous, and endopod with appendix interna. Male and female Plp1-2 smaller than Plp3-5. Plp3-5 biramous, foliaceous, and with appendices internae in both sexes. Uropodal exopod with a secondary setal lobe.

Type genus. — *Calliax* De Saint Laurent, 1973.

Genera included. — *Calliax* De Saint Laurent, 1973; *Paraglypturus* Türkay & Sakai, 1995.

Remarks. — The type genus, *Eucalliax* Manning & Felder, 1991 is considered to be a junior synonym of *Calliax* as discussed in the remarks under *Eucalliax* and *Calliax* De Saint Laurent, 1973 (see Sakai, 1999c: 110); *Callianassa quadracuta*, the type species of *Eucalliax*, is very similar to *Calliax lobata*, the type species of the genus *Calliax*, in the shape of the male Plp2 and Mxp3. Article 40.1 of the Code says: "When the name of a type genus of a nominal family-group taxon is considered to be a junior synonym of the name of another nominal genus, the family-group name is not to be replaced on that account alone". Therefore, Eucalliacinae must continue to be used (L. B. Holthuis, in litt., April 2002).

Genus **Calliax** De Saint Laurent, 1973

Calliax De Saint Laurent, 1973: 514; Manning, 1987: 397; Sakai, 1987a: 306; Sakai, 1988: 61; Manning & Felder, 1991: 783, figs. 3, 15f-j; Poore, 1994: 101; Sakai, 1999c: 109; Davie, 2002: 459.
Eucalliax Manning & Felder, 1991: 781, figs. 7, 15a-e; Poore, 1994: 101. [Junior synonym of *Calliax*.]
Calliaxina Ngoc-Ho, 2003: 493.

Definition. — [Revised from Sakai, 1999c.] Carapace lacking dorsal oval; rostrum distinct or poorly developed, rostral longitudinal carina, cardiac and hepatic prominences absent, cardiac transverse sulcus present or absent. Mxp3 ischium-merus subpediform, propodus subquadrate, and dactylus expanded in breadth; exopod present or absent. P1 unequal or subequal in size, dissimilar in shape, larger cheliped without meral hook. P2 chelate. P3 propodus subquadrate. P4 simple. P5 chelate. Male Plp1 present or not, when present, uniramous, biarticulate, distal segment elongate, rounded or chelate distally; male Plp2 biramous, with appendix interna, or with both appendix interna and appendix masculina. Female Plp1 uniramous, bi- or triarticulate, female Plp2 biramous, bearing appendix interna. Plp3-5 foliaceous and with appendices internae in both sexes. Uropodal endopod oval or triangular in shape, much longer than telson, and without a yellow-transparent circular structure; exopod without lateral notch.

Remarks. — The genus *Calliax* De Saint Laurent, 1993 is defined by Manning & Felder (1991) as: "Cornea terminal elongate. Mxp3 without exopod, ischium-merus subpediform. Chelipeds unequal, major without meral hook. Plp1 uniramous and Plp2 biramous in both sexes; Plp3-5 biramous in both sexes, type of appendices internae unknown. Uropodal exopod apparently with lateral notch or incision". Davie (2002: 459) considered three Australian species, *Calliax aequimana* (Baker, 1907), *Calliax bulimba* (Poore & Griffin, 1979), and *Calliax tooradin* (Poore & Griffin, 1979) as belonging in the genus *Callianassa*. However, in the present work those species are not included in *Callianassa*, i.e., Callianassidae s. str., but in the subfamily Eucalliacinae

Manning & Felder, 1991, because in the Australian species the carapace lacks a dorsal oval as in the Eucalliacinae. In *Calliax*, the uropodal exopod has a lateral notch or incision only in the type species of the genus *Calliax*, *C. lobata*, but not in *Calliax jonesi* from the Bahamas, *Calliax tooradin* from Australia, and *Calliax aequimana* from the south of Western Australia. Mxp3 bears a distinct exopod in *Paraglypturus calderus*, but a rudimentary one in *Calliax punica*, *Calliax bulimba*, *Calliax tooradin*, *Calliax novaebritanniae*, *Calliax sakaii*, and *Calliax tooradin*; the exopod is not present in *Calliax lobata*, *Calliax quadracuta*, *Calliax aequimana*, *Calliax bulimba*, *Calliax mcilhennyi*, *Calliax jonesi*, *Calliax cearaensis*, and *Calliax doerjesti*. Cardiac transverse sulcus(-i) is/are known in *Calliax aequimana*, *Calliax jonesi*, *Calliax mcilhennyi*, *Calliax bulimba*, *Calliax novaebritanniae*, *Calliax punica*, and *Calliax sakaii*.

The genus *Calliaxina* Ngoc-Ho, 2003 was established for *Calliax punica* De Saint Laurent & Manning, 1982, and included the species *Calliax novaebritanniae* (Borradaile, 1900) and *Calliax sakaii* (De Saint Laurent & Le Loeuff, 1979) (cf. Ngoc-Ho, 2003). Ngoc-Ho (2003) compared with *Paraglypturus calderus* Türkay & Sakai, a species of a different genus, and mentioned that *Calliax punica* is different from *P. calderus* in its features. However, the type species of the genus *Calliax* is *Calliax lobata* (De Gaillande & Lagardère, 1966) so her comparison is not relevant to justify a new genus *Calliaxina*: therefore, *Calliaxina* is not accepted.

Type species. — *Callianassa lobata* De Gaillande & Lagardère, 1966. Gender of the generic name, *Calliax*, feminine.

Species included. — East Atlantic and Mediterranean species: *Calliax lobata* (De Gaillande & Lagardère, 1966); *C. punica* (De Saint Laurent & Manning, 1982). West Atlantic species: *C. cearaensis* (Rodrigues & Manning, 1992); *C. doerjesti* Sakai, 1999; *C. jonesi* Heard, 1989; *C. mcilhennyi* (Felder & Manning, 1994); *C. quadracuta* (Biffar, 1970). Indo-West Pacific species: *C. andamanica* Sakai, 2002; *C. aequimana* (Baker, 1907); *C. bulimba* (Poore & Griffin, 1979); *C. novaebritanniae* (Borradaile, 1900); *C. sakaii* (De Saint Laurent & Le Loeuff, 1979); *C. tooradin* (Poore & Griffin, 1979); *C. sp. Sakai, 2002.

East Atlantic and Mediterranean species

Calliax lobata (De Gaillande & Lagardère, 1966)

Callianassa (*Callichirus*) *lobata* De Gaillande & Lagardère, 1966: 259, pls. 1-4.

Calliax lobata; De Saint Laurent, 1973: 514; De Saint Laurent & Bozic, 1976: 28, figs. 7, 15, 23, 27, 34; Thessalou-Legaki, 1986: 183; Manning, 1987: 397; Števčić, 1990: 218; Manning & Felder, 1991: 783, figs. 3, 15f-j; Sakai, 1999c: 110, fig. 27a-c; d'Udekem d'Acoz, 1999: 155; Tudge et al., 2000: 145.

Diagnosis. — Mxp3 ischium-merus subrectangular, merus concave on distomesial margin, exopod absent, male Plp1 uniramous, two-segmented, lanceolate distally, male Plp2 biramous, endopod with appendix interna and appendix masculina, telson subovoid, rounded on posterior margin, lacking a median spine (De Gaillande & Lagardère, 1966: 259, pls. 1b, 3a, 3d).

Type locality. — Port Miou, Toulon, France, Mediterranean; 2-8 m.

Distribution. — Port Miou, Toulon, France, Mediterranean; 2-8 m (De Gaillande & Lagardère, 1966); Patraikos Gulf, Greece, 15 m depth, sandy mud (Thessalou-Legaki, 1985, 1986).

Calliax punica De Saint Laurent & Manning, 1982

Callianassa subterranea; Ortmann, 1891: 56 (partim). [Not *Callianassa subterranea* (Montagu, 1808).]
Calliax sp. De Saint Laurent & Bozic, 1976: 29, figs. 8, 16, 24, 35. [Locality: Salammbô, Tunisia.]
Calliax punica De Saint Laurent & Manning, 1982: 211-224, figs. 1-6; Thessalou-Legaki, 1986: 183; d'Udekem d'Acoz, 1996: 54; d'Udekem d'Acoz, 1999: 155; Tudge et al., 2000: 145. [Type locality: Salammbô, Tunisia.]
Paraglypturus punica; Sakai, 1999c: 122; d'Udekem d'Acoz, 2003, fig. on website.
Calliaxina punica; Ngoc-Ho, 2003: 496, figs. 19, 20.

Material examined. — ZLUA, 2 males (Tl/Cl 48.0/11.5; 30.0/7.5), 2 females (57.0/14.0; 50.0/12.0), Salamina, Saronikos Gulf, Greece, 0.5-1.0 m depth, yabby pump, leg. et det. M. Thessalou-Legaki, 14.iv.1997.

Diagnosis. — Mxp3 ischium-merus subovoid, merus concave on distomesial margin, with distinct exopod; male Plp1 uniramous, two-segmented, chelate with trilobed teeth distally, male Plp2 biramous, from the endopod branches distomesially an appendix interna bearing long setae distally, and from the appendix interna laterally branches a rod-like, half-length appendix masculina; telson subrectangular, truncate on posterior margin, lacking a median spine.

Remarks. — The present species is included in the genus *Calliax*, because the carapace lacks the rostral carina; the cardiac prominence is slightly swollen, and the transverse cardiac sulcus is distinct. The Mxp3 exopod is distinct, extending to the level of the proximal one-third of the lateral margin of the Mxp3 merus.

Type locality. — Gulf of Tunisia.

Distribution. — Salammbô, Tunisia (De Saint Laurent & Bozic, 1976); Port of Carthago, Tunisia (De Saint Laurent & Manning, 1982); Sardinia and Naples, Italy (De Saint Laurent & Manning, 1982); southern Euboikos Gulf, Messolongi lagoon, Salamina, Saronikos Gulf, Crete, Greece (Thessalou-Legaki, 1986).

West Atlantic species

Calliax cearaensis (Rodrigues & Manning, 1992)

Eucalliax cearaensis Rodrigues & Manning, 1992a: 327, fig. 2; Tudge et al., 2000: 145.
Calliax cearaensis; Sakai, 1999c: 112.

Diagnosis. — Mxp3 ischium-merus subrectangular; male Plp1 uniramous, two-segmented, curved distally; male Plp2 biramous, endopod with demarcated appendix masculina, laterally with small appendix interna; telson subrectangular, posterior margin concave, lacking a median spine (Rodrigues & Manning, 1992a: 327, fig. 2c, j, s, t).

Type locality. — Barro de Cear Fortaleza, right bank of river mouth (3°45'S 38°35'W), Brazil.

Distribution. — Barro de Cear Fortaleza, Brazil.

Calliax doerjesti Sakai, 1999

Calliax doerjesti Sakai, 1999c: 112, figs. 27d-g, 28, 29a-f.

Diagnosis. — Mxp3 ischium-merus subrectangular, merus obliquely declined on distomesial margin; male Plp1 uniramous, two-segmented, hooked distally; male Plp2 biramous, endopod with distinctly demarcated appendix masculina, distally with small appendix interna; telson subrectangular, posterior margin broadly convex, lacking a median spine (Sakai, 1999c: 112, figs. 27g, 28, 29e).

Type locality. — Uncertain: either Georgia, Ogeechee River, or Virgin Islands, U.S.A.

Distribution. — Georgia, Ogeechee River or Virgin Islands, U.S.A.

Calliax jonesi Heard, 1989

Calliax jonesi Heard, 1989: 129, figs. 1-5; Sakai, 1999c: 116.
Eucalliax jonesi; Tudge et al., 2000: 145.

Diagnosis. — Mxp3 ischium-merus subrectangular, merus obliquely declined on distomesial margin; male Plp1 uniramous, two-segmented, chelate distally; male Plp2 biramous, endopod with distinctly demarcated appendix masculina, proximally with small appendix interna; telson subrectangular, posterior margin weakly trilobate, lacking a median spine (Heard, 1989: 129, figs. 3E, 4F, 5a, c).

Remarks. — Heard (1989) mentioned "dorsal oval and cervical groove indistinctly defined". However, examination of his fig. 3A, D confirms that the dorsal oval is not developed as in the other species of *Calliax*. The presence of a transverse cardiac sulcus is an important characteristic of this species.

Type locality. — 25°44'N 79°15'W, Bimini Harbor, Bahamas, 3-5 m.

Distribution. — Bahamas.

Calliax mcilhennyi (Felder & Manning, 1994)

Eucalliax mcilhennyi Felder & Manning, 1994: 341, figs. 1-6; Tudge et al., 2000: 145.
Calliax mcilhennyi; Sakai, 1999c: 116.

Diagnosis. — Mxp3 ischium-merus subovoid, merus obliquely rounded on distomesial margin; male Plp1 uniramous, two-segmented, chelate distally; male Plp2 biramous, endopod with distinctly demarcated appendix masculina, proximally with small appendix interna; telson subrectangular, posterior margin broadly concave, lacking a median spine (Felder & Manning, 1994: 341, figs. 1-6).

Type locality. — Indian River lagoon, St. Lucie County (27°27.7'N 80°18.7'W), Florida, sandflat with sparse seagrass, south side of Fort Pierce Inlet.

Distribution. — Santa Marta, Colombia; Indian River lagoon, St. Lucie County, Florida.

Calliax quadracuta (Biffar, 1970)

Callianassa quadracuta Biffar, 1970: 40, fig. 2; Biffar, 1971a: 694, figs. 17, 18. [Type species of *Eucalliax* Manning & Felder, 1991.]

Calliax quadracuta; De Saint Laurent & Manning, 1982: 222; Manning, 1987: 397; Sakai, 1999c: 117, fig. 30a-c.
Eucalliax quadracuta; Manning & Felder, 1991: 781; Tudge et al., 2000: 145.

Diagnosis. — Mxp3 ischium-merus subrectangular, merus obliquely rounded on distomesial margin; male Plp1 uniramous, two-segmented, simply lobed distally; male Plp2 biramous, endopod with distinctly demarcated appendix masculina, proximally with small appendix interna; telson subrectangular, posterior margin broadly convex, bearing a median swelling (Biffar, 1970: 40, fig. 2f, i, k, m; Sakai, 1999c: 117, fig. 30b).

Remarks. — Biffar (1970) mentioned that the present species has a "Short median ridge narrowing anteriorly, continuous with rostrum (in carapace)". However, as shown by Sakai (1999: 117, fig. 30a, b) there is no recognizable rostral carina. In consequence, there is no reason to recognize the genus *Eucalliax* (refer to Sakai, 1999c: 117), and *Eucalliax* must, therefore, be considered a junior synonym of *Calliax*.

Type locality. — Venezuela, Cumaná.

Distribution. — Gulf of Uraba, Colombia, Caribbean coast; Cumaná, Venezuela.

Indo-West Pacific species

Calliax andamanica Sakai, 2002

Calliax andamanica Sakai, 2002: 463, figs. 1A-G, 2A-D.

Diagnosis. — Mxp3 ischium-merus subrectangular, merus rounded at distomesial angle and straight on mesial margin; male Plp1 uniramous, two-segmented, bilobed distally; male Plp2 biramous, endopod with distinctly demarcated appendix masculina, laterally with slender appendix interna; telson subsquare, posterior margin convex (Sakai, 2002: 463, figs. 1D, C, 2A).

Type locality. — Andaman Sea 9°30.364'N 97°57.346'E, 42.5 m, sand with shell fragments.

Distribution. — Andaman Sea (7°44.638'N 98°16.496'E; 7°45.002'N 98°15.103'E; 9°30.364'N 97°57.346'E), 30.5-70.0 m, coarse and fine sand, mud, sandy mud, with shell fragments.

Calliax aequimana (Baker, 1907)

Callianassa aequimana Baker, 1907: 182-185, pl. 24 figs. 1-8; Hale, 1927: 87, fig. 83; Poore &
   Griffin, 1979: 245, figs. 12, 13.
Callianassa (*Callichirus*) *aequimana*; De Man, 1928b: 28, 93, 114.
Callianassa (*Callichirus*) *Novae-britanniae*; De Man, 1928b: 29 (partim). [Not *Callianassa no-
   vae-britanniae* Borradaile, 1900.]
Callianassa (*Callichirus*) *novae-britanniae* var.; De Man, 1928b: 114.
Calliax aequimana; De Saint Laurent & Manning, 1982: 222; Sakai, 1988: 222; Sakai, 1999c:
   118, fig. 31a-e; Tudge et al., 2000: 145; Davie, 2002: 459.

Diagnosis. — Mxp3 ischium-merus subovoid, merus angled at distomesial
corner and broadly convex on mesial margin; male Plp1 uniramous, two-segmented, simply lobed distally with a notch laterally; male Plp2 biramous,
endopod with distinctly demarcated appendix masculina, laterally with slender
appendix interna; telson subrectangular, posterior margin broadly convex
(Poore & Griffin, 1979: 245, figs. 12c, 13f, i, j).

Type locality. — Kingston, western side of South Australia.

Distribution. — Central E. coast, Queensland to Kingston, southern coast of
Australia; Maldive Archipelago, Goifurfekendu, Goidu. Estuarine, littoral to
sublittoral, mud flats; to 9 m depth.

Calliax bulimba (Poore & Griffin, 1979)

Callianassa bulimba Poore & Griffin, 1979: 257, fig. 21.
Calliax bulimba; De Saint Laurent & Manning, 1982: 222; Sakai, 1988: 61; Sakai, 1999c: 119,
   fig. 32a-c; Tudge et al., 2000: 145; Davie, 2002: 459.

Diagnosis. — Mxp3 ischium-merus subovoid, merus rounded on distomesial margin; male Plp1 uniramous, two-segmented, simply lobed and curved
distally; male Plp2 biramous, endopod with distinctly demarcated appendix
masculina, laterally with slender appendix interna; telson subrectangular, posterior margin broadly straight (Poore & Griffin, 1979: 257, fig. 21f, g, h; Sakai, 1999c: 119, fig. 32c, f).

Type locality. — Mud Island, Moreton Bay, Queensland Australia.

Distribution. — Mud Island, Moreton Bay, Queensland Australia; sublittoral.

Calliax novaebritanniae (Borradaile, 1900)

Callianassa novae-britanniae Borradaile, 1900: 419, pl. 39 fig. 14a-d.
Callianassa (*Callichirus*) *novae-britanniae*; Borradaile, 1903: 547; Borradaile, 1904: 753.

Callianassa (*Callichirus*) *Novae-britanniae*; De Man, 1928a: 48 (partim); De Man, 1928b: 29 (partim).
Callianassa (*Callichirus*) *Novae-britanniae* var. ?; De Man, 1928a: 49, pl. 12 fig. 20-20g; De Man, 1928b: 29.
Callianassa (*Callichirus*) *novae-britanniae*; De Man, 1928b: 92-93 (partim). [Not: p. 114 = *Calliax aequimana*.]
Calliax novaebritanniae; De Saint Laurent & Manning, 1982: 211-224, figs. 1c, 2b, 6c; Tudge et al., 2000: 145.
Paraglypturus novaebritanniae; Sakai, 1999c: 123, fig. 32d-f.

Material examined. — ZMUC 146, 1 female (Tl/Cl 32/7.5, P1-2 on right side detached, P1-4 on left side, P3 on right side absent), Amboina [= Ambon], Danish Exped. Kei-Islands, 1922, under stone at low tide, 11.ii.1922.

Diagnosis. — Mxp3 ischium-merus subovoid, merus angled at distomesial corner and broadly rounded on mesial margin; male Plp1 uniramous, two-segmented, chelate distally; male Plp2 biramous; telson subrectangular, posterior margin broadly convex (De Man, 1928a: 49; Sakai, 1999c: 123, fig. 32c, f).

Remarks. — The linea anomurica, the cardiac transverse carina, and the Mxp3 exopod are considered as the generic characters.

Type locality. — New Britain, Papua New Guinea.

Distribution. — New Britain, Papua New Guinea; Ambon (= Amboina), Indonesia.

Calliax sakaii (De Saint Laurent & Le Loeuff, 1979)

Callianassa (*Callichirus*) *novaebritanniae*; Sakai, 1966: 161, figs. 1-4. [Non Borradaile, 1900.]
Calliax sakaii De Saint Laurent & Le Loeuff, 1979: 95; De Saint Laurent & Manning, 1982: 211-224, figs. 1, 2, 6; Sakai, 1987a: 306; Tudge et al., 2000: 145.
Paraglypturus sakaii; Sakai, 1999c: 124, fig. 33d-e.

Diagnosis. — Mxp3 ischium-merus subovoid, merus angled at distomesial corner and broadly rounded on mesial margin; male Plp1 uniramous, two-segmented, chelate distally; male Plp2 biramous, endopod with distinctly demarcated appendix masculina, laterally with slender appendix interna; telson subrectangular, posterior margin broadly convex (Sakai, 1966: 161, figs. 1c, 2g, 4a, c).

Type locality. — Tomioka, Amakusa, Japan, intertidal under *Zostera* vegetation.

Distribution. — Tomioka, Amakusa, Kumamoto, Japan.

Calliax tooradin (Poore & Griffin, 1979)

Callianassa tooradin Poore & Griffin, 1979: 275, fig. 36.
Calliax tooradin; Sakai, 1988: 61; Tudge et al., 2000: 145; Davie, 2002: 459.
Paraglypturus tooradin; Sakai, 1999c: 124, fig. 33a-c.

Diagnosis. — Mxp3 ischium-merus subrectangular, merus obliquely declined on distomesial angle and broadly convex on mesial margin; male Plp1 absent; male Plp2 biramous, endopod with laterally demarcated, distinct appendix masculina; telson subrectangular, posterior margin broadly rounded with short, straight median part (Poore & Griffin, 1979: 245, figs. 12c, 13f, i, j).

Type locality. — Australia, Victoria, Crib Point, Western Port, fine sand sediment, 5 m.

Distribution. — Western Port, Victoria, Australia (Poore & Griffin, 1979); sandy bottom, sublittoral.

Calliax sp. Sakai, 2002

Calliax sp. Sakai, 2002: 467, fig. 3A-I.

Distribution. — Andaman Sea, 6°45.961'N 99°20.968'E; 7°30.229'N 98°29.080'E; 38.0-60.4 m depth.

Genus **Paraglypturus** Türkay & Sakai, 1995

Paraglypturus Türkay & Sakai, 1995: 26.

Definition. — [Revised from Sakai, 1999c: 122.] Carapace lacks dorsal oval, rostrum poorly developed, without rostral longitudinal carina, cardiac and hepatic prominences, and hepatic transverse sulcus. Mxp3 ischium-merus subpediform and without meral spine; propodus broadened, subquadrate; and dactylus ovate; exopod distinct. P1 unequal, larger cheliped without meral hook. P2 chelate. P3 propodus subquadrate. P4 subchelate. P5 chelate. Male Plp1 uniramous, biarticulate, distal segment chelate distally; male Plp2 biramous, foliaceous, with appendix interna and appendix masculina. Female Plp1 uniramous, biarticulate; female Plp2 foliaceous, biramous, and with appendix interna. Plp3-5 foliaceous, with appendices internae in both sexes; uropodal endopod oval, much longer than telson, and with a yellow-transparent circular structure; exopod with or without lateral notch or incision.

Remarks. — This genus is closely similar to *Calliax*, but the type species, *Paraglypturus calderus* is characteristic and different from all other genera of the family Callianassidae in having a yellow-transparent circular structure on the uropodal endopod, whereas it is almost the same with *Calliax* in the relative length of the A1-2 peduncles, the form of the Mxp3 dactylus, and the absence of a ventral hook on the merus of the male larger cheliped. *Paraglypturus punica* (De Saint Laurent & Manning, 1982), *Paraglypturus novaebri-tanniae* (Borradaile, 1900), *Paraglypturus sakaii* (De Saint Laurent & Le Loeuff, 1979), and *Paraglypturus tooradin* (Poore & Griffin, 1979) as referred to in my previous paper (Sakai, 1999c: 109), are to be transferred to *Calliax*, because those species only differ from the other *Calliax* species by the presence or absence of the Mxp3 exopod, and they bear no yellow-transparent circular structure on the uropodal endopod as does *Paraglypturus calderus*.

Type species. — *Paraglypturus calderus* Türkay & Sakai, 1995, by monotypy and original designation. Gender of the generic name, *Paraglypturus*, masculine.

Species included. — *Paraglypturus calderus* Türkay & Sakai, 1995.

Indo-West Pacific species

Paraglypturus calderus Türkay & Sakai, 1995

Paraglypturus calderus Türkay & Sakai, 1995: 27, figs. 2-6.

Diagnosis. — Mxp3 ischium-merus subrectangular, merus convex on distomesial angle and broadly convex on mesial margin, bearing a long exopod; male Plp1 uniramous, two-segmented, bilobed distally; male Plp2 biramous, endopod laterally with appendix masculina, holding proximally small appendix interna; telson subtrapezoid, posterior margin broadly convex (Türkay & Sakai, 1995: 27, figs. 3f, 4h, 6a, b).

Type locality. — North rim of Esmeralda Caldera, Marianas, 14°58'3N 145°15'14E, 63 m depth.

Distribution. — North rim of Esmeralda Caldera, Marianas; 63-114 m depth; 50-78.5°C.

Subfamily CALLIAPAGUROPINAE Sakai, 1999

Calliapaguropinae Sakai, 1999c: 7.
Callichirinae; Tudge et al., 2000: 129; Ngoc-Ho, 2002: 540.

Definition. — Rostrum narrowly elongate, lacking rostral carina. Carapace without dorsal oval (in the sense of Ngoc-Ho, 2002, with faint dorsal oval), and cardiac prominence and hepatic sulcus absent. Eyestalks cylindrical and set apart. Abdominal somite 6 lacking lateral projections. A2 peduncle remarkably thick; A2 scaphocerite developed as a small process. Mxp3 ischium-merus suboperculiform, bearing meral teeth; propodus broadened; dactylus digitiform. P1 chelate, unequal and dissimilar, lacking meral hook in both chelipeds. P2 chelate. P3 propodus broadened. P4-5 chelate. Male and female Plp1 slender and uniramous. Male Plp 2 blade-like and biramous, with appendix interna and appendix masculina; female Plp2 blade-like and biramous, with appendix interna. Male and female Plp1-2 smaller than Plp3-5. Plp 3-5 biramous, foliaceous, and with appendix interna in both sexes. Uropodal exopod with a secondary setal lobe.

Remarks. — This subfamily is aberrant in the family Callianassidae, because the eyestalks are cylindrical and set apart, and the antennal peduncle is remarkably thick (fig. 41). Recently, Ngoc-Ho (2002: 539) followed Tudge's (2000) classification of taxa, and proposed that the genus *Calliapagurops* De Saint Laurent, 1973 should be placed in the Callichirinae Manning & Felder, 1991 rather than in the Calliapaguropinae Sakai, 1999. The reason would be, that *Calliapagurops charcoti* De Saint Laurent, 1973, the type species of the genus *Calliapagurops*, and *C. foresti* Ngoc-Ho, 2002, share with *Corallianassa articulata* (Rathbun, 1906) (removed to *Callianassa*) and all its congeners, *Corallianassa borradailei* (De Man, 1928) (syn. of *Glypturus coutierei*), *Corallianassa collaroy* (Poore & Griffin, 1979) (removed to *Glypturus*), and *Corallianassa longiventris* (A. Milne-Edwards, 1870) (removed to *Glypturus*) the presence of a long rostral spine on the carapace, elongate cylindrical eyestalks with terminal corneas, peduncle of antenna 2 longer than that of antenna 1, maxilliped 3 operculiform, pereiopods 1 with numerous spines on lower border of ischium and merus, abdominal somite 2 longer than somite 6, pleopod 1 uniramous, pleopod 2 biramous in both sexes, and telson wider than long. Ngoc-Ho (2002: 548), however, differentiated those two *Calliapagurops* species from other callianassids for the following reasons: (1) Long cylindrical eyestalks with terminal corneas, longer than in any other callianassids but approached by species of *Corallianassa* (removed to *Glypturus*). (2) Antenna 2 peduncles approximately 2.0 times as long as those of antenna 1 and much thicker. It is here contended that it is certain that in *Calliapagurops charcoti* and *C. foresti* the eyestalks are cylindrically elongate, set apart, and without orbit (Sakai, 1999c: 7-9). However, in *Corallianassa articulata*,

C. borradailei (fig. 28C, D as a senior synonym of *C. coutierei*), *C. collaroy* (fig. 29A, B), and *C. longiventris*, the eyestalks are never cylindrical but subglobose, and certainly different from the species of *Corallianassa*, though the antenna 2 peduncles are indeed much thicker as she cited, whereas Ngoc-Ho (2002) did not pay attention to the fact that the thickness of the antennal 2 peduncles is characteristic within the Callianassidae. Re-examination of the eyestalks shows, that in *Corallianassa articulata* (cf. Sakai, 1999c, fig. 15), *C. xutha* (fig. 28A, B), and *C. longiventris*, they are globose with a distal cornea, touching each other closely, and in *C. collaroy* (fig. 29A, B) they are subglobose with a mesiodistal cornea, and also touching each other closely. The antennal peduncles in *Calliapagurops charcoti* and *C. foresti* are extremely thick (fig. 41A, B) in comparison with those of other callianassid species. As a result, the differences that Ngoc-Ho (2002) mentioned in the form of the eyestalks and the antenna 2 peduncles, are not relevant, so the subfamily Calliapaguropinae can be reconfirmed by the characters of the elongate eyestalks and the thick A2 peduncle.

Type genus. — *Calliapagurops* De Saint Laurent, 1973.

Genus included. — *Calliapagurops* De Saint Laurent, 1973.

Genus **Calliapagurops** De Saint Laurent, 1973

Calliapagurops De Saint Laurent, 1973: 515; Sakai, 1999c: 8; Ngoc-Ho, 2002: 540.

Type species. — *Calliapagurops charcoti* De Saint Laurent, 1973, by original designation. Gender of the generic name, *Calliapagurops*, masculine.

Species included. — *Calliapagurops charcoti* De Saint Laurent, 1973; *C. foresti* Ngoc-Ho, 2002.

East Atlantic species

Calliapagurops charcoti De Saint Laurent, 1973

Calliapagurops charcoti De Saint Laurent, 1973: 515; Sakai, 1999c: 8, fig. 1a-e; d'Udekem d'Acoz, 1999: 155; Tudge et al., 2000: 143.

Diagnosis. — Mxp3 ischium-merus suboperculiform, merus rounded with three teeth on distomesial margin, lacking exopod; male Plp1-2 unknown; telson unknown (Sakai, 1999c: 8, fig. 1b).

Type locality. — Off Azores Is., near Flores Islands, "Biaçores" Exped., 1971, R/V "Jean Charcot", Sta. 109, 39°33'N 31°17'W, 230-190 m depth.

Distribution. — Off Azores Is., near Flores Islands, 39°33'N 31°17'W.

Indo-West Pacific species

Calliapagurops foresti Ngoc-Ho, 2002
(fig. 41)

Calliapagurops foresti Ngoc-Ho, 2002: 539, figs. 1-3.

Material examined. — MNHN Th 1399, paratype, 1 female (Tl/Cl 30.5/8.5), Philippines, 14°00.8'N 120°17.8'E, MUSORSTOM 3, Sta. 102, 192 m depth, 1.vi.1985.

Diagnosis. — Mxp3 ischium-merus subovoid, merus rounded, with 6-10 teeth on distomesial margin, lacking exopod; male Plp1 uniramous, two-segmented, pointed distally; male Plp2 biramous, endopod with distally attached, small appendix masculina; telson subovoid, posterior margin broadly concave (Ngoc-Ho, 2002: 539, figs. 1C, H, G, 2E).

Type locality. — Philippines, MUSORSTOM 2, Sta. 71, 14°00.1'N 120°17.8'E, 192 m depth.

Distribution. — Philippines, 14°00.1'N 120°17'E to 14°00.8'N 120°17.8'E, 192 m depth.

Subfamily ANACALLIACINAE Manning & Felder, 1991

Anacalliinae Manning & Felder, 1991: 786.

Definition. — Rostrum developed, triangular, bearing rostral carina. Carapace with dorsal oval, cardiac prominence present, and hepatic sulcus absent. Eyestalks dorsoventrally flattened and contiguous. Abdominal somite 6 lacking lateral projections. A2 scaphocerite developed, sharp. Mxp3 ischium-merus pediform, lacking distal meral spine, propodus oblong, and dactylus digitiform. P1 chelate, unequal in size and dissimilar in shape, lacking meral hook in both chelipeds. P2 chelate. P3 propodus broadened. P4 subchelate. P5 chelate. Male Plp1 uniramous, distal article chelate or blade-like in shape. Female Plp1 biarticulate, uniramous. Male Plp2 slender and biramous, bearing both

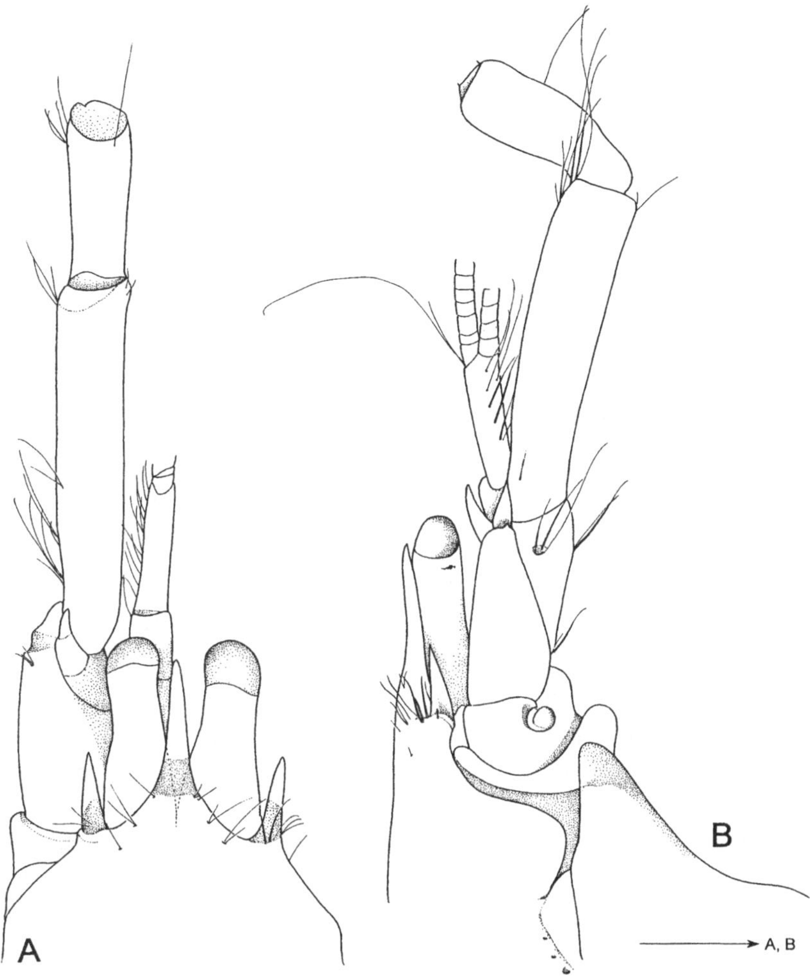

Fig. 41. *Calliapagurops foresti* (Ngoc-Ho, 2002). A, anterior part of carapace with eyes, antennal and antennular peduncles, dorsal view; B, anterior part of carapace, lateral view. MNHN Th 1399, paratype, 1 female, 14°00.8'N 120°17.8'E. Scale 1 mm.

appendix interna and appendix masculina. Female Plp2 biramous, with appendix interna. Uropodal exopod lacking a secondary setal lobe, but bearing lateral notch.

Remarks. — Manning & Felder (1991: 786) included *Anacalliax* in the family Ctenochelidae, but *Anacalliax* differs from *Ctenocheles* because in *Anacalliax* the carapace oval is present, the P3 propodus is broadened, and the male Plp2 is elongate, with a distinct appendix masculina and a small appendix interna, whereas in *Ctenocheles* the carapace lacks an oval area, the P3 propodus is oblong, and the male Plp2 is foliaceous, bearing an elongate appendix masculina and a small appendix interna. Thus, *Anacalliax* is to be separated from the Ctenochelidae and treated as a new subfamily, Anacalliacinae in the Callianassidae. The pediform Mxp3 in *Anacalliax* is similar to that of *Podocallichirus*.

Type genus. — *Anacalliax* De Saint Laurent, 1973.

Genus included. — *Anacalliax* De Saint Laurent, 1973.

Genus **Anacalliax** De Saint Laurent, 1973

Anacalliax De Saint Laurent, 1973: 515; Manning, 1987: 397; Manning & Felder, 1991: 786, figs. 2-3, 17; Sakai, 1999c: 126.

Definition. — [Revised from Sakai, 1999c.] Carapace with dorsal oval well defined; rostral carina and cardiac prominence present, but transverse cardiac sulcus absent. A2 with distinct scaphocerite. Mxp3 without exopod; ischium-merus pediform, lacking distal meral spine, propodus oblong, and dactylus digitiform. P1 unequal, larger cheliped without meral hook. P2 chelate. P3 propodus broadened. P4 subchelate. P5 chelate. Male Plp1 uniramous, biarticulate, distal segment chelate distally; Plp2 biramous, pediform, and with appendix interna and masculina. Plp3-5 similar, but larger than and different from Plp2, and bearing stubby appendices internae, embedded in margin of endopod. Uropodal exopod lacking dorsal plate, but distally thickened and with strong indentation.

Remarks. — *Anacalliax* was once included in the subfamily Eucalliacinae by Sakai (1999c) but is now transferred to the subfamily Anacalliacinae Manning & Felder, 1991.

Type species. — *Callianassa argentinensis* Biffar, 1971b, by original designation and monotypy. Gender of the generic name, *Anacalliax*, feminine.

Species included. — East Atlantic species: *Anacalliax pixii* (Kensley, 1975). West Atlantic species: *A. agassizi* (Biffar, 1971); *A. argentinensis* (Biffar, 1971).

East Atlantic species

Anacalliax pixii (Kensley, 1975)

Callianassa pixii Kensley, 1975: 53, figs. 4A-H, 5A-K; Tudge et al., 2000: 143.
Anacalliax pixii; Sakai, 1999c: 127.

Diagnosis. — Mxp3 ischium-merus pediform; male Plp1 uniramous, two-segmented, distal segment scythe-shaped; male Plp2 biramous, endopod pediform, provided distally with appendix masculina and appendix interna, exopod slender and short; telson subrectangular, posterior margin broadly concave (Kensley, 1975: 53, figs. 4B, H, 5G, H).

Type locality. — Kowie River estuary, Cape Province, South Africa.

Distribution. — South Africa, Kowie River estuary, Cape Province (Kensley, 1975).

West Atlantic species

Anacalliax agassizi (Biffar, 1971)

Callianassa agassizi Biffar, 1971b: 233, fig. 3.
Anacalliax agassizi; Manning, 1987: 397; Manning & Felder, 1991: 787 (list); Sakai, 1999c: 127; Tudge et al., 2000: 142.

Diagnosis. — Mxp3 ischium-merus pediform; male Plp1 uniramous, two-segmented, distal segment chelate; male Plp2 biramous, endopod pediform, with distally attached appendix masculina and appendix interna, exopod slender and lanceolate; telson trapezoid, posterior margin concave (Biffar, 1971b: 233, fig. 3f, k, l, m).

Type locality. — Cumaná, Venezuela.

Distribution. — Cumaná, Venezuela.

Anacalliax argentinensis (Biffar, 1971)

Callianassa argentinensis Biffar, 1971b: 229, fig. 2.
Anacalliax argentinensis; Manning, 1987: 397 (list); Manning & Felder, 1991: 787 (list), figs. 2, 3, 17; Sakai, 1999c: 127; Tudge et al., 2000: 142 (list).

Diagnosis. — Mxp3 ischium-merus pediform; male Plp1 uniramous, two-segmented, distal segment broadened distally, with two rounded lobes; male Plp2 biramous, endopod pediform, with distally attached appendix masculina and appendix interna, exopod blade-shaped; telson subsquare, posterior margin medially notched (Biffar, 1971b: 229, fig. 2g, k, i, m).

Type locality. — Sandy beach on the north coast of Isla del Rey, Río Deseado, Province of Santa Cruz, Argentina, 47°42'S 66°W.

Distribution. — Argentina: Province of Santa Cruz, Río Deseado; Province of Río Negro, Golfo San Catías; Province of Buenos Aires, Riacho Jabali, Bahia San Blas.

Subfamily LIPKECALLIANASSINAE n. subfam.

Definition. — Rostrum developed, sharp, of small size, lacking rostral carina. Carapace without dorsal oval, cardiac prominence present, but transverse hepatic sulcus absent. Eyestalks dorsoventrally flattened and contiguous. Abdominal somite 6 lacking lateral projections. A2 scaphocerite developed as a small process. Mxp3 ischium-merus subpediform, bearing mesiodistal meral tooth; propodus oblong; dactylus digitiform. P1 uncertain. P2 chelate. P3 propodus oblong. Male Plp1 slender and uniramous. Male Plp2-5 and female Plp1-5 unknown. Uropodal exopod lacking secondary setal lobe and lateral notch.

Remarks. — The new subfamily, Lipkecallianassinae, is very characteristic: the carapace lacks a rostral carina; the rostrum is elongate; and the telson is largely concave on the posterior margin, bearing a median spine; the uropodal exopod lacks a secondary setal lobe; the P3 propodus is rectangular; and P4 is lengthened, overreaching P2.

Type genus. — *Lipkecallianassa* Sakai, 2002.

Genus included. — *Lipkecallianassa* Sakai, 2002.

Genus **Lipkecallianassa** Sakai, 2002

Lipkecallianassa Sakai, 2002: 477.

Description. — Carapace without dorsal oval but with linea thalassinica. A1 peduncle shorter than A2 peduncle. Mxp3 ischium-merus subpediform, laterally parallel, merus forming a mesiodistal tooth, propodus rectangular. P3 propodus rectangular. P4 lengthened, overreaching P2. Plp3 foliaceous, biramous, with a fused appendix interna. Uropodal exopod lacking a secondary setal lobe.

Remarks. — *Lipkecallianassa* is characterized by the oblong P3 propodus and the uropodal exopod lacking a secondary setal lobe, as in the family Ctenochelidae, so this genus could have been transferred to the Ctenochelidae based on that character. However, the genus was originally included in the Callianassinae, because the carapace of the only known specimen is damaged, thus making it difficult to discern its proper shape and to assign it to a subfamily of its own. Yet, because of its separate position, it is now set apart from the other subfamilies of the Callianassidae.

Type species. — *Lipkecallianassa abyssa* Sakai, 2002, by original designation and monotypy. The gender of the generic name, *Lipkecallianassa*, is feminine.

Species included. — *Lipkecallianassa abyssa* Sakai, 2002.

Lipkecallianassa abyssa Sakai, 2002

Lipkecallianassa abyssa Sakai, 2002: 477, figs. 9A-D, 10A-H.

Diagnosis. — Mxp3 ischium-merus pediform, merus with a triangular distomedian spine, exopod absent; P4 elongated; male Plp1-2 unknown; telson subrectangular, posterior margin rounded, medially notched with a median spine (Sakai, 2002: 477, figs. 9A, D, 10A, B).

Type locality. — Andaman Sea, 7°545.179'N 97°19.907'E, 493.0 m.

Distribution. — Andaman Sea, 7°545.179'N 97°19.907'E.

Subfamily BATHYCALLIACINAE Sakai & Türkay, 1999

Bathycalliacinae Sakai & Türkay, 1999: 204.

Definition. — Rostrum developed, of small size, bearing a low rostral carina. Carapace with a dorsal oval with low dorsal carina; cardiac prominence and two transverse hepatic sulci present. Eyestalks dorsoventrally flattened and contiguous. Abdominal somite 6 lacking lateral projections. A2 scaphocerite developed as a small process. Mxp3 ischium-merus subpediform, lacking meral tooth; propodus broadened; dactylus ovate. P1 chelate, unequal in size and dissimilar in shape, lacking meral hook in both chelipeds. P2 chelate. P3 propodus broadened. Male Plp1 slender and uniramous, distal article consisting of multiarticulate flagellum. Male Plp2 biramous, with appendix interna (cf. Sakai, 1999: 208, shown erroneously as appendix masculina), no appendix masculina. Female Plp1-2 unknown. Plp3-5 foliaceous, biramous, bearing appendices internae. Uropodal exopod lacking secondary setal lobe and lateral notch.

Remarks. — The only included species, *Bathycalliax geomar*, is characteristic, because no dorsal oval is present, but a dorsal ridge on the carapace is present from the posterior half of the gastric region to the posterior margin. Epipods are present on Mxp3-P4.

Type genus. — *Bathycalliax* Sakai & Türkay, 1999.

Genus included. — *Bathycalliax* Sakai & Türkay, 1999.

Genus **Bathycalliax** Sakai & Türkay, 1999

Bathycalliax Sakai & Türkay, 1999: 204.

Diagnosis. — [Revised from Sakai & Türkay, 1999.] Carapace without dorsal oval and with a narrow, longitudinal dorsomedian carina running from the posterior half of the gastric region through a conspicuous cardiac prominence to the posterior margin; cardiac prominence with a mid-pit; cervical groove and two transverse cardiac sulci present. Abdominal somite 6 without lateral projections. Eyestalks triangular, without corneas, and touching each other, dorsal surfaces dorsoventrally depressed. Maxilla 2 scaphognathite without a posterior long seta. Mxp3 ischium-merus subpediform, propodus and dactylus broadened, exopod rudimentary. P1 unequal and dissimilar. P2 chelate. P3 propodus broadened. P4-5 subchelate. Epipods are present from Mxp3 to P4, respectively. Uropodal exopod oval, forming an anterodorsal plate in its anterior third. Male Plp2 with appendix interna.

Type species. — *Bathycalliax geomar* Sakai & Türkay, 1999, by original designation and monotypy. The gender of the generic name, *Bathycalliax*, is feminine.

Species included. — East Pacific species: *Bathycalliax geomar* Sakai & Türkay, 1999.

East Pacific species

Bathycalliax geomar Sakai & Türkay, 1999

Bathycalliax geomar Sakai & Türkay, 1999: 204, figs. 1-3, tab. 1.

Diagnosis. — Cornea absent. Mxp3 ischium-merus subrectangular, merus rounded on distomesial angle, exopod rudimentary; male Plp1 uniramous, consisting of proximal segment and flagellum; male Plp2 biramous, endopod with distally demarcated lobe of appendix masculina and with proximally attached appendix interna; telson subpentagonal, posterior margin truncate (Sakai & Türkay, 1999: 203, figs. 1f, g, 3).

Type locality. — Aleutian subduction-zone off Oregon, U.S.A., 44°40.146'N 125°06.685'W, 625 m.

Distribution. — Mauritania; Aleutian subduction-zone off Oregon, U.S.A., 44°40.146'N 125°06.685'W-44°40.193'N 125°06.605'W, 625-627 m.

Subfamily PARACALLIACINAE n. subfam.

Definition. — Rostrum triangular, of small size, bearing rostral carina. Carapace lacking dorsal oval; cardiac prominence present, but transverse hepatic sulcus absent. Eyestalks prolonged, dorsoventrally flattened and contiguous. Abdominal somite 6 with lateral projections. A2 scaphocerite developed as a small process. Mxp3 ischium-merus pediform, bearing meral spine; propodus oblong; dactylus digitiform. P1 chelate, unequal in size, dissimilar in shape, lacking proximal meral hook in both chelipeds. P2 chelate. P3 propodus oblong. Male Plp1-2 unknown. Female Plp1 two-segmented, distal segment foliaceous. Female Plp2 foliaceous, with appendix interna. Plp3-5 foliaceous, with appendices internae (De Saint Laurent & Le Loeuff, 1979: 87). Uropodal exopod lacking lateral notch or incision, lacking a secondary setal lobe.

Remarks. — The present subfamily is similar to the subfamilies Eucalliacinae, Lipkecallianassinae, and Pseudogourretiinae as well as to the families Ctenochelidae and Gourretiidae, by the character that the carapace lacks the dorsal oval. However, the Paracalliacinae are distinguishable, because the eyestalks are elongate, and have their dorsal surface concave, without cornea; and the P3 propodus is oblong. The Paracalliacinae are similar to the Ctenochelidae in the oblique form of the P3 propodus.

Type genus. — *Paracalliax* De Saint Laurent, 1979.

Genus included. — *Paracalliax* De Saint Laurent, 1979.

Genus **Paracalliax** De Saint Laurent, 1979

Paracalliax De Saint Laurent, 1979: 1396; De Saint Laurent & Le Loeuff, 1979: 47 (key), 84; Manning & Felder, 1991: 785, figs. 1, 7; Poore, 1994: 103 (key).

Diagnosis. — [Revised from Manning & Felder, 1991: 785.] Carapace lacking dorsal oval, but low rostral carina and cardiac prominence present; rostral spine absent, but without transverse cardiac sulcus. Dorsal surface of eye concave. Mxp3 ischium-merus pediform and with distal meral spine, propodus oblong, and dactylus digitiform; exopod present. Larger cheliped lacking proximal meral hook. P3 propodus oblong. Sixth abdominal somite with rounded lateral projections. Uropodal exopod lacking lateral notch or incision.

Type species. — *Paracalliax bollorei* De Saint Laurent, 1979, by original designation and monotypy. Gender of the generic name, *Paracalliax*, feminine.

Species included. — East Atlantic species: *Paracalliax bollorei* De Saint Laurent, 1979.

East Atlantic species

Paracalliax bollorei De Saint Laurent, 1979

Paracalliax bollorei De Saint Laurent, 1979: 1396; De Saint Laurent & Le Loeuff, 1979: 86, figs. 26a-i, 27a-c, 28a-h; Manning & Felder, 1991: 785 (list), figs. 1, 7; Tudge et al., 2000: 142.

Diagnosis. — Cornea absent. Mxp3 ischium-merus pediform, merus with a distomedian spine, exopod present; male Plp1-2 unknown; telson subsquare, posterior margin truncate with a median spine (De Saint Laurent & Le Loeuff, 1979: 86, figs. 26c, 28f).

Type locality. — Banc d'Arguin, Mauritania, 20-100 m.

Distribution. — Banc d'Arguin, Mauritania (De Saint Laurent, 1979; De Saint Laurent & Le Loeuff, 1979).

FAMILY GOURRETIIDAE SAKAI, 1999

Gourretinae Sakai, 1999a: 95; Ngoc-Ho, 2003: 498.
Ctenochelidae; [partim] Ngoc-Ho, 2003: 498.
Gourretiidae; Sakai, 2004: 556.

Definition. — Rostrum developed but short, unarmed laterally, and lacking rostral carina. Carapace without dorsal oval; linea thalassinica present. Abdominal somites 3-5 dorsolaterally with a tuft of setae. Abdominal somite 6 lacking lateral projections. Eyestalks dorsoventally flattened and contiguous. A2 scaphocerite an obtuse process of small size, or present as a strong spine (in *Laurentgourretia*). Maxilla 2 without posterior seta. Mxp3 subpediform; propodus oblong and dactylus digitiform. P1 chelate, dissimilar. Larger cheliped with a meral hook, palm oblong, fingers elongate, not pectinate; smaller cheliped usually with meral hook (not in *Gourretia lahouensis*), fingers elongate and not pectinate. P3 propodus broadened in a heel shape. Male Plp1 usually chelate; female Plp1 uniramous, bi- or triarticulate, distal segment simple; male Plp2 biramous, with or without appendix interna and appendix masculina; female Plp2 biramous, with appendix interna. Plp3-5 with appendices internae in both sexes. Uropodal exopod with or without lateral notch or incision.

Subfamilies included. — Gourretiinae Sakai, 1999; Callianopsinae Manning & Felder, 1991; Pseudogourretiinae nov. subfam.

Subfamily GOURRETIINAE Sakai, 1999

Gourretinae Sakai, 1999a: 95; Ngoc-Ho, 2003: 498.
Ctenochelidae; [partim] Ngoc-Ho, 2003: 498.
Gourretiinae; Sakai, 2004: 556.

Definition. — Rostrum developed, of small size, unarmed laterally, and lacking rostral carina. Carapace without dorsal oval; cardiac prominence and transverse hepatic sulcus absent; vertical row of setae rarely present on branchial region. Eyestalks dorsoventrally flattened and contiguous. Abdominal somite 6 lacking or bearing (in *Dawsonius* [see Note on p. 245]) lateral projections. A2 scaphocerite present as a small, obtuse process. Mxp3 ischium-merus subpediform, merus with mesial spine; propodus oblong; dactylus digitiform. P1 chelate, unequal in size, dissimilar in shape; larger cheliped with a ventroproximal meral hook, palm oblong, fingers elongate, not pectinate; smaller cheliped usually with ventroproximal meral hook (not in *Gourretia lahouensis*), fingers elongate, not pectinate. P2 chelate. P3 propodus broadened in a heeled form. P4 simple. P5 chelate. Male Plp1 uniramous, biarticulate, distal segment usually chelate; male Plp2 biramous, foliaceous, endopod with or without appendix interna, and with or without appendix masculina. Female Plp1 uniramous, bi- or triarticulate, distal segtment simple; female Plp2 biramous, foliaceous, endopod with appendix interna. Plp2 smaller than Plp3-5 in both sexes. Uropodal exopod with or without lateral notch, and without a secondary setal lobe.

Remarks. — The genera *Dawsonius* and *Gourretia* were included in the Ctenochelinae by Manning & Felder (1991: 784), but *Gourretia* is different from *Ctenocheles*, *Dawsonius*, and *Paracalliax* in that the carapace has no rostral carina. This led Sakai (1999a) to separate *Gourretia* from the Ctenochelinae and assign it to a separate subfamily, the Gourretiinae. In addition, in *Gourretia* the P3 propodus is broadened in a heel-shaped form, and the male Plp1 is chelate distally, whereas in *Ctenocheles* the carapace bears a rostral carina, the P3 is oblong, and the male Plp1 is not chelate distally but simple, consisting of 3-4 segments.

In the Ctenochelinae, a rostral carina and a hepatic prominence are present, the scaphocerite is distinct, the P3 propodus is oblong, the male Plp1 is uniramous, 3-4-segmented, and the male Plp2 is foliaceous, but smaller than Plp3-5, and bears an appendix interna and an elongate appendix masculina. In the Gourretiinae, a rostral carina and a hepatic prominence are absent, the scaphocerite is rudimentary, the P3 propodus is broadened in a heel-shaped form, the male Plp1 is uniramous, chelate distally, and the Plp2 is smaller than Plp3-5 and bears an appendix interna and a thick appendix masculina.

Type genus. — *Gourretia* De Saint Laurent, 1973.

Genera included. — *Gourretia* De Saint Laurent, 1973; *Laurentgourretia* Sakai, 2004; *Paragourretia* Sakai, 2004.

Genus **Gourretia** De Saint Laurent, 1973

Gourretia De Saint Laurent, 1973: 514; Le Loeuff & Intès, 1974: 26; Poore & Griffin, 1979: 278; De Saint Laurent & Le Loeuff, 1979: 48 (key), 78; Manning, 1987: 398 (list); Manning & Felder, 1991: 785, fig. 3; Poore, 1994: 103 (key); Davie, 2002: 464; Sakai, 2002: 468; De Saint Laurent, 2003: 498; Sakai, 2004: 562.

Diagnosis. — [Revised from Manning & Felder, 1991: 785.] Carapace lacking dorsal oval and rostral carina; rostral spine usually present; transverse cardiac sulcus absent; cardiac prominence present or absent. A2 with rudimentary scaphocerite. Dorsal surface of eye flattened. Mxp3 ischium-merus subpediform, usually with distolateral meral spine; propodus oblong and dactylus digitiform; exopod present. Larger cheliped with proximal meral hook. P3 propodus broadened. P4 simple. P5 chelate. Abdominal somite 6 lacking acute lateral projections. Male Plp1 uniramous, biarticulate, distal segment distally chelate. Male Plp2 biramous, foliaceous, and with appendices interna and masculina. Female Plp1 uniramous, biarticulate, distal segment simple. Female Plp2 biramous, with appendix interna. Uropodal exopod usually lacking lateral notch or incision.

Type species. — *Callianassa denticulata* Lutze, 1937, by original designation and monotypy. Gender of the generic name, *Gourretia*, feminine.

Species included. — East Atlantic and Mediterranean species: *Gourretia barracuda* Le Loeuff & Intès, 1974; *G. denticulata* (Lutze, 1937); *G. lahouensis* Le Loeuff & Intès, 1974; *G. loeuffintesi* sp. nov. West Atlantic species: *G. biffari* Blanco Rambla & Liñero Arana, 1994; *G. laresi* Blanco Rambla & Liñero Arana, 1994. Indo-West Pacific species: *G. coolibah* Poore & Griffin, 1979; *G. crosnieri* Ngoc-Ho, 1991; *G. manihinae* Sakai, 1984; *G. nosybeensis* Sakai, 2004.

East Atlantic and Mediterranean species

Gourretia barracuda Le Loeuff & Intès, 1974

Gourretia barracuda Le Loeuff & Intès, 1974: 30, fig. 6a-t; De Saint Laurent & Le Loeuff, 1979: 80, fig. 24a; Manning & Felder, 1991: 785 (list); Tudge et al., 2000: 142.

Diagnosis. — Mxp3 ischium-merus pediform, merus armed with a distomedian spine, exopod present; male Plp1 uniramous, two-segmented, distal segment pointed distally; male Plp2 biramous, endopod distally with appendix masculina and laterally attached appendix interna; telson subtriangular, posterior margin rounded (Le Loeuff & Intès, 1974: 30, fig. 6j, q, r, t).

Type locality. — Abidjan, Ivory Coast, 250-100 m.

Distribution. — Abidjan, Ivory Coast (Le Loeuff & Intès, 1974; De Saint Laurent & Le Loeuff, 1979).

Gourretia denticulata (Lutze, 1937)

Callianassa subterranea var. *minor* Gourret, 1887: 1034; Gourret, 1888: 96, pl. 8 figs. 1-15.
? *Callianassa* (*Cheramus*) *subterranea* var. *minor*; Borradaile, 1903: 546 (list).
Callianassa (*Cheramus*) *subterranea* var. *minor*; Pesta, 1918: 205 (partim).
Callianassa (*Cheramus*) *minor*; De Man, 1928b: 26 (list), 92, 100 (key).
? *Callianassa subterranea* var. *minor*; Balss, 1936: 16, fig. 15.
Callianassa denticulata Lutze, 1937: 6, figs. 1-7; Lutze, 1938: 170; Vatova, 1949, tabs. 5, 20,
     31.
Callianassa stebbingi; Gottlieb, 1953: 440. [Not *Callianassa stebbingi* Borradaile, 1903.]
Callianassa minor; Holthuis & Gottlieb, 1958: 56, figs. 11-12; Harmelin, 1964: 68, 95; Picard,
     1965: 59, 104 (list); Števčić, 1969a: 128 (list); Štjepčević & Parenzan, 1980: 47.
Gourretia minor; De Saint Laurent, 1973: 514; De Saint Laurent & Bozic, 1976: 16 (key), 27,
     figs. 6, 14, 22, 37, 41, 48; Števčić, 1976: 102; Števčić, 1979: 128 (list); Manning & Števčić,
     1982: 296; Števčić, 1985: 314.
Gourretia serrata De Saint Laurent & Le Loeuff, 1979: 79, 80 (key), fig. 24c; Thessalou-Legaki
     & Zenetos, 1985: 311; Thessalou-Legaki, 1986: 182, 184 (list); Dounas et al., 1993: 49.
Gourretia denticulata; Lewinsohn & Holthuis, 1986: 24; Števčić, 1990: 218; Manning & Felder,
     1991: 785 (list), fig. 3; Dworschak, 1992: 210; Koukouras et al., 1992: 223; López de la Rosa
     et al., 1998: 394, fig. 1; Sakai, 1999c: 128; d'Udekem d'Acoz, 1999: 156; Tudge et al., 2000:
     142; Ngoc-Ho, 2003: 499, fig. 21; Sakai & Türkay (in press). [A junior synonym of the pre-
     occupied name *Callianassa subterranea* var. *minor* Gourret, 1887.]

Not *Gourretia minor*; Le Loeuff & Intès, 1974: 26, fig. 4a-k. (= *Gourretia* sp. De Saint Laurent
     & Le Loeuff, 1979 = *Gourretia loeuffintesi* sp. nov.).

Material examined. — SMF 28053, 1 male (Tl/Cl 20.0/4.4, left minor cheliped detached,
lacking left P4), in front of Sotto Castello, Limski-Canal, Istria, Croatia, 45°08.020'N
013°39.030'E, 29 m depth, soft mud, R/V "Burin", 1.ix.1999; SMF 28336, 1 female (31.0/7.0),
around Marseille, France, 1.8 m depth, sediment with *Posidonia*, leg. A. Willsie, 18.v.1983;
SMF 28785, anterior part of carapace with appendages, including both chelipeds, same data and
collector as SMF 28336, 28.v.1983; ZLUA CE 2b, 1 male (16.0/4.0), 1 juv. (Cl 1.5 mm, telson
missing), Aegean Sea, Greece, leg. M. Thessalou-Legaki; ZLT P4500, 1 male (15.5/3.6), Olym-
piada, Strymonikos Gulf, Aegean Sea, Sta. 146, leg. A. Koukouras, 16.v.1977; ZLUA TH2b, 1
male (16.0/4.0), 1 juv. (Cl 1.5 mm, telson missing), Theologos Bay, N. Evoikos Gulf, Greece,
15 m depth, sand, Van Veen grab, leg. M. Thessalou-Legaki, 15.vii.1983; ZLUA R1d, 1 female
(14.0/4.0), N. Rhodes Is., Greece, 70 m depth, coarse coralligenous substrate, Van Veen grab,
leg. M. Thessalou-Legaki, 05.viii.1983.

Diagnosis. — Mxp3 ischium-merus pediform, merus armed with a distomedian spine, exopod present; male Plp1 uniramous, two-segmented, distal segment chelate distally; male Plp2 biramous, endopod with distally demarcated
appendix masculina and laterally attached appendix interna; telson pentagonal,
posterior margin rounded.

Remarks. — In comparing the figures of *Gourretia minor* (cf. Le Loeuff &
Intès, 1974: 26, fig. 4a-k) from Guinea to the neotypes from the Mediterranean

Sea, it is observed that in the Guinean male material the crista dentata of the Mxp3 ischium is armed with a row of eight denticles; the cutting edge of the larger P1 dactylus is smooth, and the tip of the cutting edge of the smaller P1 cheliped is simply pointed, whereas in the Mediterranean male specimen the crista dentata is based on a broader plate, armed with a row of 13 denticles, the cutting edge of the larger P1 dactylus is armed with a row of fine denticles, and the tip of the cutting edge of the smaller P1 cheliped is hook-like. In view of those differences, the Guinean material is to be separated from the Mediterranean material as a new species, *Gourretia loeuffintesi* sp. nov. (see below).

Type locality. — Limski-Canal, Istria, Croatia, Adriatic Sea, 29 m depth.

Distribution. — East Atlantic Ocean: Cadiz Bay, Spain (López de la Rosa et al., 1998). Mediterranean: Gulf of Marseille, Ionian Sea, Malta, and along the coast of Israel; Baie de Kotor (De Saint Laurent & Bozic, 1976); Piran Gulf (Manning & Števčić, 1982). Adriatic Sea (Lutze, 1937; Dworschak, 1992). Tyrrhenian Sea, Ischia (Italy) (Dworschak, 1992); Ionian Sea, Aegean Sea, N. Euboikos Gulf, Patraikos Gulf, and Rhodes, 5-15 m depth, in pure sand (Thessalou-Legaki, 1986; Koukouras et al., 1992). Cyprus (Lewinsohn & Holthuis, 1986). From 2.5 to 146 m depth.

Gourretia lahouensis Le Loeuff & Intès, 1974

Gourretia lahouensis Le Loeuff & Intès, 1974: 28, fig. 5a-v; De Saint Laurent & Le Loeuff, 1979: 79 (key), 80; Manning & Felder, 1991: 785 (list); Tudge et al., 2000: 142.

Diagnosis. — Mxp3 ischium-merus pediform, merus armed with a distomedian spine, exopod present; male Plp1 uniramous, two-segmented, and chelate distally; male Plp2 biramous, endopod with laterally demarcated appendix masculina and laterally attached appendix interna; telson subtrapezoid, narrowed in width in posterior two-thirds, posterior margin broadly convex (Le Loeuff & Intès, 1974: 28, fig. 5j, p, r, r', s).

Type locality. — Grand Lahou, Ivory Coast, 5°7.4'N 5°4.5'W, 15 m.

Distribution. — Ivory Coast, Grand Lahou (Le Loeuff & Intès, 1974).

Gourretia loeuffintesi sp. nov.

Gourretia minor; Le Loeuff & Intès, 1974: 26, fig. 4a-k.
Gourretia sp. De Saint Laurent & Le Loeuff, 1979: 81, fig. 24b.

Diagnosis. — Mxp3 ischium-merus pediform, merus armed with a distomedian spine, ischium with crista dentata bearing 8 teeth; male Plp1 uniramous, two-segmented, distal segment chelate distally; male Plp2 biramous, endopod distally with appendix masculina and laterally attached appendix interna; telson subtriangular, posterior margin convergent in posterior half, showing an obtuse tip (Le Loeuff & Intès, 1974: 26, fig. 4g, h, i).

Remarks. — *Gourretia minor*, Le Loeuff & Intès, 1974 from the Guinea Bay, 6°11.5'N 2°12.5'W, is different from *G. denticulata* from the Mediterranean in the form of the P3 propodus. In *G. denticulata* it is oblong, whereas in *G. minor* from the Guinea Bay it is oval, so herewith this form is named *G. loeuffintesi* sp. nov.

Type locality. — Guinea Bay, Benin, 6°11.5'N 2°12.5'W, 39 m.

Distribution. — Benin, 6°11.5'N 2°12.5'W (De Saint Laurent & Le Loeuff, 1979).

West Atlantic species

Gourretia biffari Blanco Rambla & Liñero Arana, 1994

Gourretia biffari Blanco Rambla & Liñero Arana, 1994: 22, figs. 6, 7; Tudge et al., 2000: 142.

Diagnosis. — Mxp3 ischium-merus pediform, merus provided with a distomedian spine, exopod present; male Plp1-2 undescribed; telson pentagonal, posterior margin rounded (Blanco Rambla & Liñero Arana, 1994: 22, figs. 6, 7).

Type locality. — Northwest of Barcelona, Venezuela, 10°15'30"N 64°42'30"W.

Distribution. — Northwest of Barcelona, Venezuela, 10°15'30"N 64°42'30"W; 50 m depth, muddy bottom.

Gourretia laresi Blanco Rambla & Liñero Arana, 1994

Gourretia laresi Blanco Rambla & Liñero Arana, 1994: 20, fig. 8; Tudge et al., 2000: 142.

Diagnosis. — Mxp3 ischium-merus pediform, merus armed with a strong distomedian spine, exopod present; male Plp1 uniramous, two-segmented, distal segment bilobed distally; male Plp2 biramous, endopod distally with appendix masculina and laterally attached appendix interna; telson trapezoid, posterior margin rounded (Blanco Rambla & Liñero Arana, 1994: 20, fig. 8c, j, h, i).

Type locality. — Northwest of Chimana Islands, Venezuela, 10°18'40"N 64°47'40"W; 71 m depth, clay bottom.

Distribution. — Northwest of Chimana Islands, Venezuela, 10°18'40"N 64°47'40"W.

Indo-West Pacific species

Gourretia coolibah Poore & Griffin, 1979

Gourretia coolibah Poore & Griffin, 1979: 278, figs. 38-39; Manning & Felder, 1991: 785 (list); Tudge et al., 2000: 142; Davie, 2002: 465.

Diagnosis. — Mxp3 ischium-merus pediform, merus lacking a distomedian spine, exopod present; male Plp1-2 unknown; telson pentagonal, posterior margin rounded (Poore & Griffin, 1979: 278, figs. 38c, 39f).

Type locality. — 350 km ENE of Troughton Is., Joseph Bonaparte Gulf, Western Australia, 58 m.

Distribution. — Australia: 350 km ENE of Troughton Is., Western Australia (Poore & Griffin, 1979).

Gourretia crosnieri Ngoc-Ho, 1991

Gourretia crosnieri Ngoc-Ho, 1991: 294, fig. 6; Tudge et al., 2000: 142.

Diagnosis. — Mxp3 ischium-merus pediform, merus lacking a distomedian spine, exopod present; male Plp1-2 unknown; telson subpentagonal, posterior margin rounded (Ngoc-Ho, 1991: 294, fig. 6e, n).

Type locality. — Prony Bay, Ile Ouen, New Caledonia, 29 m.

Distribution. — New Caledonia: Prony Bay, Ile Ouen (Ngoc-Ho, 1991).

Gourretia manihinae Sakai, 1984

Gourretia manihinae Sakai, 1984: 101, figs. 4-5; Tudge et al., 2000: 142.

Diagnosis. — Mxp3 ischium-merus pediform, merus with a distomedian spine, exopod present; male Plp1-2 unknown; telson subtriangular, posterior two-thirds rounded (Sakai, 1984: 101, figs. 4C, 5A).

Type locality. — Pangani Bay, Tanzania, 5°29.2'S 39°03.2'E, 35 m.

Distribution. — Tanzania: Pangani Bay (Sakai, 1984).

Gourretia nosybeensis Sakai, 2004

Gourretia nosybeensis Sakai, 2004: 563-568, figs. 4-6.

Type locality. — Nosy Bé and near Ile d'Ampasindava, Madagascar, 47.00 m.

Genus **Laurentgourretia** Sakai, 2004.

Laurentgourretia Sakai, 2004: 557.

Diagnosis. — Carapace lacking dorsal oval and rostral carina; rostral spine distinct. Transverse cardiac sulcus and cardiac prominence both absent. Scaphocerite strong. Mxp3 ischium-merus subpediform, merus with three mesial spines. Exopod absent. P3 propodus ovoid. P4 simple. P5 subchelate. Abdominal somite 6 lacking acute lateral projections. Male Plp1-2 unknown. Female Plp1 uniramous, biarticulate, distal segment simple. Female Plp2 biramous, with appendix interna. Uropodal exopod bearing lateral notch or incision.

Type species. — *Laurentgourretia rhopalommata* Sakai, 2004, by monotypy and original designation. Gender of the generic name, *Laurentgourretia*, feminine.

Species included. — *Laurentgourretia rhopalommata* Sakai, 2004.

Laurentgourretia rhopalommata Sakai, 2004

Laurentgourretia rhopalommata Sakai, 2004: 557-562, figs. 1-3.

Type locality. — Nosy Mitsio, Madagascar, 12°38.400'S 048°33.200'E; 50.00 m depth.

Distribution. — Known only from the type locality.

Genus **Paragourretia** Sakai, 2004

Paragourretia Sakai, 2004: 568.

Diagnosis. — Carapace large and thick, lacking dorsal oval and rostral carina; rostral spine absent; transverse cardiac sulcus incomplete; cardiac prominence present; row of transverse setae present on anterior branchial region. Dorsal surface of eye flattened. A2 scaphocerite obtuse. Maxilla 2 scaphognathite without posterior long whip. Mxp3 ischium-merus subpediform, usually with one distolateral meral spine; propodus oblong and dactylus digitiform; exopod present. Larger cheliped with proximal meral hook. P3 simple, propodus oblong. P4 subchelate. P5 chelate. Abdominal somite 6 lacking acute lateral projections. Male Plp1 uniramous, biarticulate, distal segment distally chelate. Male Plp2 biramous, foliaceous, and usually with appendix interna and appendix masculina. Female Plp1 uniramous, biarticulate, distal segment simple. Female Plp2 biramous, endopod with appendix interna. Uropodal exopod with lateral notch or incision.

Type species. — *Gourretia phuketensis* Sakai, 2002, by original designation and monotypy. Gender of the generic name, *Paragourretia*, feminine.

Species included. — *Paragourretia aungtonyai* (Sakai, 2002); *P. phuketensis* (Sakai, 2002).

Paragourretia aungtonyai (Sakai, 2002)

Gourretia aungtonyai Sakai, 2002: 476, figs. 6A-H, 7A-F, 8A-G.
Paragourretia aungtonyai; Sakai, 2004: 569.

Diagnosis. — Mxp3 ischium-merus pediform, merus lacking a distomedian spine, exopod present; male Plp1 absent (probably artifact); male Plp2 biramous, endopod distally with appendix interna; telson subtriangular (artifact), posterior margin ovate (Le Loeuff & Intès, 1974: 30, fig. 6j, q, r, t).

Remarks. — The specimens are heavily damaged, so the features reported are sometimes deformed, as indicated.

Type locality. — Andaman Sea, 9°30.351'N 97°57.168'E, 60.7 m, sandy mud, fine sand, and shell fragments.

Distribution. — Andaman Sea, 9°30.351'N, 97°57.168'E;   7°29.921'N, 99°00.977'E, 20.5-60.7 m.

Paragourretia phuketensis (Sakai, 2002)

Gourretia phuketensis Sakai, 2002: 469, figs. 4A-H, 5A-H.
Paragourretia phuketensis; Sakai, 2004: 569-574, figs. 7-8.

Diagnosis. — Mxp3 ischium-merus pediform, merus with a distomedian spine, exopod present; male Plp1 uniramous, two-segmented; male Plp2 biramous, endopod laterally with appendix masculina and laterally attached appendix interna; telson subpentagonal, posterior margin truncate (Sakai, 2002: 469, figs. 4E, 5B, G).

Type locality. — Andaman Sea, 8°29.993'N 98°06.162'E, 42.0 m, muddy sand.

Distribution. — Andaman Sea, 6°45.961'N 99°20.968'E; 8°29.993'N 98°06.162'E; 38.0-42.0 m.

Subfamily CALLIANOPSINAE Manning & Felder, 1991

Callianopsinae Manning & Felder, 1991: 787; Hopkins & Feldmann, 1997: 237.

Definition. — [Revised from Manning & Felder, 1991: 787; Hopkins & Feldmann, 1997: 237.] Rostrum developed, of small size, bearing a low rostral carina. Carapace with low dorsal oval; cardiac prominence present, but hepatic sulcus absent. Eyestalks dorsoventrally flattened and contiguous. Abdominal somite 6 bearing lateral projections. A2 scaphocerite developed as a small process. Mxp3 ischium-merus pediform; propodus broadened; dactylus ovate. P1 chelate, unequal in size, dissimilar in shape; larger cheliped with ventroproximal meral hook, palm oblong, fingers elongate, not pectinate; smaller cheliped usually without ventroproximal meral hook (not in *Gourretia lahouensis*), fingers elongate and not pectinate. P2 chelate. P3 propodus broadened. P4 simple. P5 subchelate. Male Plp1 uniramous, biarticulate; male Plp2 biramous, foliaceous, endopod with bifurcate appendix interna (Hopkins & Feldmann, 1997: 238) or appendix interna and appendix masculina [in my sense: the tip with setae in Hopkins & Feldmann, 1997, fig. 1k]. Plp3-5 similar, with finger-like appendices internae. Uropodal exopod lacking a secondary setal lobe and lateral notch.

Remarks. — The Callianopsinae as defined by Manning & Felder (1991) have the Plp2-5 similar, but different from and larger than Plp1. Hopkins & Feldmann (1997: 237) pointed out that "Both De Saint Laurent (1973) and Manning and Felder (1991) reported that the second through fifth pleopods were similar in *Callianopsis*, the only genus in the subfamily Callianopsinae. The second through fifth pairs of pleopods are all foliaceous, and the third through fifth pairs are almost identical. The second pleopods differ from the third through fifth in the shape of both the endopods and exopods, as discussed below, and the second pleopods are also sexually dimorphic. The second pleopod in the male possesses an appendix masculina, while that of the female possesses a smaller appendix interna". This morphology is the same as observed in the type species of *Ctenocheles*, *Cten. balssi* (cf. Sakai, 1999a: 87).

Manning & Felder (1991: 787) located the Callianopsinae in the family Ctenochelidae. However, this subfamily is conspicuously different from Ctenochelinae, because in the Ctenochelinae the carapace characteristically lacks a dorsal oval, the P3 propodus is oblong, and the male Plp2 is foliaceous, whereas in the Callianopsinae the carapace bears a dorsal oval, the P3 propodus is broadened, and the male Plp2 is foliaceous but narrow; nevertheless, the uropodal exopod lacks the secondary setal lobe in Callianopsinae as in Ctenochelinae. So, in all, the Callianopsinae should be given subfamily status in the new family Gourretiidae, instead of being referred to the Ctenochelidae.

Type genus. — *Callianopsis* De Saint Laurent, 1973.

Genus included. — *Callianopsis* De Saint Laurent, 1973. [See also p. 245.]

Genus **Callianopsis** De Saint Laurent, 1973

Callianopsis De Saint Laurent, 1973: 515; Hopkins & Feldmann, 1997: 237.
Dawsonius Manning & Felder, 1991: 785, figs. 4, 16; Poore, 1994: 103 (key). [Type species: *Callianassa latispina* Dawson, 1967.]

Diagnosis. — [Revised from Manning & Felder, 1991: 789.] Carapace with dorsal oval, bearing low rostral carina, rostral spine, and hepatic prominence. Dorsal surface of eye concave. Mxp3 without exopod; ischium-merus pediform, not armed with distal meral spine; propodus and dactylus broadened, ovate. P1 unequal and dissimilar. P2 chelate. P3 propodus oblong. P4 simple. P5 subchelate. Abdominal somite 6 with acute lateral projections. Uropodal exopod lacking lateral notch or incision and a secondary setal lobe.

Remarks. — I was unable to examine the type specimens of *Dawsonius latispina*, so the validity of this genus is uncertain. The presence or absence of the Mxp3 exopod is a good character to separate *Ctenocheles* and *Dawsonius*, though an Mxp3 exopod is shown in *Dawsonius latispina* by Biffar (1971a, fig. 11b), whereas an Mxp3 exopod was not shown by Dawson (1967, fig. 1f, h) and Manning & Felder (1991, fig. 16c, d). The genus *Dawsonius* resembles

Gourretia, as Manning & Felder (1991: 785) indicated, so it is necessary to re-examine the type series of *D. latispina*. However, the denial of a loan request has prevented the author from further investigating the validity of Manning & Felder's (1991) genus. *Dawsonius* is here synonymized with *Callianopsis*, because an Mxp3 exopod has been shown not to be present by Dawson (1967, fig. 1f, h) and Manning & Felder (1991, fig. 16c, d); carapace with a rounded, median oval area (Dawson, 1967, as referred to below in the remarks on *Callianopsis latispina*); there is an indication of a low mid-dorsal carina; the cardiac prominence (Manning & Felder, 1999, fig. 16b) is present; and abdominal somite 6 is armed with anterolateral projections.

Type species. — *Callianassa goniophthalma* Rathbun, 1902, by original designation and monotypy. Gender of the generic name, *Callianopsis*, feminine.

Species included. — West Atlantic species: *Callianopsis latispina* (Dawson, 1967). East Pacific species: *Callianopsis goniophthalma* (Rathbun, 1902). Indo-West Pacific species: *C. caecigena* (Alcock & Anderson, 1894).

West Atlantic species

Callianopsis latispina (Dawson, 1967)[*])

Callianassa latispina Dawson, 1967: 190, fig. 1; Biffar, 1971a: 651 (key), 654 (key), 679, figs. 11-12; Herper, 1975: 619; Rabalais et al., 1981: 105, fig. 3g-k.
Gourretia latispina; Manning, 1987: 398 (list); Williams et al., 1989: 28.
Dawsonius latispina; Manning & Felder, 1991: 785 (list), figs. 4, 16; Tudge et al., 2000: 142.

Diagnosis. — Carapace bearing a dorsal oval and with an indication of a low mid-dorsal carina. Mxp3 ischium-merus pediform, merus with a distomedian spine, without exopod. Abdominal somite 6 with anterolateral projections. Male Plp1 uniramous, two-segmented, distal segment protruded distally; male Plp2 biramous, endopod distally with appendix interna and laterally attached appendix masculina. Telson subsquare, posterior margin rounded (Biffar, 1971a: 679, figs. 11b, i, 12 b, e).

Remarks. — Dawson (1967: 190) described: "Carapace dorsolaterally rounded, median "oval area" somewhat elevated and with an indication of a low mid-dorsal carina", though careful examination of the figure shows that the carapace does not reveal the presence of a dorsal oval (Manning & Felder, 1991, fig. 16b).

---

[*]) See p. 245.

Type locality. — Off Grand Isle, Louisiana, 13.5 m.

Distribution. — Louisiana: Grand Isle (Dawson, 1967; Biffar, 1971a); Texas: off Galveston and Corpus Christi Bay (Herper, 1975), between Port Aransas and Port Isabel (Rabalais et al., 1981); off southWestern Florida (Biffar, 1971a); Honduras, off Trujillo (Biffar, 1971a); 9-51 m.

East Pacific species

Callianopsis goniophthalma (Rathbun, 1902)

Callianassa goniophthalma Rathbun, 1902: 886; Rathbun, 1904: 154, pl. 8; Schmitt, 1921: 121, fig. 82; Balss, 1925: 211; Stevens, 1928: 342, fig. 19; Williams et al., 1989: 28.
Callianassa (Calliactites) goniophthalma; Borradaile, 1903: 545; De Man, 1928b: 25 (list), 95, 96.
Callianopsis goniophthalma; Manning & Felder, 1991: 789 (list), figs. 7, 18; Hendrickx, 1995: 390 (list); Hopkins & Feldmann, 1997: 238, figs. 1-4; Tudge et al., 2000: 142.

Diagnosis. — Mxp3 ischium-merus subrectangular, merus obliquely declined to a distomesial tooth and convex on mesial margin, lacking exopod; male Plp1 uniramous, two-segmented, distal segment scythe-shaped; male Plp2 biramous, endopod ovate, laterally with small process from which branch appendix masculina with distal setae and appendix interna, exopod ovate; telson subovoid, posterior margin broadly concave (Hopkins & Feldmann, 1997: 238, figs. 1f, j, k, m).

Type locality. — Off Point Conception, California, "Albatross" sta. 3198, 513 m depth.

Distribution. — Alaska: Clarence Strait (Rathbun, 1902, Hopkins & Feldmann, 1997), Funter Bay (Stevens, 1928), Yes Bay (Hopkins & Feldmann, 1997). California: off Point Conception (Rathbun, 1902; Hopkins & Feldmann, 1997), off Harris Point, San Miguel Island (Schmitt, 1921). Baja California: San Cristobal Bay, Mexico (Hopkins & Feldmann, 1997); 351-595 m.

Indo-West Pacific species

Callianopsis caecigena (Alcock & Anderson, 1894)

Callianassa caecigena Alcock & Anderson, 1894: 163; Alcock & Anderson, 1896, pl. 26 fig. 2, 2a-b; Alcock, 1901: 198; Balss, 1925: 211.
(?) *Callianassa (Calliactites) coecigena*; Borradaile, 1903: 545.
Callianassa (Calliactites) caecigena; De Man, 1928: 25 (list), 96 (key).
Callianopsis coecigena; De Saint Laurent & Le Loeuff, 1979: 95.

Diagnosis. — Mxp3 ischium-merus subrectangular, male Plp1-2 unknown, telson subtrapezoid in posterior half, convex on posterior margin (Alcock & Anderson, 1894: 163; Alcock & Anderson, 1896, pl. 26 fig. 2a).

Remarks. — Borradaile included the present species, *Callianassa caecigena* together with (?) *Callianassa* (*Calliactites*) *goniophthalma* Rathbun, 1900 in the subgenus *Calliactites* Borradaile, 1903 (synonym of *Callianidea*). *Calliactites caecigena* is similar to *Calliactites goniophthalma* in the form of the carapace, bearing a cardiac prominence (Alcock & Anderson, 1896, pl. 26 fig. 2), the larger and smaller chelipeds, abdominal somite 6, uropods, and telson, but differs in that the abdominal somites 2-5 in *C. caecigena* have lateral spines.

Type locality. — Off Ceylon, Bay of Bengal, 365-690 m.

Distribution. — Off Sri Lanka (Alcock & Anderson, 1894); 365-690 m.

Subfamily PSEUDOGOURRETIINAE n. subfam.

Definition. — Rostrum triangular, lacking rostral carina. Carapace without dorsal oval; cardiac prominence present but hepatic sulcus absent; pleurobranchs present. Eyestalks dorsoventrally flattened and contiguous. Abdominal somite 6 unknown. A2 scaphocerite developed as a small process. Mxp3 ischium-merus oblong, bearing mesiodistal meral tooth; propodus oblong; dactylus digitiform. P1-2 unknown. P3 propodus oblong. Male Plp1 uniramous, biarticulate, distal segment chelate. Male Plp2-5 and female Plp1-5 unknown. Uropodal exopod unknown.

Remarks. — This subfamily is distinguished from the other subfamilies by the presence of pleurobranchs on P2-4. The morphological characters are similar to those of the Gourretiinae, as the rostrum is short, triangular, and lacks a rostral carina; the eyestalks are dorsoventrally flattened; the Mxp3 ischium-merus is oblong, with a mesiodistal meral tooth, the propodus and dactylus are slender, and an exopod is present.

Type genus. — *Pseudogourretia* n. gen.

Genus included. — *Pseudogourretia* n. gen.

Genus **Pseudogourretia** n. gen.

Diagnosis. — Carapace lacking dorsal oval; rostral carina absent; rostral spine and cardiac prominence present, but no transverse cardiac sulcus present;

linea thalassinica present. Dorsal surface of eye flattened, cornea not developed. Mxp3 ischium-merus oblong, and with mesiodistal meral tooth; propodus and dactylus slender; exopod present. P3 propodus oblong. Male Plp1 chelate. Pleurobranchs present on P2-4.

Remarks. — It is difficult to place this new genus in the family Callianassidae, because three pleurobranchs are present on P2-4. It also resembles the family Gourretiidae, because there is no rostral carina on the carapace, and the maxilla 2 scaphognathite lacks a posterior seta. On the other hand, the type species of *Pseudogourretia*, *P. portsudanensis*, lacks pleopods. *Pseudogourretia portsudanensis* is also similar to *Gourretia*, in that the male Plp1 is chelate, but it differs from *Gourretia*, because in *Pseudogourretia* the P3 is oblong as in *Ctenocheles*, while in *Gourretia* the P3 propodus is heel-shaped.

Etymology. — From the Greek, pseudo, false, and the generic name *Gourretia*. Gender feminine.

Type species. — *Pseudogourretia portsudanensis* sp. nov. by present designation and monotypy.

Species included. — Indo-West Pacific species: *Pseudogourretia portsudanensis* sp. nov.

Indo-West Pacific species

Pseudogourretia portsudanensis sp. nov.
(fig. 42)

Material. — SMF 28105, holotype, 1 male (Cl 8.2, lacking abdominal somite 2 to telson, lacking chelipeds), Red Sea, off Port Sudan, Me5-148 Ku, 19°43.3'N 37°40.5'E-19°44.5'N 37°40.2'E, 517-583 m depth, R/V "Meteor", 20.ii.1987.

Diagnosis. — Mxp3 ischium-merus pediform, merus with a distomedian spine, exopod present; male Plp1 uniramous, two-segmented, chelate distally; male Plp2 unknown; telson unknown.

Description of male holotype. — Rostrum (fig. 42A-B) triangular in dorsal view and distally pointed. Carapace lacking dorsal oval, smooth, without rostral carina, and with an obtuse anterolateral angle; cervical groove located in posterior third of carapace. Linea thalassinica extends at full length. Cardiac prominence present.

Eyestalks (fig. 42A) oval, longer than broad, convex and directed downward distally from rostrum on dorsal surface; tip obtuse, longer than distal end of

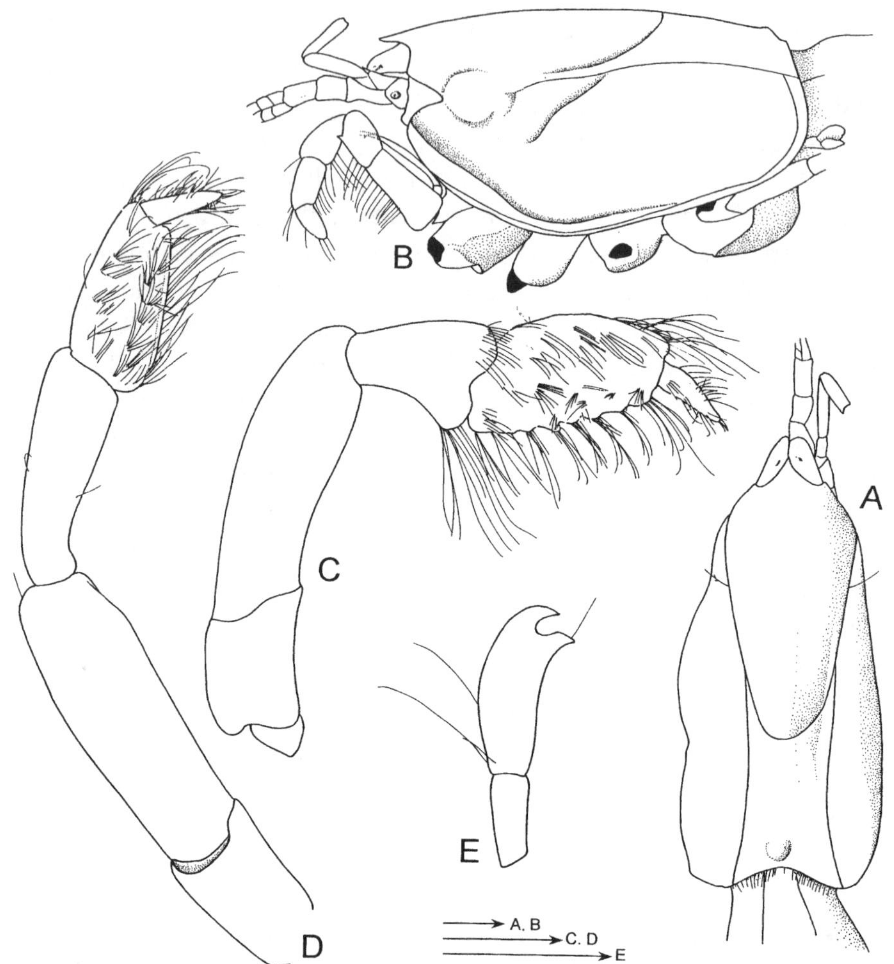

Fig. 42. *Pseudogourretia portsudanensis* sp. nov. A, carapace, dorsal view; B, same, lateral view; C, P3, lateral view; D, P4, lateral view; E, male Plp1. A-E, holotype, SMF 28105, 1 male (Cl 8.2, lacking abdominal somite 2-telson, chelipeds), Red Sea, off Port Sudan, Me5-148 Ku (19°43.3'N 37°40.5'E-19°44.5'N 37°40.2'E), 517-583 m depth, R/V "Meteor". Scales 1 mm.

antennular basal article; cornea small, located medially, pigmented black in alcohol specimen.

A1 peduncle (fig. 42A-C) shorter than A2 peduncle, terminal article about as long as penultimate. A2 scaphocerite distinct, narrow-triangular in form; terminal article slightly shorter than penultimate. Mxp3 (fig. 42B) with exopod, overreaching distal end of ischium; merus-ischium of endopod rectangular; ischium rectangular, 2.0 times as long as broad; crista dentata with a

TABLE X

Branchial formula of *Pseudogourretia portsudanensis* sp. nov.

	Maxillipeds			Pereiopods				
	1	2	3	1	2	3	4	5
Exopods	1	1	1	–	–	–	–	–
Epipods	1	1	–	–	–	–	–	–
Podobranchs	–	1	–	–	–	–	–	–
Arthrobranchs	–	–	2	2	2	2	2	–
Pleurobranchs	–	–	–	–	1	1	1	–

row of 13 stout denticles, of which proximal 4 are reduced in size and fused; merus oblong, 1.3 times as broad as long and about half length of ischium, distal margin convex, bearing a sharp tooth at mesiodistal angle; carpus triangular, 1.5 times as long as broad and longer than merus; propodus subrectangular, about as long as carpus; dactylus digitiform, 0.8 times as long as propodus and distally obtuse. Maxilla 2 scaphognathite without a posterior seta. Branchial formula as shown in table X.

Chelipeds missing. P2 chelate. P3 (fig. 42C) simple; ischium 1.2-1.3 times as long as broad; merus 2.0 times as long as ischium and three times as long as broad, carpus about half length of merus, convex at distoventral part; propodus oblong, 1.2 times as long as carpus and 1.5 times as long as broad, setose on mesial surface, dorsal margin entirely convex and ventral margin denticulate in a straight line; dactylus half length of propodus, narrowly triangular with a stout seta.

P4 (fig. 42D) simple; ischium elongate, about 1.5 times as long as broad, merus 2.3 times as long as ischium and three times as long as broad; carpus 0.7 times as long as merus; propodus oblong, 0.8 times as long as carpus and 2.0 times as long as broad, roundishly angled at posteroventral corner; dactylus narrow, sickle-shaped, and shorter than half length of propodus.

P5 chelate; ischium short; merus 2.5 times as long as ischium; carpus 0.7 times as long as merus; propodus about as long as merus, forming a posteroventral projection; dactylus half as long as propodus, curved towards external side of distoventral projection of propodus.

Abdominal somites 2-6 and tail-fan lacking; somite 1 short, smooth, glabrous dorsally; Plp1 (fig. 42E) uniramous, 2-segmented, distal segment chelate.

Remarks. — Unfortunately, the holotype is a damaged male specimen lacking abdominal somites 2-6 and the tail-fan. However, because it exhibits characters significant at the generic level, and as it is unlikely that more material will be collected in the foreseeable future, I have decided it important to describe the species now rather than to wait for better specimens.

Etymology. — The species is named after the type locality, Port Sudan. The specific name is an adjective agreeing in gender with the (feminine) generic name.

Type locality. — Port Sudan (19°43.3'N 37°40.5'E-19°44.5'N 37°40.2'E), Red Sea, 517-583 m.

FAMILY CTENOCHELIDAE MANNING & FELDER, 1991

Ctenochelidae Manning & Felder, 1991: 784; Poore, 1994: 103; Hendrickx, 1995a: 387 (key), fig. 8; Tudge et al., 2000: 135; Davie, 2002: 463.

Definition. — Rostrum developed but short, unarmed laterally, and usually bearing denticulate rostral carina. Carapace without dorsal oval; linea thalassinica present. Abdominal somites 3-5 dorsolaterally with a tuft of setae. Eyestalks dorsoventally flattened and contiguous. A2 scaphocerite developed as a sharp spine. Maxilla 2 without posterior seta. Mxp3 pediform or subpediform; propodus oblong, dactylus digitiform. P1 chelate, dissimilar. Larger cheliped without a proximal meral hook, palm subglobular, fingers elongate, pectinate; smaller cheliped without proximal meral hook, fingers elongate and not pectinate. P3 propodus oblong. Male Plp1-2 smaller than Plp3-5; female Plp1 smaller than Plp2-5 and of more slender shape; male Plp2 with appendix interna and appendix masculina; female Plp2 with finger-like appendix interna. Plp3-5 with appendices internae in both sexes. Uropodal exopod with lateral notch or incision.

Additional diagnosis. — Propyloric ossicle of gastric mill highly protruded ventrally, posterior surface bearing a low median longitudinal carina. Lateral tooth thickened anteriorly to form a molar protrusion, and posterior to it a smooth lower median carina, extended backward to a secondary molar protrusion (Sakai, 2005).

Subfamily included. — Ctenochelinae Manning & Felder, 1991.

Subfamily CTENOCHELINAE Manning & Felder, 1991

Ctenochelinae Manning & Felder, 1991: 784; Sakai, 1999a: 87.

Definition. — Rostrum narrowly protruded, unarmed laterally, and bearing rostral carina. Carapace without dorsal oval; cardiac prominence and transverse hepatic sulcus present or absent. Eyestalks dorsoventrally flattened and contiguous. Abdominal somite 6 lacking lateral projections. A2 scaphocerite developed, sharp. Mxp3 ischium-merus subpediform, merus with mesial spine; propodus oblong; dactylus digitiform. P1 chelate, unequal in size, dissimilar in shape; larger cheliped without a ventroproximal meral hook, palm subglobular, fingers elongate, pectinate; smaller cheliped without ventroproximal meral hook, fingers elongate and not pectinate. P2 chelate. P3 propodus oblong or broadened. Male Plp1 blade-like and uniramous. Male Plp2 foliaceous, biramous, endopod bearing appendix interna and appendix masculina, being conspicuously smaller than Plp3-5. Female Plp1 slender and uniramous. Female Plp2-5 foliaceous, biramous, endopods with appendices internae. Plp2 slightly smaller than Plp3-5 in size and slightly more slender in shape. Uropodal exopod lacking a secondary setal lobe, but with lateral notch.

Remarks. — Manning & Felder's (1991) taxa and Tudge et al.'s (2000) clades seem to give a basis to Poore's (1994: 95-96) assertion that the Ctenochelidae and Callianassidae are to be considered as two families, for the reason that in the Ctenochelidae Plp2 is similar to Plp3-5, with lanceolate rami; the P3 propodus is linear or weakly ovate; and the uropodal exopod is simply ovate. In Callianassidae, by contrast, the Plp2 is reduced and sexually modified; Plp3-5 are provided with broad, interacting rami; the P3 propodus is armed with a proximal heel on the lower margin; and the uropodal exopod is provided with a secondary, setose lobe. He (Poore, 1994) also placed the Upogebiidae in a separate family, considering that in Ctenochelidae and Callianassidae the eyestalks are flattened, while in the Upogebiidae they would be cylindrical. It has turned out, after a close examination of the material, that in Callianassidae the eyestalks are variable in form: flattened, subglobose, or cylindrical, whereas in the Upogebiidae they are not cylindrical, but subglobose, whence Poore (1994) erroneously described them as cylindrical in the Upogebiidae. The family Ctenochelidae was once proposed at subfamily level, as Ctenochelinae (cf. Sakai, 1999a). However, owing to Manning & Felder's (1991) analysis, the Ctenochelidae should indeed be placed at the level of a family, because it is correctly observed that (1) in *Ctenocheles balssi* Kishinouye, 1926, the type species of *Ctenocheles*, the male Plp2 is much reduced, and shows dissimilarity in shape to Plp3-5, as in *Eucallichirus*, where Plp2 is sexually modified as recently revealed by Matsuzawa & Hayashi (1997) and Sakai (1999a), and the female Plp2 is distinctly reduced in size, not so much sexually modified in size and shape as in males, but its appendices internae are different in shape (cf. *Ctenocheles collini*, see fig. 44D, E), and Plp3-5 are foliaceous and biramous as in other callianassoids. (2) The P3 propodus is not linear or weakly ovate in *Ctenocheles balssi*, but subquadrate as recently examined, and this form is also observed in *Callianassa profunda* Biffar, 1973, *C. longicauda* Sakai, 1967, and *Lipkecallianassa abyssa* Sakai, 2002. (3) In *Ctenocheles* the uropodal exopod is not provided with a secondary setose lobe,

TABLE XI

Comparison of morphological characters in the family Ctenochelidae and in taxa that show some degree of resemblance, i.e., the family Gourretiidae as a whole and its subfamily Pseudogourretiinae

(Sub)families	Ctenochelidae	Pseudogourretiinae	Gourretiidae
Genera	*Ctenocheles* Kishinouye, 1926	*Pseudogourretia* n. gen.	*Gourretia* De Saint Laurent, 1973 *Laurentgourretia* Sakai, 2004 *Paragourretia* Sakai, 2004
Rostral spine	triangular	lacking	triangular
Rostral carina	present	absent	absent
Cardiac prominence	present, with a mid-pit	present, without mid-pit	present or absent, when present, with or without a mid-pit
Cardiac transverse sulcus	absent	absent	absent or present
Pleurobranchs	absent	P2-4 present	absent
Mxp3 merus	with mesiodistal spine	with mesiodistal tooth	with or without mesiodistal spine
P3 propodus	oblong	oblong	broadened in a heeled form
Male Plp1	uniramous, three- or four-segmented	uniramous, chelate distally	uniramous, chelate distally

but this character state is also found in such callianassid species as *Callianassa propinqua* De Man, 1905, *C. longicauda* Sakai, 1967, *C. brachytelson* Sakai, 2002, *C. nigroculata* Sakai, 2002, as it is in the Ctenochelidae. As a result, it is reasonable that the genus *Ctenocheles* is to be included in the family Ctenochelidae, by the characters of the absence of the long posterior seta on the scaphognathite of maxilla 2, and the different size and shape of Plp1-2 and Plp3-5, respectively, in both sexes.

Type genus. — *Ctenocheles* Kishinouye, 1926.

Genera included. — *Ctenocheles* Kishinouye, 1926.

Genus **Ctenocheles** Kishinouye, 1926

?*Pentacheles* Balss, 1914: 75.

Thaumastocheles; Doflein, 1906: 522 (partim).

Ctenocheles Kishinouye, 1926: 63; Powell, 1949b: 369; Holthuis, 1967: 377; De Saint Laurent, 1973: 514; Le Loeuff & Intès, 1974: 24; Poore & Griffin, 1979: 277; De Saint Laurent & Le Loeuff, 1979: 47 (key), 81; Sakai, 1987a: 306 (list); Manning, 1987: 397 (list); Coelho & Ramos-Porto, 1987: 29 (key), 31; Manning & Felder, 1991: 784, figs. 2, 7; Poore, 1994: 103 (key); Sakai, 1999a: 88; Davie, 2002: 464.

Diagnosis. — [Revised from Manning & Felder, 1991: 784.] Carapace lacking dorsal oval; rostral carina present, rostral spine present, and cardiac prominence usually present, but transverse cardiac sulcus absent. Dorsal surface of eye flattened. Mxp3 usually without exopod, distal margin of merus usually with spine. Mxp3 propodus and dactylus slender. Larger cheliped with or without proximal meral hook, palm subglobular, fingers elongate, pectinate. P1 chelate, dissimilar. P2 chelate. P3 propodus oblong. P4 simple. P5 subchelate. Male Plp1 uniramous, biarticulate; male Plp2 biramous, foliaceous, and with appendices interna and masculina. Female Plp1 uniramous, biarticulate; Plp2 biramous, and with appendix interna. Plp3-5 different from Plp1-2 in size and shape. Uropodal exopod with lateral incision, lacking secondary setose lobe.

Remarks. — One character of *Ctenocheles* was described by Manning & Felder (1991: 784) as: "Plp2-5 are similar, different from and larger than Plp1". Poore (1994) and Tudge et al. (2000), continued the use of this character, but, as shown for the type species, *C. balssi*, by Sakai (1999: 86), Plp3-5 are similar, but Plp2 is different from Plp3-5 in size and shape. The species of *Ctenocheles* usually bear no Mxp3 exopod, but a rudimentary exopod is observed in *Ctenocheles serrifrons* Le Loeuff & Intès, 1974.

Type species. — *Ctenocheles balssi* Kishinouye, 1926, by monotypy. Gender of the generic name, *Ctenocheles*, masculine.

Species included. — East Atlantic species: *Ctenocheles serrifrons* Le Loeuff & Intès, 1974; *C.* sp. De Saint Laurent & Le Loeuff, 1979. West Atlantic species: *C. holthuisi* Rodrigues, 1978; *C. leviceps* Rabalais, 1979; *C.* [sp.] A. Holthuis, 1967; *C.* [sp.] B. Holthuis, 1967. Indo-West Pacific species: *C. balssi* Kishinouye, 1926; *C. collini* Ward, 1945; *C. maorianus* Powell, 1949.

East Atlantic species

Ctenocheles serrifrons Le Loeuff & Intès, 1974

Ctenocheles sp. Crosnier, 1969: 536, fig. 18.
Ctenocheles serrifrons Le Loeuff & Intès, 1974: 24, fig. 3a-u; De Saint Laurent & Le Loeuff, 1979: 83; Manning & Felder, 1991: 784 (list), fig. 2; Matsuzawa & Hayashi, 1997: 44 (in key); Sakai, 1999a: 88 (list); Tudge et al., 2000: 142.

Diagnosis. — Mxp3 ischium-merus pediform, merus armed with a distomedian spine; bearing small exopod; male Plp1 uniramous, two-segmented, distal segment obtuse distally; male Plp2 biramous, endopod ovate, laterally with small appendix interna; telson subsquare, posterior margin broadly convex (Le Loeuff & Intès, 1974: 24, fig. 3d, g, n, t).

Type locality. — Ivory Coast, 5°2'N 5°4.5'W, 50 m.

Distribution. — Ivory Coast (Le Loeuff & Intès, 1974).

Ctenocheles sp. De Saint Laurent & Le Loeuff, 1979

Ctenocheles sp. De Saint Laurent & Le Loeuff, 1979: 83, fig. 25; Sakai, 1999a: 88 (list).

Distribution. — Gabon (De Saint Laurent & Le Loeuff, 1979); Benin (De Saint Laurent & Le Loeuff, 1979); 48-110 m.

West Atlantic species

Ctenocheles holthuisi Rodrigues, 1978

Ctenocheles holthuisi Rodrigues, 1978: 113, figs. 1-21; Coelho & Ramos-Porto, 1987: 31; Manning, 1987: 397 (list); Manning & Felder, 1991: 784 (list); Matsuzawa & Hayashi, 1997: 45 (in key); Sakai, 1999a: 88 (list); Tudge et al., 2000: 142.

Diagnosis. — Mxp3 ischium-merus pediform, merus armed with a strong distomedian spine; male Plp1-2 unknown; telson pentagonal, posterior margin convex (Rodrigues, 1978: 113, figs. 1-21).

Type locality. — Off mouth of Rio São Francisco, Brazil, 10°37'09"S 36°14'00"W, 75 m.

Distribution. — Brazil: off mouth of Rio São Francisco (Rodrigues, 1978; Coelho & Ramos-Porto, 1987).

Ctenocheles leviceps Rabalais, 1979

Ctenocheles leviceps Rabalais, 1979: 295, figs. 2-29; Rabalais et al., 1981: 100; Manning, 1987: 397 (list); Williams et al., 1989: 28; Manning & Felder, 1991: 784 (list), fig. 7; Matsuzawa & Hayashi, 1997: 45 (in key); Sakai, 1999a: 88 (list); Tudge et al., 2000: 142.

Diagnosis. — Mxp3 ischium-merus pediform, merus lacking a distomedian spine, exopod present; male Plp1 uniramous, two-segmented, distal segment obtuse distally; male Plp2 biramous, endopod lanceolate, with laterally attached small appendix interna bearing minute hooks; telson subsquare, posterior margin convex, slightly concave medially (Rabalais, 1979: 295, figs. 13, 28, 25, 26, 27, 28).

Type locality. — Port Aransas, Texas, U.S.A., 27°38'N 96°41'W.

Distribution. — Texas: Port Aransas (Rabalais, 1979); Gulf of Mexico (Rabalais et al., 1981); 10-49 m.

Ctenocheles [sp.] A. Holthuis, 1967

Ctenocheles [sp.] A. Holthuis, 1967: 379, figs. 1, 2a; Manning, 1987: 397 (list); Manning & Felder, 1991: 784 (list); Sakai, 1999a: 88 (list).

Distribution. — Straits of Florida, near Bimini Group (Holthuis, 1967); off north coast of Panama (Holthuis, 1967); 109-406 m.

Ctenocheles [sp.] B. Holthuis, 1967

Ctenocheles [sp.] B. Holthuis, 1967: 382, fig. 2b; Manning, 1987: 398 (list); Manning & Felder, 1991: 784 (list); Sakai, 1999a: 88 (list).

Distribution. — Colombia: off Cartagena and off Cordoba State (Holthuis, 1967).

Indo-West Pacific species

Ctenocheles balssi Kishinouye, 1926

Thaumastocheles Doflein, 1906: 522.
? *Pentacheles* nov. sp.? Balss, 1914: 75, fig. 43.
Ctenocheles balssi Kishinouye, 1926: 63, fig. 1; Yokoya, 1933: 55; Makarov, 1938: 76, fig. 29; Holthuis, 1967: 377; Suzuki, 1979: 296, pl. 18 fig. 234; Sakai, 1987a: 306 (list); Manning & Felder, 1991: 784 (list); Noguchi & Akamine, 1992: 25, fig. 1; Liu & Zhong, 1994: 562 (list); Matsuzawa & Hayashi, 1997: 39, 44 (in key), figs. 1-3; Sakai, 1999a: 88, figs. 1a-e, 2a-g, 3a-g; Tudge et al., 2000: 142.
Ctenocheles Balssi; De Man, 1928b: 25 (list).

Diagnosis. — Mxp3 ischium-merus pediform, merus provided with a distomedian spine, exopod absent; male Plp1 uniramous, 4-segmented, distal segment scythe-shaped distally; male Plp2 biramous, endopod lanceolate, laterally with small appendix interna and appendix masculina; telson subquadrate, posterior margin convex (Sakai, 1999a: 88, figs. 1d, 2a, 3a, b, c).

Type locality. — Ohsu near Kashiwasaki, Niigata Prefecture, Japan.

Distribution. — Japan: Ohsu near Kashiwazaki, Niigata Pref. (Kishinouye, 1926), Wakasa Bay (Yokoya, 1933), Okinose Bank, Sagami Bay (Holthuis, 1967), Nezumigaseki, Yamagata Pref. (Suzuki, 1979), Igarashihama, Niigata Pref. (Noguchi & Akamine, 1992), off None, Muroto Peninsula (Matsuzawa & Hayashi, 1997; Sakai, 1999a); 30-200 m. East China Sea to South China Sea (Liu & Zhong, 1994: 562).

Ctenocheles collini Ward, 1945
(figs. 43-44)

Ctenocheles collini Ward, 1945: 134, pl. 13; Holthuis, 1967: 377; Poore & Griffin, 1979: 277, fig. 37; Manning & Felder, 1991: 784 (list); Matsuzawa & Hayashi, 1997: 45 (in key); Sakai, 1999a: 88 (list); Tudge et al., 2000: 142.

Material examined. — QMB 5953, 1 ovig. female (Tl/Cl 17.7/4.6, including rostrum); QMB 5951, 1 female (18.5/4.5), Mud Island, Moreton Bay, Queensland, leg. V. F. Collin.

Diagnosis. — Carapace with a sharp rostral spine extending to halfway the dorsal oval as a smooth carina (fig. 43A). Mxp3 ischium-merus pediform, merus about same length as ischium, bearing a strong subterminal tooth on mesial margin, exopod reaching to distal end of merus; P3 propodus broadened, posterior margin tapering to dactylus and with high posterior angle (fig. 43C). Telson subpentagonal, posterior margin convex with shallow median concavity. Uropodal exopod lacking secondary setose lobe (fig. 43D). Male Plp1-2 unknown. Female Plp1 (fig. 44A) uniramous, 2-segmented, distal segment foliaceous; Plp2 (fig. 44B) uniramous, foliaceous, obviously smaller than Plp3-5, bearing an elongate appendix interna (fig. 44C). Plp3 (fig. 44D) to Plp5 biramous, foliaceous, and with stubby appendices internae (fig. 44E).

Type locality. — Moreton Bay, Queensland, Australia; 15-40 m.

Distribution. — Moreton Bay, Queensland, Australia (Ward, 1945; Poore & Griffin, 1979).

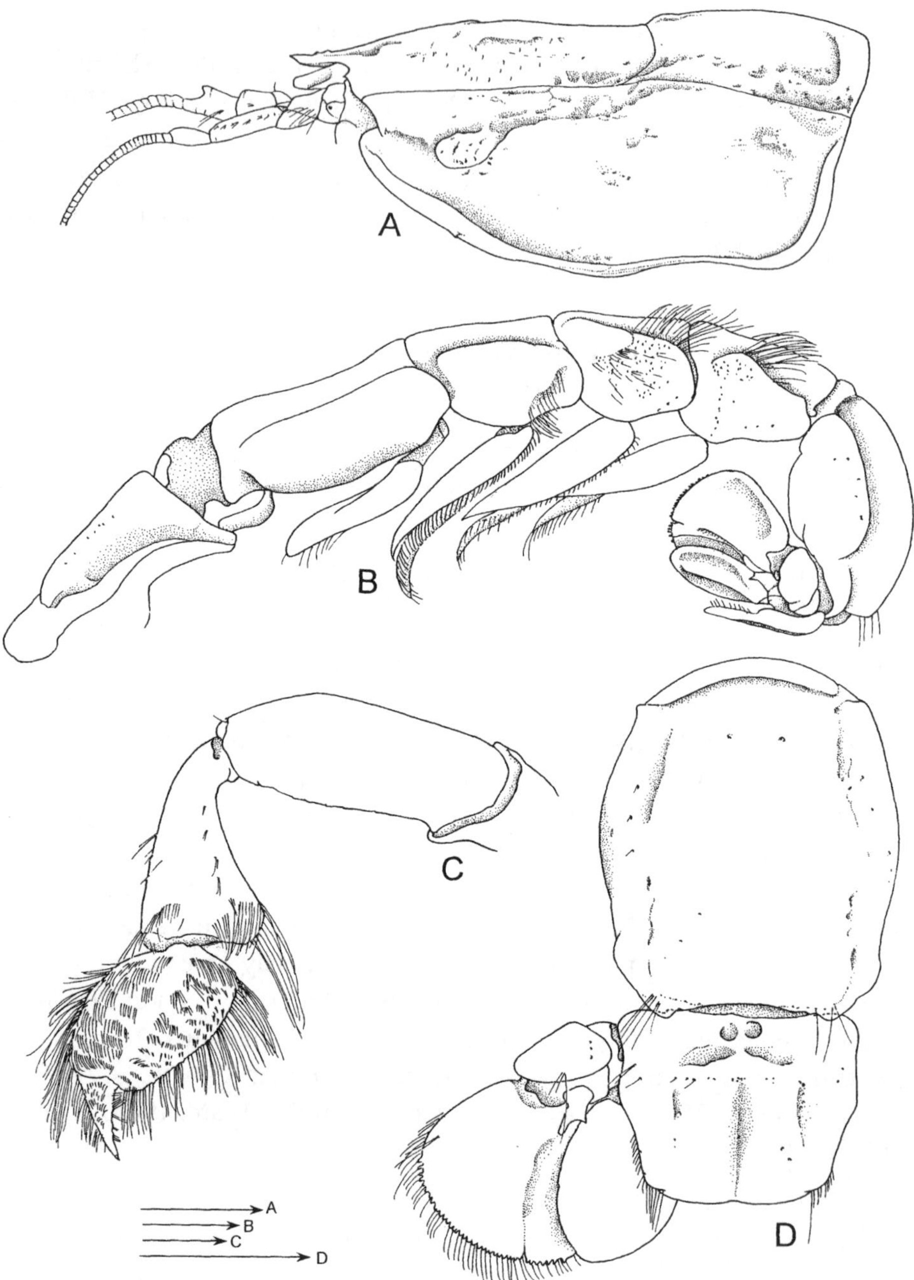

Fig. 43. *Ctenocheles collini* Ward, 1945. QMB 5953, 1 ovig. female (Tl/Cl 17.7/4.6 mm including rostrum), Mud Island, Moreton Bay, Queensland, leg. V. F. Collin. A, carapace, lateral view; B, abdomen and tail-fan, lateral view; C, P3, lateral view; D, abdominal somite 6 and left part of tail fan in dorsal view. Scales 1 mm.

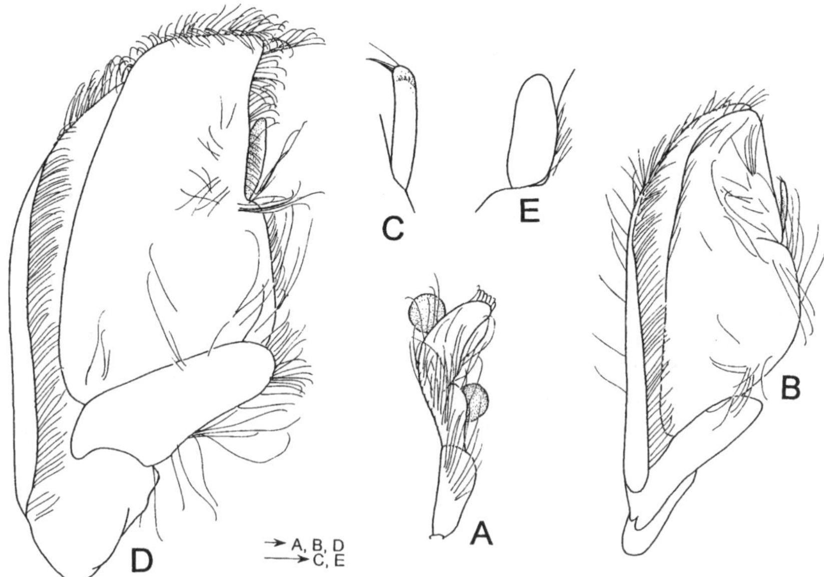

Fig. 44. *Ctenocheles collini* Ward, 1945. QMB 5953, 1 ovig. female (Tl/Cl 17.7/4.6 mm including rostrum), Mud Island, Moreton Bay, Queensland. A, female Plp1; B, female Plp2; C, appendix interna of female Plp2; D, female Plp4; E, appendix interna of female Plp4. Scales 1 mm.

Ctenocheles maorianus Powell, 1949

Ctenocheles maorianus Powell, 1949a: 408; Powell, 1949b: 369, pl. 68 figs. 3-7; Dell, 1956: 149; Holthuis, 1967: 378; Manning & Felder, 1991: 784 (list); Dworschak, 1992: 209, figs. 12a-b, 13a-c; Matsuzawa & Hayashi, 1997: 45 (in key); Sakai, 1999a: 88 (list); Tudge et al., 2000: 142.

Remarks. — Powell (1949a) mentioned that the male holotype is closely similar to *C. collini* Ward, 1945, but differs from that species, because the telson in *C. maorianus* bears two prominent, wide-spaced, calcareous spines.

Type locality. — Hauraki Gulf, New Zealand.

Distribution. — New Zealand: Hauraki Gulf, Tiritiri Island, off Plate Island, Bay of Plenty, and Tasman Bay (Powell, 1949a); off Cape Runaway, off Kaikoura, Marlborough, Castlecliffian (upper Pleistocene) near Castlecliff (Holthuis, 1967); Little Town (Dworschak, 1992); 35-73 m.

ACKNOWLEDGEMENTS

I am most grateful to Dr. T. Wolff of the Zoological Museum, University of Copenhagen, for allowing me to examine many historical specimens preserved in the Museum, by which I could pursue the present work. I am indebted to Dr. M. Türkay, Forschungsinstitut Senckenberg, Frankfurt am Main, who entrusted me with his important material collected from off Port Sudan, Red Sea by the R/V "Meteor" in 1987, from the Canary Islands, Atlantic Ocean by the R/V "Valdivia" in 1975, and from the Adriatic Sea by the Excursions of Frankfurt University in 1993, 1999, and 2001, and who also suggested me that I should examine the old Rovinj collections preserved in the Stazione Idrobio-logica di Chioggia, Dipartimento di Biologia, Università degli Studi di Padova. Thanks are also due to Dr. P. Clark, Ms. M. Lowe, and science photographer Mr. Philip R. Hurst, the Natural History Museum, London, who kindly sent me their digital photos of two old species from A. White's collection, *Callianassa abdominalis* White, 1847 and *Callianassa carinaedorsis* White, 1847 [both are in fact *Scytoleptus* sp., Axiidae, det. M. De Saint Laurent in June 1977], as well as the holotype of *Cheramus orientalis* Bate, 1888.

I am also heartily thankful to Dr. J. Carel von Vaupel Klein, University of Leiden, and editor of Crustaceana, for making the inevitable arrangements for the publication of this monograph; to Dr. Lipke B. Holthuis and Dr. C. H. J. M. Fransen of the National Museum of Natural History, Leiden, for giving me much useful advice on nomenclatorial problems; to Dr. Niel L. Bruce, New Zealand and Dr. P. Davie, Queensland Museum, Brisbane, Australia, for reading my original manuscripts for improvement; my thanks are also extended to Dr. Peter C. Dworschak, Naturhistorisches Museum, Vienna, who gave me further suggestions to improve my manuscript.

I wish to thank the following individuals for making necessary specimens available for study: Dr. J. C. Yaldwyn, National Museum of New Zealand, Wellington; Dr. Peter. J. F. Davie, Queensland Museum, South Brisbane; the late Dr. J. Haig, Allan Hancock Foundation, Southern California University, who sent me her collection for reference; Dr. M. Türkay and Mr. A. Allspach, Forschungsinstitut Senckenberg, Frankfurt am Main, who entrusted me with their important material; Dr. N. Ngoc-Ho, Muséum national d'Histoire naturelle, Paris; Dr. M. Takeda, National Science Museum, Tokyo; Prof. S. Casellato, Dipartimento di Biologia, Università degli Studi di Padova, and Prof. S. Vito, Stazione Idrobiologica di Chioggia, Dipartimento di Biologia, Università degli Studi di Padova; Dr. C. Aungtonya, Phuket Marine Biological Center, Thailand; and Dr. M. Thessalou-Legaki, Department of Zoology-Marine Biology, School of Biology, University of Athens, Athens.

NOTE ADDED IN PROOF

The genus *Dawsonius* Manning & Felder, 1991 was initially synonymized with *Callianopsis* in the above text. However, after examining the type species of *Dawsonius*, i.e., *Callianassa latispina* Dawson, 1967, it has turned out that *Dawsonius* can be considered a good genus. As a result, the number of genera in the Callianassoidea has now been raised to 21.

In an earlier stage of the investigations, the type specimen of *Dawsonius latispina* was not accessible, so the validity of this genus had to remain uncertain. However, on 17 March 2005 I received, with many thanks, the male holotype of *Callianassa latispina* Dawson, 1967 (USNM 105398), sent to me by courtesy of Dr. Rafael Lemaitre of the Smithsonian Institution, Washington, D.C. Therefore, only then I had the opportunity to examine it and study its morphology in sufficient detail.

As a result, it can now safely be stated that the holotype of *Callianassa latispina* was incorrectly described and figured by Biffar (1971a: 682, fig. 11b), as that author erroneously describes the presence of an exopod on Mxp3. In contrast, both Dawson (1967: 193) and Manning & Felder (1991: 783, fig. 16c) correctly described and figured the holotype, including the absence of an exopod on Mxp3. Thus, it is now possible to say that Dawson's species, *Callianassa latispina*, is clearly different from the genera *Gourretia* and *Ctenocheles* by the absence of the Mxp3 exopod, and in this respect similar to the genus *Callianopsis* De Saint Laurent, 1973. *Callianassa latispina* was included earlier in the genus *Dawsonius* Manning & Felder, 1991, based on the following definition: "Diagnosis. — Carapace with rostral carina and rostral spine." (Manning & Felder, 1991: 785). This obviously means (1) the presence of a dorsal oval with an indication of a low mid-dorsal carina, and (2) the presence of a spiniform rostrum. Those characters are applicable to incorporate *Dawsonius latispina* in the subfamily Callianopsinae, but they are insufficient to bring *Dawsonius latispina* under the genus *Callianopsis*. This is because in *Dawsonius latispina* the Mxp3 dactylus is digitiform, whereas in the type species of the genus *Callianopsis*, *Callianassa goniophthalma* Rathbun, 1902, this structure is oval. For that reason, the genus *Dawsonius* is to be separated from *Callianopsis* and admitted as a good genus.

Genus **Dawsonius** sensu Sakai (not Manning & Felder, 1991)

Definition. — Carapace bearing dorsal oval with an indication of a low mid-dorsal carina, and a rostral spine. Mxp3 exopod absent; merus with distoventral spine; dactylus digitiform. Chelipeds unequal. Male Plp2 with appendix masculina and appendix interna laterally fused with one another in their proximal half. Uropodal exopod lacking lateral notch.

Type species. — *Callianassa latispina* Dawson, 1967, by monotypy.

Remarks. — *Dawsonius latispina* resembles *Callianopsis goniophthalma* (Rathbun, 1902) in (1) having sharp lateral projections on abdominal somite 6; (2) bearing an indication of a low mid-dorsal carina on the carapace; (3) a rostrum shaped as a spine; (4) the Mxp3 without an exopod; and (5) the uropodal exopod lacking a lateral notch or incision. However, it differs from *Callianopsis*, because in that genus (1) the Mxp3 dactylus is ovate; and (2) the male Plp2 has a slender appendix masculina bearing an appendix interna.

Tokushima, 18 March 2005; revised 22 September 2005

K. Sakai

REFERENCES

ANONYMOUS, 1941. Annotated list of the fauna of the Grand Isle region 1928-1946. Occasional Pap. mar. Lab. Louisiana State Univ., **6**: 1-66.

ABED-NAVANDI, D. & P. C. DWORSCHAK, 1997. First record of the thalassinid *Callianassa truncata* Giard & Bonnier, 1980 in the Adriatic Sea (Crustacea: Decapoda: Callianassidae). Ann. naturhist. Mus. Wien, (B) **99**: 565-570, figs. 1-9.

—— & ——, 1998. First record of the thalassinids *Callianassa acanthura* Caroli, 1946 and *Upogebia mediterranea* Noel, 1992 and of the hermit crab *Paguristes streaensis* Pastore, 1984 in the Adriatic Sea. Ann. naturhist. Mus. Wien, (B) **100**: 605-612, figs. 1-8, tab. 1.

ABELE, L. G. & W. KIM, 1986. An illustrated guide to the marine decapod crustaceans of Florida. State of Florida, Department of Environmental Regulation, Technical Series, **8**(1): i-xvii, 1-326 (Part 1), 327-760 (Part 2), pls. (Tallahassee, Florida).

ABU-HILAL, A., M. BADRAN & J. DE VAUGELAS, 1988. Distribution of trace elements in *Callichirus laurae* burrows and nearby sediments in the Gulf of Aqaba, Jordan (Red Sea). Mar. environm. Res., **25**: 233-248, 4 figs.

ADEMA, J. P. H. M., F. CREUTZBERG & G. J. VAN NOORT, 1982. Notes on the occurrence of some poorly known Decapoda (Crustacea) in the southern North Sea. Zool. Bijdr., Leiden, **28**: 9-32.

ADENSAMER, T., 1898. Decapoden gesammelt auf S.M. Schiff "Pola" in den Jahren 1890-1894. Berichte der Commission für Erforschung des östlichen Mittelmeeres, **22** (Zoologische Ergebnisse, 11): 597-628, fig. 1. (Denkschr. Akad. Wiss. Wien).

AGASSIZ, L., 1846. Nomenclatoris zoologici Index universalis, continens Nomina systematica Classium, Ordinum, Familiarum et Generum Animalium omnium, tam viventium quam fossilium, secundum Ordinem alphabeticum unicum disposita, adjectis Homonymiis Plantarum: i-x, 1-1135. (Jent & Gassmann, Soloduri).

ALCOCK, A., 1900. Illustrations of the zoology of the royal Indian marine surveying steamer "Investigator", **7**, pls. 36-45. (Calcutta).

——, 1901. A descriptive catalogue of the Indian deep-sea Crustacea Macrura and Anomura, in the Indian Museum. Being a revised account of the deep-sea species collected by the royal Indian marine survey ship "Investigator": 1-286, i-iv, pls. 1-3. (Calcutta).

ALCOCK, A. & A. R. S. ANDERSON, 1894. Natural history notes from H.M. Indian marine survey steamer "Investigator", (2) **14**. An account of a recent collection of deep-sea Crustacea from the Bay of Bengal and Laccadive Sea. Journ. Asiatic Soc. Bengal, **63**(2)(3): 141-185, pl. 9.

—— & ——, 1899. Natural history notes from H.M. royal Indian marine survey ship "Investigator", Commander T.H. Heming, R.N., commanding, (3) **2**. An account of the deep-sea Crustacea dredged during the surveying-season of 1897-98. Ann. Mag. nat. Hist., (7) **3**: 1-27, 278-292, pls.

ALLEN, J. A., 1967. The fauna of the Clyde Sea. Crustacea: Euphausiacea and Decapoda with an illustrated key to the British species: 1-116. (Scottish Marine Biological Association, Millport).

ANDALORO, F., A. CEFALI, R. MILLEMACI & C. M. SCUDERI, 1979. I Crostacei Decapodi dello stagnone di Marsala: note ambientali. Mem. Biol. mar. Oceanogr., **9**(3): 85-90, figs. 1-2.

ATHANASSOPOULOS, G. T., 1917. Quelques éléments de recherches hydrobiologiques en Grèce. Bull. Sta. Hydrobiol. mar., Greece, **2**(1): 1-40.

BĂCESCU, M., 1949. Données sur la faune carcinologique de la Mer Noire le long de la côte bulgare. Trav. Sta. Biol. mar. Varna, **14**: 7-10, figs.

— —, 1967. Crustacea Decapoda. Fauna Republicii soc. Romania, (Crustacea) **4**(9): 1-351, figs. 1-141, pls. 1, 2. [In Romanian.]

BAKER, W. H., 1907. Notes on South Australian decapod Crustacea. Part V. Trans. Roy. Soc. South Australia, **31**: 173-191, pls. 23-25.

BALSS, H. [V.], 1914. Ostasiatische Decapoden II. Die Natantia und Reptantia. In: F. DOFLEIN, Beiträge zur Naturgeschichte Ostasiens. Abh. bayerische Akad. Wiss., (Suppl.) **2**(10): 1-101, figs. 1-50, pl. 1.

— —, 1915. Anomuren, Dromiaceen und Oxystomen. Die Decapoden des Roten Meeres. II. Expeditionen S.M. Schiff "Pola" in das Rote Meer. Nördliche und Südliche Hälfte 1895/96-1897/98. Zoologische Ergebnisse, **31**. Berichte der Kommission für ozeanographische Forschungen. Denkschr. Akad. Wiss. Wien, (Math.-naturwiss. Kl.) **92**(10): 1-20, figs. 1-9.

— —, 1916. Crustacea II: Decapoda Macrura und Anomura (außer Fam. Paguridae). In: W. MICHAELSEN, Beiträge zur Kenntnis der Meeresfauna Westafrikas, **2**: 13-46, figs. 1-16.

— —, 1924. Westindische Decapoden. Zool. Anz., **61**: 177-182, figs. 1-5.

— —, 1925. Macrura der Deutschen Tiefsee-Expedition. 1. Palinura, Astacura und Thalassinidea. Wissenschaftliche Ergebnisse der Deutschen Tiefsee-Expedition auf dem Dampfer "Valdivia" 1808-1810, **20**(4): 209-213 (Thalassinidea), 16 text-figs. pls. 18-19.

— —, 1933. Über einige systematisch interessante indopazifische Dekapoden. Mitt. zool. Mus. Berlin, **19**: 84-97, figs. 1-9, pls. 1-2.

— —, 1957. Decapoda. VII. Systematik. In: H. G. BRONN, Klassen und Ordnungen des Tierreichs, **5**(1)(7)(12): 1505-1672, text-figs. 1131-1199.

BARNARD, K. H., 1946. Description of new species of South African decapod Crustacea, with notes on synonymy and new records. Ann. Mag. nat. Hist., (11) **13**(102): 361-392.

— —, 1950. Descriptive catalogue of South African decapod Crustacea. Ann. South African Mus., **38**: 1-837, figs. 1-154.

BATE, C. S., 1888. Report on the Crustacea Macrura collected by H.M.S. "Challenger" during the years 1873-1876. Rep. scient. Res. Challenger Exped., (Zool.) **24**: i-xc, 1-942, figs. 1-76, pls. 1-150.

BEAUBRUN, P.-C., 1979. Crustacés Décapodes marcheurs des côtes marocaines (sections des Astacidea Eryonidea Palinura Thalassinidea). Bull. Inst. Sci., Rabat, **3**: 1-110, figs. 1-75.

BEHRE, E. H., 1950. Annotated list of the fauna of the Grand Isle Region 1928-1946. Occ. Pap. mar. Lab. Louisiana State Univ., **6**: 1-66.

BELL, T., 1844-1853. A history of British stalk-eyed Crustacea: i-lxv, 1-386, figs. 1-174. (London). [Dates of publication: part 1, pp. 1-48, 1 October 1844; part 2, pp. 49-96, 2 December 1844; part 3, pp. 97-144, 1 May 1845; part 4, pp. 145-192, December 1845; part 5, pp. 193-240, December 1846; part 6, pp. 241-288, December 1847; part 7, pp. 289-336, 1851; part 8, pp. 337-386, 1852; see Gordon (1960: 191) for details.]

BERKENBUSCH, K. & A. A. ROWDEN, 1998. Population dynamics of the burrowing ghost shrimp *Callianassa filholi* on an intertidal sandflat in New Zealand (Decapoda; Thalassinidea). Ophelia, **49**(1): 55-69, illustr.

— — & — —, 2000. Intraspecific burrow plasticity of an intertidal population of *Callianassa filholi* (Crustacea: Decapoda: Thalassinidea) in relation to environmental conditions. New Zealand Journ. mar. freshw. Res., **34**(3): 397-408.

BIFFAR, T. A., 1970. Three new species of callianassid shrimp (Decapoda, Thalassinidea) from the western Atlantic. Proc. biol. Soc. Washington, **83**(3): 35-50, figs. 1-3.

— —, 1971a. The genus *Callianassa* (Crustacea, Decapoda, Thalassinidea) in south Florida, with keys to the western Atlantic species. Bull. mar. Sci. Univ. Miami, **21**(3): 637-715, figs. 1-22.

— —, 1971b. New species of *Callianassa* (Decapoda, Thalassinidea) from the western Atlantic. Crustaceana, **21**(3): 225-236, figs. 1-3.

— —, 1972. A study of the eastern Pacific representatives of the genus *Callianassa* (Crustacea, Decapoda, Callianassidae): i-xv, 1-276. (Ph.D. Thesis, University of Miami, Coral Gables, Florida). [Not referred to as a publication.]

— —, 1973. The taxonomic status of *Callianassa occidentalis* Bate, 1888, and *C. Batei* Borradaile, 1903 (Decapoda, Callianassidae). Biological results of the University of Miami Deep-Sea Expedition 98. Crustaceana, **24**(2): 224-230, figs. 1-2.

BLANCO RAMBLA, J. P. & R. LEMAITRE, 1999. *Neocallichirus raymanningi*, a new species of ghost shrimp from the northeastern coast of Venezuela. Proc. biol. Soc. Washington, **112**(4): 768-777, figs. 1-4.

BLANCO RAMBLA, J. P. & I. LIÑERO ARANA, 1994. New records and new species of ghost shrimps (Crustacea: Thalassinidea) from Venezuela. Bull. mar. Sci., **55**: 16-29.

BLANCO RAMBLA, J. P., I. LIÑERO ARANA & M. L. BELTRÁN LARES, 1995. A new callianassid (Decapoda: Thalassinidea) from the southern Caribbean Sea. Proc. biol. Soc. Washington, **108**(1): 102-106, figs. 1-3.

BLOHM, A., 1913. Die Dekapoden der Nord- und Ostsee (mit Ausnahme der Natantia Boas): 1-115, tabs., 4 maps. (Heider Anzeiger, Kiel).

BOAS, J. E. V., 1880. Studier over Decapodernes Slaegtskabsforhold. (Recherches sur les affinités des Crustacés Décapodes). Kong. Danske Selsk. Skr., (6) **1**: 25-210, pls. 1-7.

BODENHEIMER, F. S., 1937. Prodromus Faunae Palestinae. Essai sur les éléments zoogéographiques et historiques du sud-ouest du sous-règne paléarctique. Mém. Inst. Egypte, **33**: i-iii, 1-286, figs. 1-4.

BOGDANOS, C. & J. SATSMADJIS, 1983. The macrozoobenthos of an Aegean embayment. Thalassographica, **6**: 77-105.

BOONE, L., 1927. Crustacea from tropical east American seas. Scientific results of the first oceanographic expedition of the "Pawnee" 1925. Bull. Bingham oceanogr. Coll., **1**(2): 1-147, figs. 1-33.

BORRADAILE, L. A., 1900. On the Stomatopoda and Macrura brought by Dr. Willey from the South Seas. In: A. WILLEY, Zoological results based on the material from New Britain, New Guinea, Loyalty Islands and elsewhere collected during the years 1895, 1896, and 1897, **4**: 395-428, pls. 36-39. (Cambridge).

— —, 1903. On the classification of the Thalassinidea. Ann. Mag. nat. Hist., (7) **12**: 534-551, 638.

— —, 1904. Marine Crustaceans. XIII. The Hippidea, Thalassinidea and Scyllaridea. In: J. STANLEY GARDINER, The Fauna and Geography of the Maldive and Laccadive Archipelagoes, **2**: 750-754, figs. 1-6, pl. 58. [Publ. 5 July 1904, P. Clark, in litt.]

— —, 1910. Penaeidea, Stenopidea and Reptantia from the western Indian Ocean. The Percy Sladen Trust Expedition to the Indian Ocean in 1905, under the leadership of Mr. J. Stanley Gardiner, M.A., **2**(10). Trans. Linn. Soc. London, (Zool.) **13**: 257-264, pl. 16.

BOTT, R. 1955. Litorale Dekapoden, außer *Uca*. Dekapoden (Crustacea) aus El Salvador. Ergebnisse der Forschungsreise A. Zilch 1951 nach El Salvador, 18. Senckenb. Biol., **36**(1/2): 45-72, figs. 1-7, pls. 1-8.

BOURDON, R., 1965. Inventaire de la faune marine de Roscoff. Décapodes - Stomatopodes. Ed. Sta. Biol. Roscoff: 1-45.

BOUVIER, E. L., 1895. Sur une collection de Crustacés Décapodes recueillis en Basse-Californie par M. Diguet. Bull. Mus. Hist. nat., Paris, **1**: 6-8.

— —, 1901. Sur quelques Crustacés du Japon, offerts au Muséum par M. le Dr. Harmand. Bull. Mus. Hist. nat., Paris, **7**: 332-334.

— —, 1905. Sur les Thalassinidés recueillis par le "Blake" dans la mer des Antilles et le golfe du Mexique. C. R. Acad. Sci., Paris, **141**: 802-806.

— —, 1925. Les Macroures marcheurs. Reports on the results of dredging under the supervision of Alexander Agassiz in the Gulf of Mexico (1877-78), in the Caribbean Sea (1878-79), and along the Atlantic coast of the United States (1880), by the U.S. Coast Survey Steamer "Blake". Mem. Mus. comp. Zool., Harvard, **47**(5): 401-472, figs.1-28, pls. 1-11.

— —, 1940. Décapodes marcheurs. Faune de France, **37**: 1-404, figs. 1-222, pls. 1-14.

BRITTON, J. C. & B. MORTON, 1989. Shore ecology of the Gulf of Mexico: i-viii, 1-387. (University of Texas, Austin, Texas).

BRONN, H. G. & F. ROEMER, 1852. Lethaea geognostica (ed. 3), **2**(5): 1-412, i-iv.

CANO, G., 1891. Sviluppo postembrionale della *Gebia*, *Axius*, *Callianassa* e *Calliaxis*. Morfologia dei Thalassinidi. Boll. Soc. Naturalisti, Napoli, **5**(1): 5-30, tabs. 1-4.

CAROLI, E., 1921. Identificazione delle supposte larve di *Calocaris macrandreae* Bell ed *Axius stirhynchus* Leach. Pubbl. Staz. zool. Napoli, **3**: 241-252.

— —, 1940. In difesa della *Callianassa truncata* Giard et Bonnier. Boll. Zool., Union. zool. Italiano, **11**(3-4): 73-77.

— —, 1946. Una nuova *Callianassa* (*C. acanthura* n. sp.) del golfo di Napoli, con alcune considerazioni sulle forme giovanili del genere. Pubbl. Staz. zool. Napoli, **20**: 66-74, figs. 1-3.

— —, 1950. Sulla validatà del nome *Callianassa laticauda* Otto. Pubbl. Staz. zool. Napoli, **22**: 189-191.

CARUS, J. V., 1885. Prodromus Faunae Mediterraneae sive Descriptio Animalium Maris Mediterranei Incolarum quam comparata siva Rerum quatenus innotuit adiectis Locis et Nominibus vulgaribus eorumque Auctoribus in Commodum Zoologorum: i-xi, i-iv, 1-525 [Decapoda, 470-525]. (Stuttgart).

CASPERS, H., 1939. Die Bodenfauna der Helgoländer Tiefen Rinne. Helgoländer wiss. Meeresunters., **2**(1): 1-112, figs. 1-35, tabs. 1-22.

— —, 1941. Zur Systematik von *Callianassa*. Antwort an J. Lutze. Zool. Anz., **133**(7/8): 658.

CHACE, F. A., JR., 1962. The non-brachyuran decapod crustaceans of Clipperton Island. Proc. U.S. natn. Mus., **113** (3466): 605-635, figs. 1-7.

CHACE, F. A., JR., J. J. MCDERMOTT, P. A. MCLAUGHLIN & R. B. MANNING, 1986. Order Decapoda (shrimps, lobsters and crabs). In: W. STERRER & C. SCHOEPFER-STERRER (eds.), Marine fauna and flora of Bermuda. A systematic guide to the identification of marine organisms: 312-358, pls. 101-118, col. pls. 9-11. (New York).

CHILTON, C., 1907. Notes on the Callianassidae of New Zealand. Trans. Proc. New Zealand Inst., **39**: 456-464, pl. 16.

— —, 1911. The Crustacea of the Kermadec Islands. Trans. Proc. New Zealand Inst., **43**: 544-573, figs. 1-4.

CHRISTIANSEN, M. E., 1972. Crustacea Decapoda. Tifotkreps: 1-71, figs. 1-91. (Universitetsforlaget, Oslo).

— —, 2000. On the occurrence of Thalassinidea (Decapoda) in Norwegian waters. Journ. Crust. Biol., (spec. no.) **2**: 230-237, 1 fig., 1 tab.

CHRISTIANSEN, M. E. & L. GREVE, 1982. First record of the thalassinid *Callianassa subterranea* (Montagu) (Crustacea, Decapoda) from the coast of Norway. Sarsia, **67**(3): 213-214.

CHRISTIANSEN, M. E. & R. STENE, 1998. Occurrence of the thalassinid *Callianassa subterranea* (Montagu) (Crustacea, Decapoda) on the coast of southern Norway. Sarsia, **83**: 75-77.

CLARK, P. F. & B. PRESSWELL, 2001. Adam White: the crustacean years. Bull. Raffles Mus., **49**(1): 149-166, tabs. 1-2.

COELHO, P. A. & M. DE A. RAMOS, 1973 [not 1972]. A constituição e a distribuição da fauna de Decápodos do litoral leste da América do sul entre as latitudes de 5°N 39'S. Trab. oceanogr. Univ. Fed. Pernambuco, Recife, **13**: 133-236.

COELHO, P. A. & M. RAMOS-PORTO, 1987. Sinopse dos Crustáceos Decápodos Brasileiros (familias Callianassidae, Callianideidae, Upogebiidae, Parapaguridae, Paguridae, Diogenidae). Trab. oceanogr. Univ. Fed. Pernambuco, Recife, **19**: 27-53.

COLOSI, G., 1923. Crostacei Decapodi della Cirenaica. Mem. R. Com. Talassogr. Italiano, **104**: 1-11.

COUTIÈRE, H., 1899. Note sur *Callianassa grandidieri* n. sp. (Voyage de M. Guillaume Grandidier à Madagascar). Bull. Mus. Hist. nat., Paris, **5**: 285-287, figs. 1-5.

CROSNIER, A., 1969. Sur quelques Crustacés Décapodes ouest-africains: description de *Pinnotheres leloeuffi* et *Pasiphaea ecarina* spp. nov. Bull. Mus. Hist. nat., Paris, (2) **41**(2): 529-543, figs. 1-36.

CZERNIAVSKY, V., 1884. Crustacea Decapoda Pontica littoralia. Materalia ad Zoogeographiam Ponticam comparatam. II. Trans. Soc. Univ., Kharkov, **13**(Suppl.): 1-268, pls.1-7.

— —, 1868. I. Materialia ad Zoogeographiam Ponticam comparatam. I. Studiosi Universitatis Charcoviensis: 16-136, pls. 1-8. [= Crustacea sinum Jaltensem incolentia et Catalogus. Crustaceorum ponticorum in Museo Zool. Acad. Petrop. etc. conserv., Petropólis: 1-120, pls. 1-8.]

DAKIN, W. J., I. BENNETT & E. C. POPE, 1952. Australian seashores. A guide for the beach-lover, the naturalist, the shore fisherman, and the student (ed. 1): i-xii, 1-372, figs. 1-23, pls. 1-99. (Sydney).

DAKIN, W. J. & A. N. COLEFAX, 1940. The plankton of the Australian coastal waters off New South Wales. Part. I. With special reference to the seasonal distribution, the phytoplankton, and the planktonic Crustacea, and in particular, the Copepoda and crustacean larvae, together with an account of the more frequent members of the groups Mysidacea, Euphausiacea, Amphipoda, Mollusca, Tunicata, Chaetognatha, and some reference to the fish eggs and fish larvae. Publ. Univ. Sydney, Dept. Zool., (Mon.) **1**: 1-215, figs. 1-303, pls. 1-4.

DANA, J. D., 1852a. Macroura. Conspectus Crustaceorum, &c. Conspectus of the Crustacea of the exploring expedition under Capt. C. Wilkes, U.S.N. Proc. Acad. nat. Sci., Philadelphia, **6**: 10-28.

— —, 1852b. Crustacea. In: United States Exploring Expedition during the Years 1838, 1839, 1840, 1841, 1842, under the Command of Charles Wilkes, U.S.N., **13** (pt. 1): i-viii, 1-1685, pl. 8; **13** (Atlas): 1-27, pls. 1-96 [1855]; **13** (pt. 2): 1-1593 [1853].

— —, 1854. Catalogue and descriptions of Crustacea collected in California by Dr. John L. Le Conte. Proc. Acad. nat. Sci., Philadelphia, **7**: 175-177.

DARNELL, R. M., 1958. Food habits of fishes and larger invertebrates of Lake Pontchartrain, Louisiana, an estuarine community. Publ. Inst. mar. Sci., Univ. Texas, **5**: 353-416, figs. 1-17.

DAVIE, P., 2002. Crustacea: Malacostraca: Phyllocarida, Eucarida (part 1). In: A. WELLS & W. W. K. HOUSTON (eds.), Zoological catalogue of Australia, **19**(3A): i-xii, 1-551. (CSIRO Publishing, Melbourne).

DAVIS, F. M., 1925. Quantitative studies on the fauna of the sea bottom, no. 2. Results of the investigations in the southern North Sea, 1921-24. Fishery Invest., London, (2) **8**(4): 1-50, figs. 1-2, tabs. 1-7, appends. 1-5.

DAWSON, C. E., 1967. *Callianassa latispina* (Decapoda, Thalassinidea), a new mud shrimp from the northern Gulf of Mexico. Crustaceana, **13**: 190-196, fig. 1.

DEKAY, J. E., 1844. Crustacea. In: Zoology of New York, or the New-York fauna; comprising detailed descriptions of all the animals hitherto observed within the state of New-York, with brief notices of those occasionally found near its borders, and accompanied by appropriate illustrations, **6**: 11-70, pls. 1-13. (Carroll and Cook, Albany).

DELL, R. R., 1956. A record of *Latreillopsis pettersi* Grand (Crustacea, Brachyura) from New Zealand, with notes on some other species of Crustacea. Rec. Dom. Mus. New Zealand, **2**(3): 147-149, fig. 1.

DESMAREST, A. G., 1825. Considérations générales sur la classe des Crustacés et description des espèces de ces animaux, qui vivent dans la mer, sur les côtes, ou dans les eaux douces de la France: i-xix, 1-446, pls. 1-56, tabs. 1-5 (F.G. Levrault, Paris and Strasbourg).

DOFLEIN, F., 1902. Ostasiatische Dekapoden. Abh. bayerischen Akad. Wiss., Munich, (2) **21**(3): 613-670, figs. 1-4(A-D), pls. 1-6.

— —, 1906. Mitteilungen über japanische Crustaceen. Zool. Anz., **30**: 521-525.

DOLGOPOL'SKAIA, M. A., 1954. Métamorphose des Crustacés de la Mer Noire. 2. Callianassidae. Trav. Sta. biol. Sevastopol, **8**: 178-213, figs. 1-14. [In Russian.]

— —, 1969. Larves de Décapodes, Macrura et Anomura. In: F. D. MORDUKHAY-BOLTOVSKOY (ed.), Manuel de détermination de la faune de la Mer Noire et de la Mer d'Azov, 2. Invertébrés libres, Crustacea. Akademija Nauk Ukrainskoi SSR, Institut Biologii Yuznykh Morey, Izdatel'stvo "Naukova Dumka", Kiev: 307-362, pls. 1-44. [In Russian.]

DOMENECH, J. L., M. M. DE LA HOZ & J. A. ORTEA, 1981. Crustáceos Decápodos de la costa Asturiana, nuevas citas y especies probables, I. Macruros. Bol. Cienc. natur. I.D.E.A., **27**: 119-157, figs. 1-36.

DOUNAS, C., I. KARAKASSIS & C. H. J. M. FRANSEN, 1993. Towards a revision of the east Atlantic species of the genus *Ebalia* Leach (Crustacea, Brachyura, Leucosiidae). Preliminary results. Première Conférence Européenne sur les Crustacés, Résumés, Paris: 49.

DWORSCHAK, P. C., 1992. The Thalassinidea in the Museum of Natural History, Vienna, with some remarks on the biology of the species. Ann. naturhist. Mus. Wien, (B) **93**: 189-238, figs. 1-18.

— —, 2000a. The burrows of *Callianassa tyrrhena* (Petagna, 1792) (Decapoda: Thalassinidea). Marine Ecol., **22**(1-2): 155-166, figs. 1-6.

— —, 2000b. On the burrows of *Lepidophthalmus louisianensis* (Schmitt, 1935) (Decapoda: Thalassinidea: Callianassidae). Senckenb. marit., **30**(3-6): 99-104.

— —, 2002. The burrows of *Callianassa candida* (Olivi, 1792) and *C. whitei* Sakai, 1999c (Crustacea: Decapoda: Thalassinidea). In: M. BRIGHT, P. C. DWORSCHAK & M. STACHOWITSCH (eds.), The Vienna school of marine biology: a tribute to Jörg Ott: 63-71. (Vienna).

DWORSCHAK, P. C., A. ANKER & D. ABED-NAVANDI, 2000. A new genus and three new species of alpheids (Decapoda: Caridea) associated with thalassinids. Ann. naturhist. Mus. Wien, (B) **102**: 301-320.

DWORSCHAK, P. C. & J. A. OTT, 1993. Decapod burrows in mangrove-channel and back-reef environments at the Atlantic Barrier Reef, Belize. Ichnos, **2**: 277-290.

DWORSCHAK, P. C. & P. PERVESLER, 1988. Burrows of *Callianassa bouvieri* Nobili, 1904 from Safaga (Egypt, Red Sea) with some remarks on the biology of the species. Senckenb. marit., **20**(1-2): 1-17, figs. 1-5, 1 tab., pls. 1-2.

EDMONDSON, C. H., 1944. Callianassidae of the central Pacific. Occ. Pap. Bernice P. Bishop Mus., Honolulu, **18**(2): 35-61, figs. 1-11.

——, 1946. Reef and shore fauna of Hawaii. Spec. Publ. Bernice P. Bishop Mus., Honolulu, **22**: 1-381, figs. 1-223.

ESTAMPADOR, E. P., 1937. A check list of Philippine crustacean decapods. Philippine Journ. Sci., **62**: 465-559.

FAURE, G., 1970. Bionomie et écologie de la macrofaune des substrats meubles des côtes charentaise. Tethys, **1**(3): 751-778.

FELDER, D. L., 1973. An annotated key to crabs and lobsters (Decapoda, Reptantia) from coastal waters of the northwestern Gulf of Mexico: i-vii, 1-103, pls. 1-12. Center for Wetland Resources, Louisiana State University, Baton Rouge, Publ. Np. LSU-SG-**73-02**.

——, 1975. Physioecological studies of Louisiana Callianassidae (Crustacea, Decapoda, Thalassinidea), **1-2**: i-ix, 1-117. (Ph.D. Thesis, Louisiana State University, Baton Rouge, Louisiana).

——, 1978. Osmotic and ionic regulation in several western Atlantic Callianassidae (Crustacea, Decapoda, Thalassinidea). Biol. Bull., Woods Hole, **154**: 409-429, figs. 1-10.

——, 1979. Respiratory adaptations of the estuarine mud shrimp, *Callianassa jamaicense* (Schmitt, 1935) (Crustacea, Decapoda, Thalassinidea). Biol. Bull., Woods Hole, **157**: 125-137.

——, 2003. Ventrally sclerotized members of *Lepidophthalmus* (Crustacea: Decapoda: Callianassidae) from the eastern Pacific. Ann. naturhist. Mus. Wien, (B) **104**: 429-443.

FELDER, D. L. & B. E. FELGENHAUER, 1993. Morphology of the midgut-hindgut juncture in the ghost shrimp *Lepidophthalmus louisianensis* (Schmitt) (Crustacea: Decapoda: Thalassinidea). Acta Zool., Stockholm, **74**(4): 263-276.

FELDER, D. L. & D. L. LOVETT, 1989. Relative growth and sexual maturation in the estuarine ghost shrimp *Callianassa louisianensis* Schmitt, 1935. Journ. Crust. Biol., **9**: 540-553, figs. 1-6.

FELDER, D. L. & R. B. MANNING, 1994. Description of the ghost shrimp *Eucalliax mcilhennyi*, new species, from south Florida, with reexamination of its known congeners (Crustacea: Decapoda: Callianassidae). Proc. biol. Soc. Washington, **107**(2): 340-353, figs. 1-6.

—— & ——, 1995. *Neocallichirus cacahuate*, a new species of ghost shrimp from the Atlantic coast of Florida, with reexamination of *N. grandimana* and *N. lemaitrei* (Crustacea: Decapoda: Callianassidae). Proc. biol. Soc. Washington, **108**(3): 477-490.

—— & ——, 1997. Ghost shrimps of the genus *Lepidophthalmus* from the Caribbean region, with description of *L. richardi*, new species, from Belize (Decapoda: Thalassinidea: Callianassidae). Journ. Crust. Biol., **17**(2): 309-311, figs. 1-7.

—— & ——, 1998. A new ghost shrimp of the genus *Lepidophthalmus* from the Pacific coast of Colombia (Decapoda: Thalassinidea: Callianassidae). Proc. biol. Soc. Washington, **111**(2): 398-408.

FELDER, D. L. & S. DE A. RODRIGUES, 1993. Reexamination of the ghost shrimp *Lepidophthalmus louisianensis* (Schmitt, 1935) from the northern Gulf of Mexico and comparison to *L. siriboia*, new species, from Brazil (Decapoda: Thalassinidea: Callianassidae). Journ. Crust. Biol., **13**(2): 357-376, figs. 1-6.

FELDER, D. L. & J. L. STATON, 1990. Relationship of burrow morphology to population structure in the estuarine ghost shrimp *Lepidophthalmus louisianensis* (Decapoda, Thalassinoidea). American Zool., **30**: 137A.

—— & ——, 2000. *Lepidophthalmus manningi*, a new ghost shrimp from the southwestern Gulf of Mexico (Decapoda: Thalassinidea: Callianassidae). Journ. Crust. Biol., **20** (Spec. No.): 170-181, figs. 1-2.

FELDER, D. L., J. L. STATON & S. DE A. RODRIGUES, 1991. Patterns of endemism in the ghost shrimp genus *Lepidophthalmus* (Crustacea, Decapoda, Callianassidae): evidence from morphology, ecology and allozymes. American Zool., **31**: 101A.

FELDER, J. M., D. L. FELDER & S. C. HAND, 1984. Larval ontogeny of osmoregulation in the thalassinid ghost shrimp *Callianassa jamaicense*. American Zool., **24**: 67A.

—— & ——, 1986. Ontogeny of osmoregulation in the estuarine ghost shrimp *Callianassa jamaicense* var. *louisianensis* Schmitt (Decapoda, Thalassinidea). Journ. exp. mar. Biol. Ecol., **99**: 91-105.

FELDMAN, K. L., D. A. ARMSTRONG, B. R. DUMBAULD, T. H. DeWITT & D. C. DOTY, 2000. Oysters, crabs, and burrowing shrimp: review of an environmental conflict over aquatic resources and pesticide use in Washington State (USA) coastal estuaries. Estuaries, **23**(2): 141-176, illustr.

FELGENHAUER, B. E. & D. L. FELDER, 1986. Histology of the posterior midgut caecum in *Callianassa jamaicense* var. American Zool., **26**: 34A.

FERRARI, L., 1981. Aportes para el conocimiento de la familia Callianassidae (Decapoda, Macrura) en el Oceano Atlántico sudoccidental. Physis, Buenos Aires, (A) **39**(97): 11-21, 1 fig., pls. 1-3.

FILHOL, H., 1886. Passage de Vénus sur le soleil. Rec. Mém., Rapp. Doc., 3(Iie): 576, [Atlas] pls. 1-55.

FISCHER, P., 1866. In: A. FALSAN & A. LOCARD, Monographie géologique du Mont d'Or Lyonnais: 435, pl. 1 fig. 2. (Paris, Lyon).

FOREST, J., 1967. Sur une collection de Crustacés Décapodes la région de Port Cesareo: description de *Portumnus pestai* sp. nov. Thalassia Salentina, **2**: 1-29, figs. 1-6, pls. 1-4.

FOREST, J. & H. GANTÈS, 1960. Sur une collection de Crustacés Décapodes marcheurs du Maroc. Bull. Mus. nat., Paris, (2) **32**(4): 346-358, figs. 1-3.

FOREST, J. & D. GUINOT, 1956. Sur une collection de Crustacés Décapodes et Stomatopodes des mers tunisiennes. Bull. sta. océanogr. Salammbô, **53**: 24-43, figs. 1-5, 1 map.

—— & ——, 1958. Sur une collection de Crustacés Décapodes des côtes d'Israël. Bull. Sea Fish. Res. Sta., Haifa, **15**: 4-16.

FOTHERINGHAM, N., 1980 [reissued without change 1985, 1988]. Beach-comber's guide to Gulf coast marine life: 1-124. (Gulf Publishing Co., Houston, Texas).

FOTHERINGHAM, N. & S. BRUNENMEISTER, 1975. Common marine invertebrates of the northwestern Gulf coast: i-ix, 1-197, figs. (Gulf Publishing Co., Houston, Texas).

—— & ——, 1989. Beach-comber's guide to Gulf coast marine life, Florida, Alabama, Mississippi, Louisiana & Texas (2nd ed.): 1-142, figs. (Gulf Publishing Co., Houston, Texas).

FOURMANOIR, P., 1955. Crustacés Macroures et Anomoures, Stomatopodes. Notes sur la faune intercôtidale des Comores. I. Naturaliste Malgache, **7**(1): 19-33, 5 text-figs.

FRANKENBERG, D., S. L. COLES & R. E. JOHANNES, 1967. The potential trophic significance of *C. major* fecal pellets. Limnol. Oceanogr., **12**(1): 113-120, fig. 1.

FROGLIA, C. & G. B. GRIPPA, 1986. Types of decapod Crustacea (annotated catalog). A catalogue of the types kept in the collections of Museo Civico de Storia Naturale di Milano. VIII. Atti Soc. italiano Sci. nat. Mus. civ. Stor. nat. Milano, **127**(3-4): 253-283, figs. 1-5, pls. 1-2.

FULTON, S. W. & F. E. GRANT, 1906. Some little known Victorian decapod Crustacea, with descriptions of new species. No. 3. Proc. Roy. Soc. Victoria, (2) **19**(1): 5-15, pls. 3-5.

GAILLANDE, D. DE, 1970. Peuplements benthiques de l'herbier de *Posidonia oceanica* (Delile), de la pelouse à *Caulerpa prolifera* Lamouroux et du large du golfe de Gabès. Tethys, **2**(2): 373-384.

GAILLANDE, D. DE & J.-P. LAGARDÈRE, 1966. Description de *Callianassa* (*Callichirus*) *lobata* nov. sp. (Crustacea Decapoda Callianassidae). Rec. Trav. Sta. mar. Endoume, **40**(56): 259-265, pls. 1-4.

GARCÍA RASO, J. E., 1983. Aportaciones al conocimiento de los Thalassinidea Latreille, 1831 (Crustacea, Decapoda) del sur de España. Inv. Pesq., Barcelona, **47**(2): 317-324, figs.

— —, 1985a. Nuevas aportaciones a la fauna de Crustáceos Decápodos de la isla Alboran (España). Bol. Soc. Portuguesa Ent., (Suppl.) **1**: 11-18.

— —, 1985b. Presencia de una población de *Brachynotus atlanticus* Forest, 1957 (Crustacea, Decapoda, Brachyura, Grapsidae) en el sur de la Peninsula Iberica. Bol. Soc. Portuguesa Ent., (Suppl.) **1**: 19-26.

GARCIA SOCIAS, LL. & C. MASSUTI JAUME, 1987. Inventari bibliogràphic dels Crustaceis Decàpodes de les Balears (Crustacea Decapoda). Boll. Soc. Hist. nat. Balears, Palma de Mallorca, **31**: 67-92.

GIARD, A. & J. BONNIER, 1890. Sur une espèce nouvelle de *Callianassa* du Golfe de Naples (*C. truncata*). Bull. scient. France Belgique, **22**: 362-366, 4 figs.

GIBBES, L. R., 1850. On the carcinological collections of the cabinets of natural history in the United States with an enumeration of the species contained therein, and description of new species. Proc. American Ass. Adv. Sci., **3rd** meeting: 167-201.

GIORDANI SOIKA, A., 1943. Su alcuni crostacei descritti nella "Zoologia Adriatica" dell'Olivi. Arch. Oceanogr. Limnol., **3**: 81-86, pl. 1.

— —, 1945. I crostacei adriatici descritta dall'abate Stefano Chieregin. Atti Ist. Veneto, (Cl. Sci. mat. nat.) **104**(2): 927-966.

GLAÇON, R., 1971. Faune et flore du littoral du Pas-de-Calais et de la Manche orientale. Éd. Inst. Biol. mar. région. Wimereux: 1-46.

GORDON, I., 1957. Eucarida. In: Plymouth marine fauna: 240-261. (Marine Biological Association of the United Kingdom, Plymouth).

GOTTLIEB, E., 1953. Decapod crustaceans in the collection of the Sea Fisheries Research Station, Caesarea Israel. Bull. Res. Counc. Israel, **2**(4): 440-441.

GOURRET, P., 1887. Sur quelques Décapodes Macroures nouveaux du golfe de Marseille. C. R. hebd. Séanc. Acad. Sci., Paris, **105**: 1033-1035.

— —, 1888. Révision des Crustacés Podophthalmes du golfe de Marseille, suivie d'un essai de classification de la classe des Crustacés. Mém. Mus. Hist. nat., Marseille, **3**(5): 1-121, pls. 1-18.

GREBENJUK, L. P., 1975. Dva novykh vida desyatinogikh rakoobraznykh nadsemeistva Thalassinidea [Two new Decapoda species of the superfamily Thalassinidea]. Zool. Zh., **54**(2): 299-304, figs.

GRIFFIS, R. B. & T. H. SUCHANEK, 1991. A model of burrow architecture and trophic modes in thalassinidean shrimp (Decapoda: Thalassinidea). Mar. Ecol. Progr. Ser., **79**: 171-183.

GRUNER, H.-E., 1993. Lehrbuch der speziellen Zoologie. I. Wirbellose Tiere: 1-1279. (Gustav Fischer Verlag, Stuttgart).

GUÉRIN-MÉNEVILLE, F. E., 1857. Crustáceos. In: R. DE LA SAGRA (ed.), Historia fisica, politica y natural de la isla de Cuba par Ramon de la Sagra, **2** (Hist. nat.) (7): 14-57.

GURNEY, R., 1944. The systematics of the crustacean genus *Callianassa*. Proc. zool. Soc., London, **114**(5): 82-90, figs. 1-19.

GUSTAFSON, G., 1934. On the Thalassinidea of the Swedish west coast. Ark. Zool., Stockholm, (A) **28**(1): 1-19, figs. 1-4.

HAAN, W. DE, 1833-1850. Crustacea. In: P. F. VON SIEBOLD (ed.), Fauna Japonica sive descriptio animalium, que in itinere per Japoniam, jussu et auspiciis superiorum, qui summum in

India Batava Imperium tenent, suscepto, annis 1823-1830 collegit, notis, observationibus et adumbrationibus illustravit: 1-243, pls. 1-55, A-Q. (Amsterdam).

HADJICHRISTOPHOROU, M. A., A. DEMETROPOULOS & T. S. BIANCHI, 1997. A species list of the sublittoral soft-bottom macrobenthos of Cyprus. Acta Adriatica, 38(1): 3-32, 1 fig., 1 tab.

HAIG, J. & D. P. ABBOTT, 1980. Macrura and Anomura: the ghost shrimps, hermit crabs, and allies. In: R. H. MORRIS et al., Intertidal invertebrates of California: 577-593, pls. 165-172. (Stanford University Press, Stanford, CA).

HAILSTONE, T. S., 1962. They're good bait! Australian nat. Hist., 14(1): 29-31, figs. 1-2.

HAILSTONE, T. S. & W. STEPHENSON, 1961. The biology of Callianassa (Trypaea) australiensis Dana, 1852 (Crustacea, Thalassinidea). Univ. Queensland Pap., (Dept. Zool.), 1: 259-285, figs. 1-15, pls. 1-3.

HALE, H. M., 1927. The crustaceans of South Australia: 1-201, figs. 1-202. (Adelaide).

HARMELIN, J.-G., 1964. Étude de l'endofaune des "mattes" d'herbiers de Posidonia oceanica Delile. Rec. Trav. St. mar. Endoume, Bull., 35(51): 43-106.

HART, J. F. L., 1982. Crabs and their relatives from British Columbia. British Columbia provinc. Mus., Handbk., 40: i-iii, 1-266, figs. 1-102.

HASWELL, W. A., 1882. Catalogue of the Australian stalk- and sessile-eyed Crustacea. Australian Museum, Sydney, 8: i-xxiv, 1-324 + 2 pp. addenda, figs. 1-8, pls. 1-4.

HAY, W. P. & C. A. SHORE, 1917. The decapod crustaceans of Beaufort, N.C., and the surrounding region. Bull. U.S. Bur. Fish., (1915-1916) 35: 369-475, figs. 1-20, pls. 25-39.

HAYWARD, P. F., M. F. ISAAC, P. MAKINGS, F. MOYSE, E. NAYLOR & G. SMALDON, 1995. Crustaceans. In: P. J. HAYWARD & J. S. RYLAND (eds.), Handbook of the marine fauna of north-west Europe: 289-461, figs. 8.1-8.64.

HEALY, A. & J. C. YALDWYN, 1970. Australian crustaceans in colour: 1-112, figs. 1-57, pls. 1-52. (A.H. & A.W. Reed, Sydney).

HEARD, R. W., 1982. Guide to common tidal marsh invertebrates of the northeastern Gulf of Mexico: 1-82, figs. 1-73. (Mississippi Alabama Sea Grant Cons. Publ. (MASGP), 79-004).

— —, 1989. Calliax jonesi, n. sp. (Decapoda: Thalassinidea: Callianassidae) from the northwestern Bahamas. Gulf Res. Repts., 8(2): 129-136.

HEARD, R. [W.] & R. B. MANNING, 1998. A new genus and species of ghost shrimp (Crustacea: Decapoda: Callianassidae) from the Atlantic Ocean. Proc. biol. Soc. Washington, 111(4): 883-888.

— — & — —, 2000. A new genus and species of ghost shrimp from Tobago, West Indies (Crustacea: Decapoda: Callianassidae). Proc. biol. Soc. Washington, 113(1): 70-76, figs. 1-5.

HEARD, R. W. & R. C. REAMES, 1979. Callianassa (Callichirus) acanthochrius [sic] (Stimpson, 1866) (Crustacea: Decapoda: Thalassinidea) from the coastal waters of Alabama. Northeast Gulf Sci., 3(1): 51-52.

HEDGPETH, J. W., 1950. Notes on the marine invertebrate fauna of salt flat areas in Aransas National Wildlife Refuge, Texas. Publs. Inst. mar. Sci., Univ. Texas, 1: 103-119, figs. 1-2, tabs. 1-2.

HELLER, C., 1863. Die Crustaceen des südlichen Europa. Crustacea Podophthalmia. Mit einer Übersicht über die horizontale Verbreitung sämtlicher europäischer Arten: i-xi, 1-336, pls. 1-10. (Vienna).

HEMMING, F., 1958. Official list of generic names in zoology, first instalment: names 1-1274, 1: i-xxxvi, 1-200. (London).

HENDRICKX, M. E., 1995. Langostas. In: W. FISCHER, F. KRUPP, W. SCHNEIDER, C. SOMMER, K. E. CARPENTER & V. H. NIEM (eds.), Guía FAO para la identificación de especies para los fines de la pesca. Pacifico Centro-Oriental, **1**: 383-415, figs.

HERNÁNDEZ-AGUILERA, J. L., 1998. On a collection of thalassinids (Crustacea: Decapoda) from the Pacific Coast of Mexico, with description of a new species of the genus *Biffarius*. Ciencias Marinas, **24**: 303-312.

HERNÁNDEZ-AGUILERA, J. L., LL. LÓPEZ SALGADO & P. SOSA HERNÁNDEZ, 1986. Crustáceos estomatópodos y decápodos de Isla Clarión. Fauna carcinologica insular de Mexico, **1**. Investigac. oceanogr. Biol., **3**(1): 183-250.

HERBST, J. F. W., 1804. Versuch einer Naturgeschichte der Krabben und Krebse nebst einer systematischen Beschreibung ihrer verschiedenen Arten, **3**(4): 1-49, pls. 59-62.

HERPER, D. E., JR., 1975. The occurrence of *Callianassa latispina*, new record (Crustacea Decapoda Thalassinidea) off the Texas coast, U.S.A. Texas Journ. Sci., **26**(3/4): 619-620.

HILTON, W. A., 1916. Crustacea from Laguna Beach. Journ. Ent. Zool. Claremont California, **8**: 65-73, pls. 1-3.

HOLME, N. A., 1961. The bottom fauna of the English Channel. Journ. mar. biol. Ass. U.K., **41**(2): 397-461, figs. 1-15.

— —, 1966. The bottom fauna of the English Channel. Part II. Journ. mar. biol. Ass. U.K., **46**(2): 401-493, figs. 1-38.

HOLMES, S. J., 1900. Synopsis of California stalk-eyed Crustacea. Occ. Pap. California Acad. Sci., **7**: 1-262, figs. 1-6, pls. 1-4.

— —, 1904. On some new or imperfectly known species of West American Crustacea. Proc. California Acad. Sci., (Zool.) **3**: 307-328. (San Francisco).

HOLTHUIS, L. B., 1947. Nomenclatorial notes on European macrurous Crustacea Decapoda. Zool. Meded., Leiden, **27**: 312-322, fig. 1.

— —, 1950. Decapoda (K IX). Natantia, Macrura Reptantia en Stomatopoda (K X). Fauna Nederland, **15**: 1-166, figs. 1-54.

— —, 1952. The Crustacea Decapoda Macrura of Chile. Rep. Lund Univ. Chile Exped. 1948-1949, **5**. Lunds Univ. Årsskr., (n.s.) (2) **47**(10): 1-109.

— —, 1953a. On the supposed validity of the specific names *Callianassa laticauda* Otto and *Callianassa pontica* Czerniavsky. Pubbl. Staz. zool., Napoli, **24**(1): 91-98, 5 figs.

— —, 1953b. Enumeration of the decapod and stomatopod Crustacea from Pacific coral islands. Atoll Res. Bull., **24**: 1-66, 2 maps.

— —, 1954a. On a collection of decapod Crustacea from the Republic of El Salvador (Central America). Zool. Verh., Leiden, **23**: 1-43, figs. 1-15, pls. 1-2.

— —, 1954b. Observationes sobre los Crustáceos Decápodos de la Republica de El Salvador. Comun. Inst. trop. Invest. cient. El Salvador, **3**(4): 159-166, figs. 1-3.

— —, 1954c. Proposed use of the plenary powers to validate the generic names *"Upogebia"* Leach, 1814, and *"Processa"* Leach, 1815 (class Crustacea, order Decapoda). Bull. zool. Nomencl., **9**(2): 334-340.

— —, 1967. Biological investigations of the deep sea. 30. A survey of the genus *Ctenocheles* (Crustacea: Decapoda: Thalassinidea), with a discussion of its zoogeography and its occurrence in the Atlantic Ocean. Bull. mar. Sci., **17**(2): 376-385, figs. 1-2.

— —, 1958. Macrura. Crustacea Decapoda from the northern Red Sea (Gulf of Aqaba and Sinai Peninsula). I. Contributions to the knowledge of the Red Sea, No. 8. Bull. Sea Fish. Res. Sta., Haifa, **17**: 1-40, figs. 1-15.

— —, 1969. Thomas Say as a carcinologist. In: Thomas Say, an account of the Crustacea of the United States. In: J. CRAMER & H. K. SWANN (eds.), Historiae Naturalis Classica, **73**: v-xv.

————, 1974. Subterranean Crustacea Decapoda Macrura collected by Mr. Botosaneanu during the 1973 Cuban-Roumanian Biospeological Expedition to Cuba. Int. Journ. Speleol., **6**: 231-242, figs. 1-3.

————, 1977. The Mediterranean decapod and stomatopod Crustacea in A. Risso's published works and manuscripts. Ann. Mus. Hist. nat. Nice, **5**: 37-88, pls. 1-7.

————, 1979. H. Milne Edwards's "Histoire naturelle des Crustacés" (1834-1840) and its dates of publication. Zool. Meded., Leiden, **53**(27): 285-296.

————, 1991. Marine lobsters of the world. Fao Fish. Syn., **125**. Fao Species Cat., **13**: 1-292, figs. 1-459.

HOLTHUIS, L. B. & E. GOTTLIEB, 1958. An annotated list of the decapod Crustacea of the Mediterranean coast of Israel, with an appendix listing the Decapoda of the eastern Mediterranean. Bull. Res. Counc. Israel, (B) **7**(1-2): 1-126, figs. 1-15, pls. 1-3, maps 1-2.

HOLTHUIS, L. B. & G. R. HEEREBOUT, 1976. De nederlandse Decapoda (garnalen, kreeften, krabben). Wet. Meded. Kon. nederlandse natuurh. Ver., **111**: 1-56, figs. 1-66.

HOPKINS, C. S. & R. M. FELDMANN, 1997. Sexual dimorphism in fossil and extant species of *Callianopsis* De Saint Laurent. Journ. Crust. Biol., **17**(2): 236-252, figs. 1-7.

HOYT, J. H. & J. WEIMER, 1963. *Callianassa major* burrows, geologic indicators of littoral and shallow neritic environment. Bull. Georgia Acad. Sci., **21**: 10, 11.

HUGHES, D. J., R. J. A. ATKINSON & A. D. ANSELL, 2000. A field test of the effects of megafaunal burrows on benthic chamber of sediment-water solute fluxes. Mar. Ecol. Progr. Ser., **195**: 189-199.

HULT, J., 1938. Crustacea Decapoda from the Galapagos Islands collected by Mr. Rolf Blomberg. Ark. Zool., Stockholm, (A) **30**(5): 1-18, figs. 1-4, pl. 1.

HUMM, H. J., 1953. Check list of the marine fauna and flora of the St. George's Sound Apalache Bay Region, Florida Gulf coast (3rd ed.). Oceanogr. Inst., Florida State Univ., Tallahassee, Contrib., **23**: 1-18.

HUXLEY, T. H., 1879. On the classification and the distribution of the crayfishes. Proc. zool. Soc. London, **1878**: 752-788, 7 figs.

ICZN (INTERNATIONAL COMMISSION ON ZOOLOGICAL NOMENCLATURE), 2000. International code of zoological nomenclature adopted by the Union of Biological Sciences: i-xxxix, 1-306. (International Trust for Zoological Nomenclature, London).

KAMITA, T., 1957. Studies on the decapod crustaceans of Corea, II. Hermit-crabs (4). Sci. Rep. Shimane Univ., **7**: 91-109, fig. 49.

KATTOULAS, M. & A. KOUKOURAS, 1974. Benthic fauna of the Evvoia coast and Evvoia Gulf. IV. Macrura Reptantia (Crustacea, Decapoda). Sci. Anns Fac. Phys. Mathem., Univ. Thessaloniki, **14**: 341-347, fig. 1.

KAZMI, Q. B. & M. A. KAZMI, 1992. A new species of a callianassid shrimp, *Neocallichirus manningi*, with a note on the genus *Neocallichirus* Sakai, 1988, not previously recorded from the Arabian Sea (Decapoda, Thalassinidea). Crustaceana, **63**(3): 296-300, figs. 1-2.

KEMP, S. W., 1915. Crustacea Decapoda. In: Fauna of the Chilka Lake, No. 3. Mem. Indian Mus., Calcutta, **5**: 199-325, figs. 1-38, pls. 12-13.

KENSLEY, B., 1974. The genus *Callianassa* (Crustacea Decapoda) from the west coast of South Africa with a key to the South African species. Ann. South African Mus., **62**(8): 265-278, figs. 1-5.

————, 1975. Records of mud-prawns (genus *Callianassa*) from South Africa and Mauritius (Crustacea, Decapoda, Thalassinidea). Ann. South African Mus., **69**(3): 47-58.

————, 2001. A new species of *Corallichirus* Manning, 1992 (Crustacea: Decapoda: Callianassidae) from Guam. Bull. biol. Soc. Washington, **10**: 328-333, figs. 1-2.

KIKUCHI, K., 1932. Decapod crustaceans of Toyama Bay: 1-23. (Toyama Kyoyuku).

KINAHAN, J. R., 1856. Remarks on the habits and distribution of marine Crustacea on the eastern shores of Port Phillip, Victoria, Australia, with descriptions of undescribed species and genera. Journ. Roy. Dublin Soc., **1**(3): 111-134, pls. 3-4.

— —, 1859. Report on Crustacea of Dublin district. Decapoda Podophthalmata. Rep. British Ass. Adv. Sci., **1858**: 262-268.

KINGSLEY, J. S., 1878. List of decapod Crustacea of the Atlantic coast, whose range embraces Fort Macon. Proc. Acad. nat. Sci., Philadelphia, **1878**: 316-330.

— —, 1899. Synopsis of North-American invertebrates. IV. Astacoid and thalassinoid Crustacea. American Natural., **33**: 819-824, figs. 1-8.

KIRK, T. W., 1879. Note on some New Zealand crustaceans. Trans. Proc. New Zealand Inst., **11**: 401-402.

KISHINOUYE, K., 1926. Two rare and remarkable forms of macrurous Crustacea from Japan. Annot. zool. Japonenses, **11**(1): 63-70, figs. 1-2.

KOBJIAKOVA, Z. I. & M. A. DOLGOPOL'SKAIA, 1969. Opredeliteli Fauny Cernogo I Azovskogo Morei [Key to the Fauna of the Black and Azov Sea], Kiev: 268-306, figs. 1-3, pls. 1-7. [In Russian.]

KUBO, I., 1949. Studies on the penaeids of Japan and its adjacent waters. Journ. Tokyo Coll. Fish., **36**(1): 1-467, figs. 1-160.

KOCATAS, A., 1981. Liste préliminaire et répartition des Crustacés Décapodes des eaux Turquies. Rapp. Comm. int. Mer Méditerranée, **27**(2): 161-162.

KOSSMANN, R., 1880. Zoologische Ergebnisse einer Reise in die Küstengebiete des Rothen Meeres, **2**(1)(3), Malacostraca. Zoologische Ergebnisse im Aufträge der Königlichen Academie der Wissenschaften zu Berlin, **1880**: 67-140, pls. 4-15.

KOUKOURAS, A., C. DOUNAS, M. TÜRKAY & E. VOULTSIADOU-KOUKOURA, 1992. Decapod crustacean fauna of the Aegean Sea: new information, check list, affinities. Senckenb. marit., **22**(3-6): 217-244.

LAGARDÈRE, J.-P., 1966. Recherches sur la biologie et l'écologie de la macrofaune des substrats meubles de la côte des Landes et de la côte basque. Bull. Cent. Etud. Rech. scient. Biarritz, **6**(2): 143-209, figs. 1-28, pls. 1-5.

— —, 1973. Distribution des Décapodes dans le sud du Golfe de Gascogne. Rev. Trav. Inst. Pêches marit., **37**(1): 77-95, figs. 1-11.

LAGERBERG, T., 1908. Sveriges decapoder. Göteborg, vet. Handl., **11**(2): i-x, 1-117, pls. 1-5.

LANCHESTER, W. F., 1900. On some malacostracous crustaceans from Malaysia in the collection of the Sarawak Museum. Ann. Mag. nat. Hist., (7) **6**: 249-265, pl. 12.

— —, 1902. On the Crustacea collected during the "Skeat" Expedition to the Malay Peninsula, together with a note on the genus *Actaeopsis*. Part 1. Brachyura, Stomatopoda, and Macrura. Proc. zool. Soc. London, **1901**(2): 534-574, pls. 33-34.

LATREILLE, P. A., 1831. Cours d'entomologie ou de l'histoire naturelle des Crustacés, des Arachnides, des Myriopodes et des Insectes. Exposition méthodique des ordres, des familles et des genres des trois premières classes. A l'usage des élèves de l'ecole du Muséum d'Histoire naturelle, Paris: i-xiii, 1-568 + 26 pp., Atlas: 1-26, 24 pls.

LEACH, W. E., 1814. Crustaceology. In: D. BREWSTER (ed.), Edinburgh Encyclopedia, **7**(2): 385-437.

— —, 1815. A tabular view of the external characters of four classes of animals, which Linné arranged under Insecta, with the distribution of the genera composing three of these classes into orders, and descriptions of several new genera and species. Trans. Linn. Soc., London, (1) **11**: 306-400.

——, 1816. Malacostraca Podophthalmata Britanniae, or descriptions of the British species of crabs, lobsters, prawns, etc. illustrated with coloured figures of all the species by James Sowerby, 54 col. pls.

LEARY, S. P., 1964 [reissued without change, 1967]. The crabs of Texas. Texas Parks and Wildlife Department, Bull., (7, Coastal Fisheries) **43**: 1-57.

LE LOEUFF, P. & A. INTÈS, 1974. Les Thalassinidea (Crustacea Decapoda) de Golfe de Guinée: systématique – écologie. Cahiers ORSTOM, (Océanogr.) **12**(1): 17-69, figs. 1-22, pls. 1-5.

LEMAITRE, R. & D. L. FELDER, 1996. A new species of ghost shrimp of the genus *Sergio* Manning & Lemaitre, 1994 (Crustacea: Decapoda: Callianassidae) from the Caribbean coast of Colombia. Proc. biol. Soc. Washington, **109**(3): 453-463.

LEMAITRE, R. & R. A. LEÓN, 1992. Crustáceos decápodos del Pacifico colombiano: lista de especies y consideraciones zoogeográficas. An. Inst. Invest. mar. Punta Betin, Santa Marta, **21**: 33-76.

LEMAITRE, R. & G. E. RAMOS, 1992. A collection of Thalassinidea (Crustacea: Decapoda) from the Pacific coast of Colombia, with description of a new species and a checklist of eastern Pacific species. Proc. biol. Soc. Washington, **105**(2): 343-358, figs. 1-5.

LEMAITRE, R. & S. DE A. RODRIGUES, 1991. *Lepidophthalmus sinuensis*: a new species of ghost shrimp (Decapoda: Thalassinidea: Callianassidae) of importance to the commercial culture of penaeid shrimps on the Caribbean coast of Colombia, with observations on its ecology. Fish. Bull., U.S., **89**: 623-630, figs. 1-4.

LENZ, H., 1911. *Callianassa turnerana* White und *C. diademata* Ortmann. SitzBer. Ges. naturf. Fr., Berlin, **1911**: 316-318, figs.1-11.

LENZ, H. & F. RICHTERS, 1881. Beitrag zur Crustaceenfauna von Madagascar. Abh. Senckenb. naturf. Ges., **12**: 421-428, pl. 1.

LEWINSOHN, C. H., 1976. Crustacea Decapoda von der Insel Rhodos, Griechenland. Zool. Meded., Leiden, **49**(17): 237-254.

LEWINSOHN, C. H. & L. B. HOLTHUIS, 1986. The Crustacea Decapoda of Cyprus. Zool. Verh., Leiden, **230**: 3-64, fig. 1.

LIU, J. Y., 1955. Economic shrimps and prawns of north China: i-iii, 1-73, figs. 1-3. pls. 1-24. (Marine Biological Institute of the Academy of Science, Peiping). [In Chinese.]

LIU, J. Y. & Z. ZHONG 1994. In: HUANG ZONG-GUO (ed.), Marine species and their distributions in China's sea: 1-764, 1-134. (China Ocean Press, Beijing). [In Chinese.]

LOCKINGTON, W. N., 1878. Remarks upon the Thalassinidea and Astacidea of the Pacific coast of North America, with description of a new species. Ann. Mag. nat. Hist., (5) **2**: 299-304.

LONGHURST, A. R., 1958. An ecological survey of the west African marine benthos. Fish. Publ. colon. Off., London, **11**: 1-102, figs. 1-11.

LÓPEZ DE LA ROSA, I., J. E. GARCÍA RASO & A. RODRÍGUEZ MARTÍN, 1998. First record of *Gourretia denticulata* (Lutze, 1937) (Crustacea, Decapoda, Thalassinidea) from the Atlantic coast of Spain. Sci. mar., **62**(4): 393-395, fig. 1.

LOVETT, D. L. & D. L. FELDER, 1984. Relative growth and sexual maturation in the estuarine ghost shrimp *Callianassa jamaicense* Schmitt. American Zool., **24**: 74A.

—— & ——, 1989. Application of regression techniques to studies of relative growth in crustaceans. Journ. Crust. Biol., **9**(4): 529-539, figs. 1, 2.

LUNZ, G. R., 1937. Notes on *Callianassa major* Say. Charleston Mus. Leafl., **10**: 1-15, figs. 1-5, pls. 1-2.

LUTZE, J., 1937. Eine neue *Callianassa*-Art aus der Adria. Note Ist. Biol. Rovigno, 2(1): 1-12, figs. 1-7, 1 map.

——, 1938. Über Systematik, Entwicklung und Oekologie von *Callianassa*. Helgoländer wiss. Meeresunters., **1**(1-3): 162-199, figs. 1-107.

——, 1941. Zur Systematik des Dekapodenkrebses *Callianassa*. Zool. Anz., **135**: 34-35.

MACGINITIE, G. E., 1934. The natural history of *Callianassa californiensis* Dana. American Midl. Natural., **15**(2): 166-176, pls. 5-6.

——, 1935. Ecological aspects of a California marine estuary. American Midl. Natural., **16**(5): 629-765.

MCNEILL, F. A., 1968. Crustacea, Decapoda and Stomatopoda. Sci. Rep. Great Barrier Reef Exped., **7**(1): 1-98, figs. 1-2, pls. 1-2.

MAKAROV, V. V., 1935. Beschreibung neuer Dekapoden-Formen aus den Meeren des Fernen Ostens. Zool. Anz., **109**: 323, figs. 1-4.

——, 1938. Décapodes Anomoures. Crustacés, **10**(3). Fauna U.S.S.R., (n.s.) **16**: i-x, 1-324, figs. 1-133, pls. 1-5. [English translation from eds. in Washington, D.C. and Israel.]

MAN, J. G. DE, 1888. Bericht über die im Indischen Archipel von Dr. J. Brock gesammelten Decapoden und Stomatopoden. Arch. f. Naturges., **53**: 215-600, pls. 1-16a.

——, 1902. Die von Herrn Professor Kükenthal im Indischen Archipel gesammelten Dekapoden und Stomatopoden. Abhand. Senckenb. naturf. Ges., **25**: 465-929, pls. 19-27.

——, 1905. Diagnoses of new species of macrurous decapod Crustacea from the "Siboga-Expedition". Tijdschr. nederlandse dierk. Ver., (2) **9**(3-4): 587-614.

——, 1911. On two new species of decapod Crustacea. Notes Leyden Mus., **33**: 223-232.

——, 1916. Description of a new species of the genus *Calianassa* Leach and of a species of the genus *Alpheus* Fabr., both from the Indian Archipelago. Zool. Meded., Leiden, **2**: 57-61, pl. 1.

——, 1928a. A contribution to the knowledge of twenty-two species and three varieties of the genus *Callianassa* (Leach). Capita zool., **2**(6): 1-56, pls. 1-12.

——, 1928b. The Thalassinidae and Callianassidae collected by the Siboga-Expedition with some remarks on the Laomediidae. The Decapoda of the Siboga-Expedition. Part VII. Siboga Exped. Mon., **39**(a6): 1-187, pls. 1-20.

MANNING, R. B., 1987. Notes on western Atlantic Callianassidae (Crustacea: Decapoda: Thalassinidea). Proc. biol. Soc. Washington, **100**(2): 386-401, figs. 1-9.

——, 1988. The status of *Callianassa hartmeyeri* Schmitt, 1935, with the description of *Corallianassa xutha* from the west coast of America (Crustacea, Decapoda, Thalassinidea). Proc. biol. Soc. Washington, **101**(4): 883-889, figs. 1-3.

——, 1992. A new genus for *Corallianassa xutha* Manning (Crustacea: Decapoda: Callianassidae). Proc. biol. Soc. Washington, **105**(3): 571-574.

——, 1993. Two new species of *Neocallichirus* from the Caribbean Sea (Crustacea: Decapoda: Callianassidae). Proc. biol. Soc. Washington, **106**(1): 106-114, figs. 1-6.

MANNING, R. B. & F. A. CHACE, JR., 1990. Decapod and stomatopod Crustacea from Ascension Island, South Atlantic Ocean. Smithon. Contr. Zool., **503**: i-v, 1-88, figs. 1-47, tabs. 1-4.

MANNING, R. B. & D. L. FELDER, 1986. The status of the callianassid genus *Callichirus* Stimpson, 1866 (Crustacea: Decapoda: Thalassinidea). Proc. biol. Soc. Washington, **99**(3): 437-443, figs. 1-3.

—— & ——, 1989. The *Pinnixa cristata* complex in the western Atlantic, with descriptions of two new species (Crustacea, Decapoda, Pinnotheridae). Smithson. Contr. Zool., **474**: i-iii, 1-26.

—— & ——, 1991. Revision of the American Callianassidae (Crustacea: Decapoda: Thalassinidea). Proc. biol. Soc. Washington, **104**: 764-792.

—— & ——, 1992. *Gilvossius*, a new genus of callianassid shrimp from the eastern United States (Crustacea: Decapoda: Thalassinidea). Bull. mar. Sci. Univ. Miami, **49**: 558-561, fig. 1.

—— & ——, 1995. Description of the ghost shrimp *Sergio mericeae*, a new species from south Florida, with reexamination of *S. guassutinga* (Crustacea: Decapoda: Callianassidae). Proc. biol. Soc. Washington, **108**(2): 266-280.

MANNING, R. B. & R. W. HEARD, 1986. Additional records for *Callianassa rathbunae* Schmitt, 1935, from Florida and the Bahamas (Crustacea: Decapoda: Callianassidae). Proc. biol. Soc. Washington, **99**(2): 347-349, 1 fig.

MANNING, R. B. & R. LEMAITRE, 1994 [not 1993]. *Sergio*, a new genus of ghost shrimp from the Americas (Crustacea: Decapoda: Callianassidae). Nauplius, Rio Grande, **1**: 39-43, fig. 1.

MANNING, R. B. & Z. ŠTEVČIĆ, 1982. Decapod fauna of the Piran Gulf. Quad. Lab. Tecnol. Pesca, Ancona, **3**(2-5): 285-304.

MANNING, R. B. & A. TAMAKI, 1998. A new genus of ghost shrimp from Japan (Crustacea: Decapoda: Callianassidae). Proc. biol. Soc. Washington, **111**(4): 889-892.

MARCUZZI, G., 1972. Le collezioni dell'ex Istituto di Biologia Marina di Rovigno conservate presso la Stazione Idrobiologica di Chioggia: 1-129, 4 figs. (Società Cooperativa Tipografica, Padova).

MARTENS, E. VON, 1869. Über einige neue Crustaceen. Monatsber. Akad. Wiss., Berlin, **1868**: 608-615.

MATSUZAWA, K. & K. I. HAYASHI, 1997. Male of *Ctenocheles balssi* (Crustacea, Decapoda, Callianassidae) from off Mutoto Peninsula, Shikoku, Japan. Journ. natn. Fish. Univ., **46**(1): 39-46, figs. 1-3.

MAYORAL, M. A., L. LÓPEZ-SERRANO & J. M. VIÉITEZ, 1994. Macrofauna bentónica intermareal de tres playas de la desembocadura del río Piedras (Huelva, España). Bol. Roy. Soc. Español Hist. nat., (Biol.) **91**(1-4): 231-240.

MAZMANIDI, N. & A. KOMAKHIDZE, 1998. On the biodiversity of the Georgian Black Sea coast. NATO ASI Ser. Partnersh., (Sub-Ser. 2, Environm.) **46**: 129-154, illustr.

MELIN, G., 1939. Paguriden und Galatheiden von Prof. Dr. Sixten Bocks Expedition nach den Bonin-Inseln 1914. Kong. Svenska vetensk. Akad. Handl., (3) **18**(2): 1-119, figs. 1-71.

MENZEL, R. W., (ed.), 1956. Annotated check-list of the marine fauna and flora of the St. George's Sound-Apalachee Bay Region, Florida Gulf coast. Contr. oceanogr. Inst. Florida State Univ., **61**: i-iv, 1-78, 1 map.

——, 1971. Checklist of the marine fauna and flora of the Apalachee Bay and the St. George's Sound area (3rd ed.). Contr. oceanogr. Inst. Florida State Univ., **61**: 1-134. (Tallahassee, Florida).

MICHEL, C., 1974. Notes on marine biology studies made in Mauritius. Bull. Mauritius Inst., **7**(2): 1-287.

MIERS, E. J., 1882. On some Crustaceans collected at the Mauritius. Proc. zool. Soc. London, **1882**: 339-342.

——, 1884. On some Crustaceans from Mauritius. Proc. zool. Soc. London, **1884**: 10-17, pl. 1.

MIKASHAVIDZE, E. V., 1981. New findings of some species of polychaetes, molluscs and crustaceans on the shelf of the south east Black Sea. Zool. Zh., **60**: 1415-1417.

MILLER, M. & G. BATT, 1973. Reef and beach life of New Zealand: 1-141. (Collins, Auckland and London).

MILNE-EDWARDS, A., 1860. Monographie des Décapodes macroures fossiles de la famille des Thalassiniens. Ann. Sci. nat. Zool., (4) **16**: 302, pls. 11-16.

——, 1861. Historie des Crustacés Podophthalmaires fossiles. Monographies des Portuniens et des Thalassiniens: 1-222, pls. 1-16.

——, 1870. Révision du genre *Callianassa* (Leach) et description de plusieurs espèces nouvelles de ce groupe. Nouv. Arch. Mus. Hist. nat., Paris, **6**: 75-102, pls. 1-2.

——, 1878. Additions à la famille des Thalassiniens. Bull. Soc. philos., Paris, (7) **3**: 110-113.

MILNE EDWARDS, H., 1837a. Histoire naturelle des Crustacés, comprenant l'anatomie, la physiologie et la classification de ces animaux, **2**: 1-531 [Thalassinidés: 303-325]; Atlas [1834, 1837a, 1840]: 1-32, pls. 1-42. (Libraire Encyclopédique de Roret Paris). [See Holthuis, 1979 for dates of publication.].

— —, 1837b. Les Crustacés. In: G. CUVIER, Le règne animal distribué d'après son organisation, pour servir de base à l'histoire naturelle des animaux et d'introduction à l'anatomie comparé (4th ed.), **17**: 1-278; Atlas, **18**, pls. 1-80. (Paris).

— —, 1838. Arachnides, Crustacés, Annélides, Cirrhipèdes. In: J. B. P. A. DE LAMARCK, Histoire naturelle des animaux sans vertébrés ... (2nd ed.), **5**: 1-699.

MIYABI, S., K. KONISHI, Y. FUKUDA & A. TAMAKI, 1998. The complete larval development of the ghost shrimp, *Callianassa japonica* Ortmann, 1891 (Decapoda: Thalassinidea: Callianassidae), reared in the laboratory. Crust. Res., Tokyo, **27**: 101-121, figs. 1-13.

MIYAKE, S., 1965. New illustrated encyclopedia of the fauna of Japan, **2**: 1-803, figs. 1-1462. (Hokyryu-kan, Tokyo).

— —, 1982. Macrura, Anomura and Stomatopoda. Japanese crustacean decapods and stomatopods in color, **1**: 1-261, pls. 1-56. (Hoikusha, Osaka).

MIYAKE, S., K. SAKAI & S. NISHIKAWA, 1962. A fauna-list of the decapod Crustacea from the coasts washed by the Tsushima Warm Current. Rec. oceanogr. Works Japan, (Spec. No.) **6**: 121-131.

MIYAZAKI, I., 1936. Habits and larval forms of some decapod Crustaceans used for fish bait. Bull. Japanese Soc. scient. Fish., **5**(1-6): 312-325.

MONCHARMONT, U., 1979. Notizie biologiche e faunistiche sui Crostacei Decapodi de Golfo di Napoli. Annu. Ist. Mus. Zool. Univ. Napoli, **23**: 1-132, i-xi.

MONOD, TH., 1927. Sur le Crustacé auquel le Cameroun doit son nom (*Callianassa turnerana* White). Bull. Mus. Hist. nat., Paris, **33**(1): 593-624.

— —, 1931. Inventaire des manuscripts de Risso conservés à la bibliothèque du Muséum d'Histoire Naturelle. Arch. Mus. Hist. nat., Paris, (6) **7**: 103-133, figs. 1-10.

— —, 1933. Sur quelques Crustacés de l'Afrique occidentale (liste des Décapodes Mauritaniens et des Xanthidés ouest-Africains). Bull. Comun. Étud. scient. Afrique occ. français, **15**(2-3): 456-548 [1-93 on offprint], figs. 1-26.

MONTAGU, G., 1808. Description of several marine animals found on the south coast of Devonshire. Trans. Linn. Soc., London, **9**: 18-114, pls. 2-8.

MORTENSEN, TH., 1923. The Danish Expedition to the Kei Islands 1922: 1-45, 3 figs., 3 double maps.

NAKAZAWA, K., 1927. Crustacea. In: Figuraro de Japanaj Bestoj: 992-1124, figs. 1910-2166. (Tokyo). [In Japanese.]

NAKAZAWA, K. & I. KUBO, 1947. Illustrated encyclopedia of the fauna of Japan, **2**: 1-1898, figs. 1-5213. (Tokyo).

NATES, S. F. & D. L. FELDER, 1999. Growth and maturation of the ghost shrimp *Lepidophthalmus sinuensis* Lemaitre and Rodrigues, 1991 (Crustacea, Decapoda, Callianassidae), a burrowing pest in penaeid shrimp culture ponds. Fish. Bull., U.S., **97**(3): 526-541, illustr.

NEUMANN, R., 1878. Systematische Uebersicht der Gattungen der Oxyrhynchen. Catalog der Podophthalmen Crustaceen des Heidelberger Museums. Beschreibung einiger neuer Arten: 1-39. (Ph.D. Thesis, Leipzig).

NEVES, A. M., 1974. Crustaceos Decapodes marinhos de Portugal continental, **2**. Macrura Reptantia. Est. Fauna Portugal, **3**: 1-20.

NGOC-HO, N., 1991. Sur quelques Callianassidae et Upogebiidae de Nouvelle-Calédonie (Crustacea, Thalassinidea). In: B. RICHER DE FORGES, Le benthos des fonds meubles des lagons de Nouvelle Calédonie, **1**: 281-311, figs. 1-11. (Paris).

——, 1994. Some Callianassidae and Upogebiidae from Australia with description of four new species (Crustacea: Decapoda: Thalassinidea). Mem. Mus. Victoria, **54**: 51-78, figs. 1-12.

——, 1995. Une espèce nouvelle de *Neocallichirus* aux îles Tuamotu, Polynésie française (Crustacea, Decapoda, Thalassinidea). Bull. Mus. Hist. nat., Paris, (4) **17**(1-2): 211-218.

——, 2002. A new species of *Calliapagurops* De Saint Laurent from the Philippines with a discussion of the taxonomic position of the genus (Thalassiniidea, Callianassidae). Crustaceana, **75**(3-4): 539-549, figs. 1-3.

——, 2003. European and Mediterranean Thalassinidea (Crustacea, Decapoda). Zoosystema, **25**(3): 439-555, 37 figs.

NICKELL, L. A., R. J. A. ATKINSON & E. H. PINN, 1998. Morphology of thalassinidean (Crustacea: Decapoda) mouthparts and pereiopods in relation to feeding, ecology and grooming. Journ. nat. Hist., London, **32**: 733-761.

NICOLET, H., 1849. Crustacaeos. In: C. GAY, Historia fisica y politica de Chile segun documentos adquiridos en exta republica durante doce años de residencia en ella y publicada bajo los auspicios del supremo gobierno, (Zool.) **3**: 115-318, pls. 1-4.

NOBILI, G., 1900. Descrizione di un nuovo *Palaemon* di Giava e osservazioni sulla *Callianassa turnerana* Wh. del Camerun. Boll. Mus. Zool. Anat. comp. Torino, **15**(379): 1-4.

——, 1904 (28 June). Diagnoses préliminaires de vingt-huit espèces nouvelles de Stomatopodes et Décapodes Macroures de la mer Rouge. Bull. Mus. Hist. nat., Paris, **10**: 228-237.

——, 1905. Diagnoses préliminaires de 34 espèces et variétés nouvelles, et de 2 genres nouveaux de Décapodes de la mer Rouge. Bull. Mus. Hist. nat., Paris, **11**: 393-411, figs. 1-2.

——, 1906a. Mission J. Bonnier et Ch. Pérez (Golfe Persique, 1901) Crustacés Décapodes et Stomatopodes. Bull. scient. France Belgique, **40**: 13-159, figs. 1-3, pls. 2-7.

——, 1906b. Faune carcinologique de la Mer Rouge, Décapodes et Stomatopodes. Ann. Sci. nat., (Zool.) **4**(1-3): 1-347, figs. 1-12, pls. 1-11.

NOBRE, A., 1936. Fauna marinha de Portugal. IV. Crustáceos Decápodes e Stomatópodes marinhos de Portugal: 1-213, 137 + 1 figs. (Porto).

NOGUCHI, M. & T. AKAMINE, 1992. Chela of *Ctenocheles balssi*. Renraku, News from Nihonkai natn. Fish. Res. Inst., **360**: 25-26. [In Japanese.]

O'CÉIDIGH, P., 1962. The marine Decapoda of the counties Galway and Clare. Proc. Roy. Irish Acad., (B) **62**(11): 163-164.

OLIVI, G., 1792. Zoologia Adriatica ossia Catalogo ragionato degli Animali del Golfo e delle Lagune di Venezia; preceduto da una Dissertazione sulla storia fisica e naturale del Golfo; e accompagnato da Memorie, ed Osservazioni di Fisica Storia naturale ed Economia: i-xxxii, 1-334, pls. 1-9.

ORTMANN, A., 1891. Die Decapoden-Krebse des Straßburger Museums, mit besonderer Berücksichtigung der von Herrn Dr. Döderlein bei Japan und den Liu-Kiu-Inseln gesammelten und z.Z. im Strassburger Museum aufbewahrten Formen. III. Die Abtheilungen der Reptantia Boas: Homaridea, Loricata und Thalassinidea. Zool. Jb., **6**: 1-58, pl. 1.

——, 1894. Crustaceen. In: R. SEMON (ed.), Zoologische Forschungsreisen in Australien und dem Malayischen Archipel, **5**. Denkschr. med.-naturw. Ges., Jena, **8**: 1-80, pls. 1-13.

——, 1899. Malacostraca (Lief. 47-62). In: A. GERSTÄCKER & A. E. ORTMANN, Die Klassen und Ordnungen der Arthropoden wissenschaftlich dargestellt in Wort und Bild. Bronn's Klassen und Ordnungen des Tierreichs, **5**(2): 1134-1232.

OTTO, A. W., 1821. Conspectus animalium quorundam maritimorum: 1-20. (Bratislava).

——, 1828. Beschreibung einiger neuen, in den Jahren 1818 und 1819, im Mittelländischen Meere gefundener Crustaceen. Nova Acta Acad. Leop. Carol., **14**: 331-354, pls. 1-22.

PARISI, B., 1915. Note su alcuni Crostacei del Mediterraneo. Monit. zool. Italiano, **26**(3): 62-66, figs. 1-2.

— —, 1917. I Decapodi giapponesi del Museo di Milano. V. Galatheidea e Reptantia. Atti Soc. Italiana Sci. nat., **56**: 1-24, figs. 1-7.

PASTORE, M., 1976. Decapoda Crustacea in the Gulf of Taranto and the Gulf of Catania with a discussion of a new species of Dromiidae (Decapoda Brachyura) in the Mediterranean Sea. Thalassia Jugoslavica, **8**(1) [1972]: 105-117, fig. 1.

PEARSE, A. S., H. J. HUM & G. W. WHARTON, 1942. Ecology of sand beaches at Beaufort, North Carolina. Ecol. Monogr., **12**(2): 135-190, figs. 1-24, tabs. 1-17.

PEARSON, J., 1905. Report on the Macrura collected by Professor Herdman at Ceylon in 1902. In: Report to the government of Ceylon on the pearl oyster fisheries of the Gulf of Manaar by W.A. Herdman. With supplementary reports upon the marine biology of Ceylon by other naturalists, **4** (suppl. Rep.) 4: 65-92, pls. 1-2.

PÉREZ SÁNCHEZ, J. M. & E. MORENO BATET, 1991. Invertebrados marinos de Canarias: 1-335. (Ediciones del Abildo Insular de Gran Canaria, Las Palmas de Gran Canaria).

PESTA, O., 1912. Die Dekapoden-Krebse der Adria in Bestimmungstabellen zusammengestellt. Arch. Naturges., (A) **78**(1): 93-126.

— —, 1918. Die Decapodenfauna der Adria. Versuch einer Monographie: i-x, 1-500, figs. 1-150, map 1.

PETAGNA, V., 1792. Institutiones entomologicae, **1**: i-xii, 1-439; **2**: 441-718, pls. 1-10. (Naples).

PETRESCU, I. & A.-M. BALASESCU, 1995. Contributions to the knowledge of decapod fauna (Crustacea) from the Romanian coast of the Black Sea. Trav. Mus. Hist. nat. "Grigore Antipa", **35**: 99-146.

PHILLIPS, J. P., 1971. Observations on the biology of mudshrimps of the genus *Callianassa* (Anomura: Thalassinidea) in Mississippi Sound. Gulf Res. Repts, Ocean Springs, Mississippi, **3**(2): 165-196, figs. 1-8.

PICARD, J., 1957. Note sur un nouveau peuplement des sables infra-littoraux: biocoenose a *Callianassa laticauda* Otto et *Kellya* (*Bornia*) *corbuloides* Philippi. Rec. Trav. Sta. mar. Endoume, **21**: 48-94.

— —, 1965. Recherches qualitatives sur les biocoenoses marines des substrats meubles dragales de la région marseillaise. Rec. Trav. Sta. mar. Endoume, Bull., **52**(36): 1-160.

PILLAI, N. K., 1954. A note on *Callianassa maxima* M. Edwards, (Decapoda). Bull. centr. Res. Inst. Univ. Travancore, (C) **5**(1): 23-26, figs. 1-5.

POHL, M. E., 1946. Ecological observations on *Callianassa major* Say at Beaufort, North Carolina. Ecology, **27**(1): 71-80, figs. 1-28.

POORE, G. C. B., 1975. Systematics and distribution of *Callianassa* (Crustacea, Decapoda, Macrura) from Port Phillip Bay, Australia, with descriptions of two new species. Pacific Sci., **29**(2): 197-209.

— —, 1994. A phylogeny of the families of Thalassinidea (Crustacea: Decapoda) with keys to families and genera. Mem. Mus. Victoria, **54**: 79-120, figs. 1-9.

— —, 2000. A new genus and species of callianassid ghost shrimp from Kyushu, Japan (Decapoda: Thalassinidea). Journ. Crust. Biol., **20** (Spec. No.) (2): 150-156, figs. 1-4.

POORE, G. C. B. & D. J. G. GRIFFIN, 1979. The Thalassinidea (Crustacea: Decapoda) of Australia. Rec. Australian Mus., **32**(6): 217-321, figs. 1-56, 2 tabs.

POORE, G. C. B. & T. H. SUCHANEK, 1988. *Glypturus motupore*, a new callianassid shrimp (Crustacea: Decapoda) from Papua New Guinea with notes on its ecology. Rec. Australian Mus., **40**(3): 197-204, figs. 1-4.

POULSEN, E. M., 1940. On the occurrence of the Thalassinidea in Danish waters. Vidensk. Medd. Danske naturh. Foren., Copenhagen, **104**: 207-239, figs. 1-12.

POUNDS, S. G., 1961. The crabs of Texas. Bull. mar. Lab. Texas Game Fish Comm., (7) **43**: 1-57.

POWELL, A. W. B., 1949a. New species of Crustacea from New Zealand of the genera *Scyllarus* and *Ctenocheles* with notes on *Lyreidus tridentatus*. Rec. Auckland Inst. Mus., **3**(6): 368-371, pl. 68.

——, 1949b. Biological primary types in the Auckland Museum, 3. Zoological supplement. Rec. Auckland Inst. Mus., **3**(6): 403-409.

RABALAIS, N. N., 1979. A new species of *Ctenocheles* (Crustacea: Decapoda: Thalassinidea) from the northwestern Gulf of Mexico. Proc. biol. Soc. Washington, **92**: 294-306.

RABALAIS, N. N., S. A. HOLT & W. FLINT, 1981. Mud shrimps (Crustacea, Decapoda, Thalassinidea) of the northwestern Gulf of Mexico. Bull. mar. Sci., **31**(1): 96-115, figs. 1-6.

RABALAIS, S. C., W. M. PULICH, JR., N. N. RABALAIS, D. L. FELDER, R. K. TINNIN & R. D. KALKE, 1989. A biological and physical characterization of the Rio Carrizal estuary, Tamaulipas, Mexico. Contr. mar. Sci. Univ. Texas, **31**: 25-37, figs. 1-2.

RAO, P. V. & K. N. R. KARTHA, 1967. On the occurrence of *Callianassa* (*Callichirus*) *audax* De Man (Crustacea Decapoda Callianassidae) on the southwest coast of India with a description of male. Mar. biol. Ass. India, Symp. Cochin, **1**: 279-284, figs. 1-2.

RATHBUN, M. J., 1900a. The decapod crustaceans of West Africa. Proc. U.S. natn. Mus., **22** (1199): 271-316, figs. 1-2.

——, 1900b. The decapod and stomatopod Crustacea. Results of the Branner-Agassiz Expedition to Brazil, **1**. Proc. Washington Acad. Sci., **2**: 133-155, pl. 8.

——, 1901. The Brachyura and Macrura of Porto Rico. Bull. U.S. Fish Comm., **20**(2): 1-137, pls. 1-2, 24 figs.

——, 1902. Descriptions of new decapod crustaceans from the west coast of North America. Proc. U.S. natn. Mus., **24**: 885-905.

——, 1904. Decapod crustaceans of the northwest coast of North America. Harriman Alaska Exp. 1904, **10**: 1-210, figs. 1-95, pls. 1-10.

——, 1906. The Brachyura and Macrura of the Hawaiian Islands. Bull. U.S. Fish Comm., **23**(3): 827-930, pls. 1-24.

——, 1920. Stalk-eyed crustaceans of the Dutch West Indies. In: J. BOEKE, Rapport betreffende een voorloopig onderzoek naar den toestand van de visscherij en de industrie van zeeproducten in de kolonie Curaçao, ingevolge het ministeriëel besluit van 22 November 1904 uitgebracht door Prof. Dr. J. Boeke Hoogleeraar aan de Rijks-Universiteit te Utrecht, **2**: 317-349, figs. 1-5.

——, 1926. The fossil stalk-eyed Crustacea of the Pacific slope of North America. Bull. U.S. natn. Mus., **138**: i-vii, 1-155, figs. 1-6, pls. 1-39.

REED, C. T., 1941. Marine life in Texas waters: i-xii, 1-88, figs. (Texas Academy Publications in Natural History, Nontechnical Series).

RETAMAL, M. A., 1975. Descripcion de una nueva especie del cenero *Callianassa* y clave para reconocer las especies Chilenas. Boln Soc. Biol. Concepción, **49**: 177-183, figs. 1-16.

REVERBERI, G., 1942a. Annotazione sulla *Parthenopea subterranea* Kossmann e sulla *Thompsonia mediterranea* Caroli e dati sulle modificazioni dei caratteri sessuali prodotte dai parassiti sui loro ospiti. Pubbl. Staz. zool. Napoli, **19**: 89-102.

——, 1942b. Sul significazione della "castrazione parassitaria". La trasformazione del sesso nei Crostacei parassitati da Bopyridi e da Rizocefali. Pubbl. Staz. zool. Napoli, **19**: 225-316.

RISSO, A., 1816. Histoire naturelle des Crustacés des environs de Nice: 1-175, pls. 1-3.

——, 1822. Mémoire sur quelques nouveaux Crustacés observés dans la mer de Nice. Journ. Phys. Chim. Hist. nat. Art, **95**: 241-248.

——, 1827. Histoire naturelle des principales productions de l'Europe méridionale et particulièrement de celles des environs de Nice et des Alpes-Maritimes, **5**: i-vii, 1-403, pls. 1-10, figs. 1-62.

RODRIGUES, S. DE A., 1971. Mud shrimps of the genus *Callianassa* Leach from the Brazilian coast (Crustacea, Decapoda). Arq. zool. São Paulo, **20**(3): 191-223, figs. 1-98.

——, 1978. *Ctenocheles holthuisi* (Decapoda, Thalassinidea), a new remarkable mud shrimp from the Atlantic Ocean. Crustaceana, **34**(2): 113-120.

——, 1983. Aspectos da biologia de Thalassinidea do Atlantico tropical Americano: 1-174. (Universidade de São Paulo, São Paulo).

——, 1984a. Desenvolvimento pós-embrionário de *Callichirus mirim* (Rodrigues, 1971) obtido em condições artificiais (Crustacea, Decapoda, Thalassinidea). Bol. Zool. Univ. São Paulo, **8**: 239-256, figs. 1-39.

——, 1984b. Crescimento relativo de *Callianidea laevicauda* Gill e *Callichirus major* (Say) (Crustacea, Thalassinidea). Ciênc. Cult., **36**(7): 914.

RODRIGUES, S. [DE] A. & W. HÖDL, 1990. Burrowing behaviour of *Callichirus major* and *C. mirim.* Begleitveröffentlichung zum wissenschaftlichen Film C 211 des ÖWF. Wiss. Film Wien **41**/April: 48-58.

RODRIGUES, S. [DE] A. & R. B. MANNING, 1992a. Two new callianassid shrimps from Brazil (Crustacea: Decapoda: Thalassinidea). Proc. biol. Soc. Washington, **105**(2): 324-330, figs. 1-2.

—— & ——, 1992b. *Poti gaucho*, a new genus and species of ghost shrimp from southern Brazil (Crustacea: Decapoda: Callianassidae). Bull. mar. Sci. Univ. Miami, **51**: 9-13, fig. 1.

ROSSIGNOL, M., 1962. *Callianassa pentagonocephala* nov. sp. (Callianassidae) et *Sicyonia foresti* nov. sp. (Penaeidae), Crustacés Décapodes, Anomoures et Macroures nouveaux du plateau continental congolais. Cah. ORSTOM, (Océanogr.) **2**: 139-145, pl. 1 a-c, 1 pl. n.n.

SAINT LAURENT, M. DE, 1973. Sur la systématique et la phylogénie des Thalassinidea: définition des familles des Callianassidae et des Upogebiidae et diagnose de cinq genres nouveaux (Crustacea Decapoda). C. R. Acad. Sci., Paris, (D) **277**: 513-516.

——, 1979. Sur la classification et la phylogénie des Thalassinidea: définitions de la superfamille des Axioidea, de la sous-famille des Thomassiniinae et deux genres nouveaux (Crustacea Decapoda). C. R. Acad. Sci., Paris, (D) **288**:1395-1397.

SAINT LAURENT, M. DE & B. BOZIC, 1976 [not 1972]. Diagnoses et tableaux de détermination des Callianasses de l'Atlantique nord oriental et de Méditerranée (Crustacea, Decapoda, Callianasidae). Proceedings of the First Colloquium Crustacea Decapoda Mediterranea. Thalassia Jugoslavica, **8**(1): 15-40, figs. 1-35.

SAINT LAURENT, M. DE & P. LE LOEUFF, 1979. Upogebiidae et Callianassidae. Crustacés Décapodes: Thalassinidea, 1. Ann. Inst. océanogr. Monaco, **55**(Suppl.): 29-101, figs. 1-28.

SAINT LAURENT, M. DE & R. B. MANNING, 1982. *Calliax punica*, espèce nouvelle de Callianassidae (Crustacea, Decapoda) des eaux Méditerranéennes. Quad. Lab. Tecnol. Pesca, Ancona, **3**(2-5): 211-224.

SAKAI, K., 1966. On *Callianassa* (*Callichirus*) *novaebritanniae* Borradaile (Thalassinidea, Crustacea) from Japan. Journ. Fac. Agric. Kyushu Univ., **14**(1): 161-171, figs. 1-4.

——, 1967. Three new species of Thalassinidea (Decapoda, Crustacea) from Japan. Res. Crust., Tokyo, **3**: 39-51, pls. 3-5.

——, 1968. On Thalassinidea (Decapoda, Crustacea). Nat. Study nat. Hist. Mus. Osaka, **14**(9): 114-115, figs. 1-9.

——, 1969. Revision of Japanese callianassids based on the variations of larger cheliped in *Callianassa petalura* Stimpson and *C. japonica* Ortmann (Decapoda: Anomura). Publs Seto mar. biol. Lab., **17**: 209-252, figs. 1-8, pls. 9-15.

— —, 1970a. A small collection of thalassinids from the waters around Tsushima Islands, Japan, including a new species of *Callianassa* (Crustacea, Anomura). Publs Seto mar. biol. Lab., **18**(1): 37-47, figs. 1-4.

— —, 1970b. Supplementary description of *Callianassa* (*Callichirus*) *tridentata* Von Martens (Crustacea, Thalassinidea). Publs Seto mar. biol. Lab., **17**: 393-401, figs. 1-3.

— —, 1983. On a new species of the genus *Callianassa* (Crustacea, Decapoda) from Thailand. Res. Crust., Tokyo, **12**: 111-115, figs. 1-2.

— —, 1984. Some thalassinideans (Decapoda: Crustacea) from Heron Is., Queensland, eastern Australia, and a new species of *Gourretia* from East Africa. Beagle, Occ. Pap. Northern Territory Mus. Arts Sci., **1**(11): 95-108.

— —, 1987a. Two new Thalassinidea (Crustacea: Decapoda) from Japan, with the biogeographical distribution of the Japanese Thalassinidea. Bull. mar. Sci. Univ. Miami, **41**(2): 296-308, figs. 1-3, 1 tab.

— —, 1987b. On *Callianassa ranongensis* (Thalassinidea: Decapoda: Crustacea) from Halmahera, Indonesia. Rep. Usa mar. biol. Inst. Kochi Univ., **9**: 45-49.

— —, 1988. A new genus and five new species of Callianassidae (Crustacea: Decapoda: Thalassinidea) from northern Australia. Beagle, Rec. Northern Territory Mus. Arts Sci., **5**: 51-69.

— —, 1992a. Axiid collections of the Zoological Museum, Copenhagen, with the description of one new genus and six new species (Axiidae, Thalassinidea, Crustacea). Zool. Scr., **21**(2): 157-180, figs. 1-18.

— —, 1992b. Notes on some species of Thalassinidea from French Polynesia. Senckenb. marit., **22**(3/6): 211-216, figs. 1-2.

— —, 1999a. Description of *Ctenocheles balssi* Kishinouye, 1926, with comments on its systematic position and establishment of a new subfamily Gourretiinae (Decapoda, Callianassidae). Crustaceana, **72**(1): 85-97, figs. 1-3.

— —, 1999b. A new species, *Callianassa poorei* sp. nov. (Decapoda: Crustacea: Callianassidae) from Tasmania. Journ. mar. biol. Ass. U.K., **79**: 373-374, 2 text-figs.

— —, 1999c. Synopsis of the family Callianassidae, with keys to subfamilies, genera and species, and the description of new taxa (Crustacea: Decapoda: Thalassinidea). Zool. Verh., Leiden, **326**: 1-152, figs. 1-33.

— —, 2000. A new species of *Neocallichirus*, *N. angelikae*, from South Australia (Decapoda: Callianassidae). Mitt. Hamburger zool. Mus. Inst., **97**: 91-98, figs. 1-3.

— —, 2001. A review of the common Japanese callianassid species, *Callianassa japonica* and *C. petalura* (Decapoda, Thalassinidea). Crustaceana, **74**(9): 937-949, figs. 1-4, 1 tab.

— —, 2002. Callianassidae (Decapoda, Thalassinidea) in Phuket, Thailand. In: N. L. BRUCE, M. BERGGREN & S. BUSSAWARIT (eds.), Proceedings of the International Workshop on the Biodiversity of Crustacea of the Andaman Sea. Phuket mar. biol. Center spec. Publ., **23**: 461-532.

— —, 2004. Dr. R. Plante's collection of the families Callianassidae and Gourretiidae (Decapoda, Thalassinidea) from Madagascar, with the description of two new genera and one new species of the Gourretiidae Sakai, 1999 (new status) and two new species of the Callianassidae Dana, 1852. Crustaceana, **77**(5): 553-601, figs. 1-23.

— —, 2005. The diphyletic nature of the infraorder Thalassinidea (Decapoda, Pleocyemata) as derived from the morphology of the gastric mill. Crustaceana, **77**(9): 1117-1129, figs. 1-6.

SAKAI, K. & M. APEL, 2002. Thalassinidea (Crustacea: Decapoda) from Socotra Archipelago, Yemen, with a new species of *Lepidophthalmus* (Callianassidae). Fauna Arabia, **19**: 273-288, figs. 1-7.

SAKAI, K. & M. TÜRKAY, 1999. A new subfamily, Bathycalliacinae n. subfam., for *Bathycalliax geomar* n. gen., n. sp. from the deep water cold seeps off Oregon, USA (Crustacea, Decapoda, Callianassidae). Senckenb. biol., **79**(2): 203-209, figs. 1-3, 1 tab.

SANKOLLI, K. N., 1971. The Thalassinoidea (Crustacea, Anomura) of Maharashtra. Journ. Bombay nat. Hist. Soc., **68**(1): 94-106, figs. 5-8.

SAY, T., 1817-1818. An account of the Crustacea of the United States [Part 5]. Journ. Acad. nat. Sci. Philadelphia, **1**(2)(1): 235-253. [See Holthuis, 1969: vi-vii, for dates of publication.]

SCHELLENBERG, A., 1928. Krebstiere oder Crustacea. II. Decapoda, Zehnfußer (14. Ordnung). In: F. DAHL (ed.), Die Tierwelt Deutschlands und der angrenzenden Meeresteile nach ihren Merkmalen und nach ihrer Lebensweise, **10**: 1-146, figs. 1-110.

SCHMITT, W. L., 1921. The marine decapod Crustacea of California with special reference to the decapod Crustacea collected by the United States Bureau of Fisheries Steamer "Albatross" in connection with the biological survey of San Francisco Bay during the years 1912-1913. Univ. California Publ. Zool., **23**: 1-470, figs. 1-165, pls. 1-50.

— —, 1924. Report on the Macrura, Anomura and Stomatopoda, collected by the Barbados Antigua Expedition from the University of Iowa in 1918. Univ. Iowa Stud. nat. Hist., **10**(4): 65-99, pls. 1-5.

— —, 1935a. Crustacea Macrura and Anomura of Porto Rico and the Virgin Islands. Sci. Survey Porto Rico Virgin Islands, **15**(2): 127-227, figs. 1-79.

— —, 1935b. Mud shrimps of the Atlantic coast of North America. Smithson. misc. Coll., **93**(2): 1-21, pls. 1-4.

— —, 1939. Decapod and other Crustacea collected on the Presidential Cruise of 1938 (with introduction and station data). Smithson. misc. Coll., **98**(6): 1-29, figs. 1-2, pls. 1-3.

SELBIE, C. M., 1914. Palinura, Astacura, and Anomura (except Paguridea). The Decapoda Reptantia of the coasts of Ireland. Part I. Scient. Invest. Fish. Branch Ireland, **1914**: 1-116, pls. 1-15.

SHIPP, L. P., 1977. The vertical and horizontal distribution of decapod larvae in relation to some environmental conditions within a salt marsh area of the north central Gulf of Mexico: 1-129. (M.Sc. Thesis, University of South Alabama, Mobile, Alabama).

SIMBOURA, N., A. ZENETOS, M.-A. PANCUCCI-PAPADOPOULOU, M. THESSALOU-LEGAKI & S. PAPASPYROU, 1998. A baseline study on benthic species distribution in two neighbouring gulfs, with and without access to bottom trawling. Marine Ecol., **19**(4): 293-309, 6 figs., 3 tabs.

SMITH, S. I., 1873. Habits and distribution of the invertebrate animals. In: A. E. VERRILL, Report upon the invertebrate animals of Vineyard Sound and the adjacent waters. Rep. U.S. Commr Fish., **1871-1872**: 295-778, pls. 1-39.

SQUIRES, H. J., 1990. Decapod Crustacea of the Atlantic coast of Canada. Canadian Bull. Fish. aquat. Sci., **221**: 1-532.

STALIO, L., 1877. Catalogo metodico e descrittivo dei Crostacei podottalmi et edriottalmi dell' Adriatico. Atti Ist. Veneto Sci., (5) **3**: 1-274.

STAMHUIS, E. J., C. E. SCHREURS & J. J. VIDELER, 1997. Burrow architecture and turbative activity of the thalassinid shrimp *Callianassa subterranea* from the central North Sea. Mar. Ecol. Progr. Ser., **151**(1-3): 155-163, illustr.

STATON, J. L. & D. L. FELDER, 1995. Genetic variation in populations of the ghost shrimp genus *Callichirus* (Crustacea: Decapoda: Thalassinoidea) in the western Atlantic and Gulf of Mexico. Bull. mar. Sci. Univ. Miami, **56**(2): 523-536.

STATON, J. L., D. L. FELDER & D. W. FOLTZ, 1988. Genetic differences between three species of thalassinid shrimp from Louisiana. American Zool., **28**: 125A.

STATON, J. L., D. W. FOLTZ & D. L. FELDER, 2000. Genetic variation and systematic diversity in the ghost shrimp genus *Lepidophthalmus* (Decapoda: Thalassinidea: Callianassidae). Journ. Crust. Biol., **20** (Spec. No.): 157-169, figs. 1-4, tabs. 1-5.

STEBBING, TH. R. R., 1893. A history of Crustacea: recent Malacostraca. The International Scientific Series, London, **74**: i-xvii, 1-466, figs. 1-32, pls. 1-19. [New York ed., 1893, D. Appleton & Co., is in International Scientific Series, **71**, with same pagination.]

— —, 1900. South African Crustacea. Mar. Invest. South Africa, **1**: 1-66, pls. 1-4.

— —, 1902. South African Crustacea. Part 2. Mar. Invest. South Africa, **2**: 1-92, pls. 5-16.

— —, 1910. General catalogue of South African Crustacea (Part 5 of S.A. Crustacea, for the Marine Investigations in South Africa). Ann. South African Mus., **6**(4): 281-593, pls. 15-22.

STEINITZ, W., 1933. Beiträge zur Kenntnis der Küstenfauna Palaestinas. II. Pubbl. Staz. zool. Napoli, **13**: 143-154, figs. 1-3.

STEPHENSON, T. A., A. STEPHENSON, G. TANDY & M. SPENDER, 1931. The structure and ecology of Low Isles and other reefs. Scient. Rep. Great Barrier Reef Exped., **3**(2): 17-112, figs. 1-15, pls. 1-27.

ŠTEVČIĆ, Z., 1969a. Lista desetonožnih[1]) rakova Jadrana. Biol. Veštn., **17**: 125-134.

— —, 1969b. Da li su dekapodi Jadrana dobro poznati? Thalassia Jugoslavica, **5**: 345-351.

— —, 1971. Beitrag zur Revision der Decapodenfauna der Umgebung von Rovinj. Thallassia Jugoslavica, **7**(2): 525-531.

— —, 1979. Cruises of the Research Vessel "Vila Velebita" in the Kvarner region of the Adriatic Sea. XIX. Crustacea Decapoda. Thalassia Jugoslavica, **15**(3/4): 279-287, 1 fig.

— —, 1985. New and rarely reported species of decapod Crustacea from the Adriatic Sea. Rapp. Comm. int. Mer. Méditerranée, **29**(5): 313-314.

— —, 1998. Marine decapod Crustacea of the Kvarner region. Nat. Hist. Mus. Rijeka, **1**: 647-660, figs. 1-3.

— —, 1990. Check-list of the Adriatic decapod Crustacea. Acta Adriatica, **31**: 183-274.

STEVENS, B. A., 1928. Callianassidae from the west coast of North America. Publ. Puget mar. biol. Sta., **6**: 315-369, figs. 1-71.

STIMPSON, W., 1856. On some Californian Crustacea. Proc. California Acad. Sci., **1**: 87-90.

— —, 1857a. Notices of new species of Crustacea of western North America; being an abstract from a paper to be published in the Journal of the Society. Proc. Boston Soc. nat. Hist., **6**: 84-89.

— —, 1857b. On the Crustacea and Echinodermata of the Pacific shores of North America. Boston Journ. nat. Hist., **6**: 444-532, pls. 18-23.

— —, 1860. Prodromus descriptionis animalium evertebratorum, quae in expeditione ad Oceanum Pacificum septentrionalem, a Republica Federata missa, Cadwaladaro Ringgold et Johanne Rodgers ducibus, observavit et descripsit. Pars 8. Crustacea Macrura. Proc. Acad. nat. Sci. Philadelphia, **1860**: 22-47.

— —, 1866: Descriptions of new genera and species of macrurous Crustacea from the coasts of North America. Proc. Chicago Acad. Sci., **1**: 46-48.

— —, 1871. Notes on North American Crustacea, in the museum of the Smithsonian Institution. No. 3. Ann. Lyc. nat. Hist., New York, **10**(3-4): 92-136.

ŠTJEPČEVIĆ, J. & P. PARENZAN, 1980. Il Golfo delle Bocche di Cattaro. Condizioni generali e biocenosi bentoniche con carta ecologia delle sue due baie inerne: di Kotor (Cattaro) e di Risan (Risano). Studia mar., **9-10**: 3-146.

---

[1]) desetonožnih – fused as desetonoznih

STRAHL, C., 1862a. Über einige neue von Hrn. F. Jagor eingesandte Thalassinen und die systematische Stellung dieser Familie. Monatsber. Kön. Akad. Wiss. Berlin, **1861**: 1055-1072, pl. 1.

— —, 1862b. On some new Thalassinae sent from the Philippines by M. Jagor, and on the systematic position of that family. Ann. Mag. nat. Hist., (3) **9**: 383-396.

STRASSER, K. M. & D. L. FELDER, 1999a. Stand as a stimulus for settlement in the ghost shrimp *Callichirus major* (Say) and *C. islagrande* (Schmitt) (Crustacea: Decapoda: Thalassinidea). Journ. exp. mar. Biol. Ecol., **239**(2): 211-222, illustr.

— — & — —, 1999b. Larval development in two populations of the ghost shrimp *Callichirus major* (Decapoda; Thalassinidea) under laboratory conditions. Journ. Crust. Biol., **19**(4): 844-878, illustr.

— — & — —, 2000. Larval development of the ghost shrimp *Callichirus islagrande* (Decapoda; Thalassinidea; Callianassidae) under laboratory conditions. Journ. Crust. Biol., **20**(1): 100-117, illustr.

SUZUKI, S., 1979. Marine invertebrates in Yamagata Prefecture: 1-370, pls. 1-22. [In Japanese.]

TAMAKI, A., B. INGOLE, K. IKEBE, K. MURAMATSU, M. TAKA & M. TANAKA, 1997. Life history of the ghost shrimp, *Callianassa japonica* Ortmann (Decapoda: Thalassinidea), on an intertidal sandflat in western Kyushu, Japan. Journ. exp. mar. Biol. Ecol., **210**: 223-250. figs. 1-14.

TAMAKI, A. & S. MIYABE, 2000. Larval abundance patterns for three species of *Nihonotrypaea* (Decapoda: Thalassinidea: Callianassidae) along an estuary-to-open-sea-gradient in western Kyushu, Japan. Journ. Crust. Biol., **20** (Spec. No.) (2): 182-191, figs. 1-5.

TAMAKI, A., H. TANOUE, J. ITOH & Y. FUKUDA, 1996. Brooding and larval developmental periods of the callianassid ghost shrimp, *Callianassa japonica* (Decapoda: Thalassinidea). Journ. mar. biol. Ass. U.K., **76**: 675-689, fig. 1, tabs. 1-6.

THALLWITZ, J., 1891. Decapoden-Studien, insbesondere basiert auf A.B. Meyer's Sammlungen im Ostindischen Archipel, nebst einer Aufzählung der Decapoden und Stomatopoden des Dresdener Museums. Abhandl. Ber. zool. Mus. Dresden, **1890/91**(3): 1-55, 1 pl.

THATJE, S., 2000. *Notiax santarita*, a new species of the Callianassidae (Decapoda, Thalassinidea) from the Beagle Channel, southernmost America. Crustaceana **73**(3): 289-299, figs. 1-6, 1 tab.

THESSALOU, M., 1979. Preliminary report on the larvae of decapod Crustacea from Pagassitikos Gulf, Greece. Rapp. Comm. int. Mer Méditerranée, **25/26**(8): 145-146.

THESSALOU-LEGAKI, M., 1985. On a population of *Callianassa subterranea* (Crustacea, Decapoda, Thalassinidea) in the N. Euboikos Gulf (Greece). Poster, IIIrd Colloquium Crustacea Decapoda Mediterranea. Invest. Pesq., Barcelona, **51**(Suppl.)(1): 457.

— —, 1986. Preminary data on the occurrence of Thalassinidea (Crustacea, Decapoda) in the Greek seas. Biol. Gallo-Hellenica, **12**: 181-187.

— —, 1987. On a population of *Callianassa subterranea* (Crustacea, Decapoda, Thalassinidea) in the N. Euboikos Gulf (Greece). Invest. Pesq., Barcelona, **51**(Suppl.)(1): 457.

— —, 1990. Advanced larval development of *Callianassa tyrrhena* (Decapoda: Thalassinidea) and the effect of environmental factors. Journ. Crust. Biol., **10**(4): 659-666.

THESSALOU-LEGAKI, M., P. KERAMBRUN & G. VERRIOPOULOS, 1997. Differentiation of physiological aspects of the burrowing shrimp *Callianassa tyrrhena* in relation to general pollution load. Journ. mar. biol. Ass. U.K., **77**: 439-450, 3 figs., 2 tabs.

THESSALOU-LEGAKI, M. & V. KIORTSIS, 1997. Estimation of the reproductive output of the burrowing shrimp *Callianassa tyrrhena*: a comparison of three different biometrical approaches. Mar. Biol., Berlin, **127**: 435-442.

THESSALOU-LEGAKI, M., A. PEPPA & M. ZACHRAKI, 1999. Facultative lecithotrophy during larval development of the burrowing shrimp *Callianassa tyrrhena* (Decapoda: Callianassidae). Mar. Biol., Berlin, **133**: 635-642.

THESSALOU-LEGAKI, M. & A. ZENETOS, 1985. Autoecological studies on the Thalassinidea (Crustacea, Decapoda) of the Patras Gulf and Ionian Sea (Greece). Rapp. Comm. int. Mer Méditerranée, **29**(5): 309-312.

THOMPSON, W., 1844. Report on the fauna of Ireland: Div. Invertebrata. Drawn up, at the request of the British Association. Rep. British Assoc. Advmt Sci., **1843**: 245-291. (London).

TIEFENBACHER, L., 1976. *Callianassa jamaicensis* Schmitt (Decapoda, Thalassinidea) an der brasilianischen Küste südlich der Amazonasmündung. Crustaceana, **30**(3): 314-316, 1 fig.

TIRMIZI, N. M., 1967. On the occurrence of *Callianassa* (*Callichirus*) *audax* De Man off west Pakistan (Decapoda, Thalassinidea). Crustaceana, **13**(2): 151-154, figs. 1-2.

——, 1970. A new species of *Callianassa* (Decapoda, Thalassinidea) from West Pakistan. Crustaceana, **19**(3): 245-250, figs. 1-3.

——, 1974. A description of *Callianassa martensi* Miers, 1884 (Decapoda, Thalassinidea) and its occurrence in the northern Arabian Sea. Crustaceana, **26**(3): 286-292, figs. 1-4.

——, 1977. A redescription of the holotype of *Callianassa mucronata* Strahl, 1862 (Decapoda, Thalassinidea). Crustaceana, **32**(1): 21-26, figs. 1-3.

TUDGE, C. C., G. C. B. POORE & R. LEMAITRE, 2000. Preliminary phylogenetic analysis of generic relationships within the Callianassidae and Ctenochelidae (Decapoda: Thalassinidea: Callianassoidea). Journ. Crust. Biol., **20** (Spec. No.) (2): 129-149, figs. 1-4, appendices 1-3.

TÜRKAY, M., 1982. On the occurrence of *Callianassa acanthura* Caroli, 1946 in the Aegean Sea (Crustacea: Decapoda: Callianassidae). Quad. Lab. Tecnol. Pesca, Ancona, **3**(2-5): 225-226.

TÜRKAY, M., G. FISCHER & V. NEUMANN, 1987. List of the marine Crustacea Decapoda of the Northern Sporades (Aegean Sea) with systematic and zoogeographic remarks. Inv. Pesq., Barcelona, **51** (Suppl. 1): 87-109.

TÜRKAY, M. & K. SAKAI, 1995. Decapod crustaceans from a volcanic hot spring in the Marianas. Senckenb. marit., **26**(1/2): 25-35, figs. 1-9, tab. 1.

UDEKEM D'ACOZ, C. D', 1986. Etude d'une collection de Crustacés Décapodes de Bretagne. Strandvlo, **5**(4): 97-130. [For the year 1985.]

——, 1989. Seconde note sur les Crustatacés Décapodes de la Bretagne. Strandvlo, **8**(4): 166-205. [For the year 1988.]

——, 1996. Contribution à la connaissance des Crustacés Décapodes helléniques II: Penaeidea, Stenopodidea, Palinuridea, Homaridea, Thalassinidea, Anomura, et note sur les Stomatopodes. Bios, Macedonia, Greece, **3**: 51-77.

——, 1999. Inventaire et distribution des Crustacés Décapodes de l'Atlantique nord-oriental, de la Méditerranée et des eaux continentales adjacentes au nord de 25°N. Patrimoines naturels (M.N.H.N./S.P.N.), **40**: 1-383.

——, 2003. A pictorial guide to the Crustacea Decapoda (shrimps, lobsters and crabs) of the eastern Atlantic, the Mediterranean Sea and the adjacent continental waters, including brief information on their distribution and ecology, as well as identification references: http://www.tmu.uit.no/crustikon/index.htm

VATOVA, A., 1949. La fauna bentonica dell'alto e medio Adriatico. Nova Thalassia, **1-3**: 1-110, tabs. 1-6, 7-27.

VANHAELEN, M.-T., 2001. Het gravend kreeftje *Callianassa tyrrhena* (Petagna) aangespoeld op het Koksijdse strand. [The burrowing crayfish *Callianassa tyrrhena* (Petagna) washed up on the beach at Koksijde.] Strandvlo, **21**(4): 147-149. [In Dutch.]

UTINOMI, H., 1956. Anomura. Coloured illustrations of seashore animals of Japan: 1-168, pls. 1-64, 1-12. (Hoikusha, Osaka).

VANHÖFFEN, E., 1911. Ueber die Krabben, denen Kamerun seinen Namen verdankt. SitzBer. Ges. naturf. Fr., Berlin, **2**: 105-110, 1 fig.

VAUGELAS, J. DE, 1990. Ecologie des Callianasses (Crustacea, Decapoda, Thalassinidea) en milieu récifal Indo-Pacifique. Conséquences du remaniement sédimentaires sur la distribution des matières humiques, des métaux traces et des radionucléides: 1-226. (Mémoire présenté à l'Université de Nice-Sophia Antipolis pour l'obtention du Diplome d'Habilitation à diriger des recherches en sciences).

VAUGELAS, J. DE & M. DE SAINT LAURENT, 1984. Preliminary observations on two types of callianassid (Crustacea, Thalassinidea) burrows. Gulf of Aqaba (Red Sea). In: M. A. H. SAAD (ed.), Proceedings of the Symposium on Coral Reef Environment of the Red Sea, Jeddah, Jan., 1984: 520-539.

VERRILL, A. E., 1922. Decapod Crustacea of Bermuda. Part 2. Macrura. Trans. Connecticut Acad. Arts Sci., **26**: 1-179, figs. 1-12, pls. 1-48.

WARD, M., 1942. Notes on the Crustacea of the Desjardins Museum, Mauritius Institute, with descriptions of new genera and species. Mauritius Inst. Bull., **2**: 49-109, pls. 5-6.

— —, 1945. A new crustacean. Mem. Queensland Mus., Brisbane, **12**: 134-135, pl. 13.

WASS, M. L., 1955. The decapod crustaceans of Alligator Harbor and adjacent inshore areas of northwestern Florida. Quart. Journ. Florida Acad. Sci., **18**: 129-176, figs. 1-13.

WEIMER, R. J. & J. H. HOYT, 1964. Burrows of *Callianassa major* Say, geologic indicators of littoral and shallow neritic environments. Journ. Paleont., **38**(4): 761-767, pls. 123-124, figs. 1-2.

WHITE, A., 1847. List of species in the collections of the British Museum: i-viii, 1-143. (British Museum, London).

— —, 1861a. Descriptions of two species of Crustacea belonging to the families Callianassidae and Squillidae. Proc. zool. Soc., London, **1861**: 42-44, pls. 6-7.

— —, 1861b. Descriptions of two species of Crustacea belonging to the families Callianassidae and Squillidae. Ann. Mag. nat. Hist., (3) **7**: 479-481.

WILLIAMS, A. B., 1965. Marine decapod crustaceans of the Carolinas. Fishery Bull., Fish Wildl. Serv., U.S., **65**(1): 1-298, figs. 1-252.

— —, 1984. Shrimps, lobsters, and crabs of the Atlantic coast of the eastern United States, Maine to Florida: i-xviii, 1-550, figs. 1-380. (Washington, D.C.).

WILLIAMS, A. B., L. G. ABELE, D. L. FELDER, H. H. HOBBS, JR., R. B. MANNING, P. A. MCLAUGHLIN & I. PÉREZ-FARFANTE, 1989. Common and scientific names of aquatic invertebrates from the United States and Canada: decapod crustaceans. American Fish. Soc. Spec. Publ., **17**: i-ivii, 1-77.

WILLIS, E. R., 1942. Some mud shrimps of the Louisiana coast. Occ. Pap. mar. Lab. Louisiana State Univ., **2**: 1-6.

WITBAARD, R. & G. C. A. DUINEVELD, 1989. Some aspects of the biology and ecology of the burrowing shrimp *Callianassa subterranea* (Montagu) (Thalassinidea) from the southern North Sea. Sarsia, **74**: 209-219, figs. 1-7.

WOLFF, T., 1964. The Galathea Expedition 1950-52. List of benthic stations from 0-400 metres, near-surface stations, and land staions. Vidensk. medd. Dansk naturh. Foren., **127**: 1-258.

— —, 1995. Carlsbergfondets Dana-ekspedition 1928-30, Forudsætninger, gennemførelse og væsentlige resultater. Præsentationer ver 8. Danske Havforskermøde, Dansk Nationalråd for Oceanologi, Fyns Amt: 185-204.

——, 2002. The Danish "Dana" Expedition 1928-30: purpose and accomplishments mainly in the Indo-Pacific. In: K. R. BENSON & P. F. REHBOCK (eds.), Oceanographic history, the Pacific and beyond: 196-203. (University of Washington Press, Seattle).

WOODWARD, H., 1868. Fourth report on the structure and classification of the fossil Crustacea. Rept. 38th Meet. British Assoc. Advmt Sci., Norwich: 72-75, pl. 2.

WYNBERG, R. P. & G. M. BRANCH, 1997. Trampling associated with bait-collection for sandprawns *Callianassa kraussi* Stebbing: effects on the biota of an intertidal sandflat. Environm. Cons., **24**(2): 139-148, illustr.

YAMAGUCHI, T. & L. B. HOLTHUIS, 2001. Kai-ka Rui Siya-sin, a collection of pictures of crabs and shrimps, donated by Kurimoto Suiken to Ph. F. von Siebold. Calanus, Bull. Aitsu mar. biol. Sta. Kumamoto Univ., (Spec. No.) **3**: 1-182.

YOKOYA, Y., 1930. Macrura of Mutsu Bay. Report on the biological survey of Mutsu Bay, No. 16. Sci. Rep. Tôhoku Imp. Univ., (4) **5**(3): 525-548, figs. 1-5, pl. 16.

——, 1933. On the distribution of decapod crustaceans inhabiting the continental shelf around Japan, chiefly based upon the materials collected by S.S. Sôyo-Maru, during the years 1923-1930. Journ. Coll. Agric. Tokyo Imp. Univ., **12**(1): 1-226, figs. 1-71, tabs. 1-4.

——, 1939. Macrura and Anomura of decapod Crustacea found in the neighbourhood of Onagawa, Miyagi-ken. Scient. Rep. Tohoku Imp. Univ., (4) **14**(2-3): 261-289, figs. 1-13.

YÜ, S. C., 1931. On some species of shrimp-shaped Anomura from north China. Bull. Fan Mem. Inst., **2**(6): 513-516, figs. 1-3.

ZARIQUIEY ALVAREZ, R., 1946. Crustáceos Decápodos mediterráneos. Manual para la clasificación de las especies que pueden capturarse en las costas mediterráneas españolas. Publ. Biol. mediterránea Inst. Español Est. Mediterráneos, **2**: 1-181, figs. 1-174, pls. 1-16.

——, 1950. Mas formas interesantes del Mediterráneo y de las costas españolas. Decápodos españoles. III. Eos, Madrid, **26**(1-2): 73-113, figs. 1-4, pls. 1-8.

——, 1968. Crustáceos Decápodos Ibéricos. Invest. Pesq., Barcelona, **32**: i-xv, 1-510, figs. 1-164.

ZARKANELLAS, A. & C. BOGDANOS, 1977. Benthic studies of a polluted area in the upper Saronikos Gulf. Thalassographica, **1**(2): 155-177.

ZAVODNIK, D., 1979. Cruises of the research vessel "Vila Velebita" in the Kvarner region of the Adriatic Sea. XXX. Benthic investigations. Thalassia Jugoslavica, **15**: 313-350.

ZEHNTNER, L., 1894. Crustacés de l'Archipel Malais. Voyage de MM. Bedot et C. Pictet dans l'Archipel Malais. Rev. Suisse Zool., **2**: 135-214, pls. 7-9.

ZIEBIS, W., S. FORSTER, M. HUETTEL & B. B. JORGENSEN, 1996 (cf. a). Complex burrows of the mud shrimp *Callianassa truncata* and their geochemical impact in the sea bed. Nature, London, **382**: 619-622.

ZIEBIS, W., M. HUETTEL & S. FORSTER, 1996 (cf. b). Impact of biogenic sediment topography on oxygen fluxes in permeable seabeds. Mar. Ecol. Progr. Ser., **140**: 227-237.

TAXONOMIC INDEX